The Social Construction of the Dutch Air Quality Clash

The Social Construction of the Dutch Air Quality Clash

How Road Expansions Bit the Dust Against Particulate Matter

Tobias Arnoldussen

eleven
international publishing

Published, sold and distributed by Eleven International Publishing
P.O. Box 85576
2508 CG The Hague
The Netherlands
Tel.: +31 70 33 070 33
Fax: +31 70 33 070 30
e-mail: sales@elevenpub.nl
www.elevenpub.com

Sold and distributed in USA and Canada
International Specialized Book Services
920 NE 58th Avenue, Suite 300
Portland, OR 97213-3786, USA
Tel.: 1-800-944-6190 (toll-free)
Fax: +1-503-280-8832
orders@isbs.com
www.isbs.com

Eleven International Publishing is an imprint of Boom uitgevers Den Haag.

ISBN 978-94-6236-639-8
ISBN 978-94-6274-456-1 (E-book)

Printed in The Netherlands

Acknowledgements

This research would not have been accomplished without the generous help of a great number of people. I like to extend my gratitude first of all to my promotor Nick Huls and my copromotor Roel Pieterman for their incessant support during every stage of this research process. Their valuable suggestions and critical commentary did not only improve the book, but edified the author as well. I would like to thank the members of the inner committee, Ellen Hey, René van Swaaningen and Adam Burgess for their thorough assessment of this study, and my colleagues at the EUR and the OMV for the stimulating discussions we have had about it. In particular I would like to thank Willem Jan Kortleven, Mireille Hildebrandt, Elaine Mak, Harm Kloosterhuis, Sanne Taekema, Wybren van den Burg and Munish Ramlal for their insightful comments.

Good friends and colleagues took their time to read parts of this manuscript and offer their suggestions for improvement. I am very grateful for their thoughtful remarks and suggestions. I wish to mention Jitte Waagen, Rob Schwitters and Jaap Hanekamp, thank you very much for your patience and for giving me the opportunity to discuss this book with you. I would like to especially mention and thank my mother Elly Leijenaar and Laura Holswilder for their help, not only did they read parts of the book, but you also assisted me in my attempts to get to grips with all the intricate possibilities of Microsoft Office.

The support staff at the Erasmus University has invested a lot of time in ordering the references. I wish to express my sincere gratitude for the work of Fausia Sual on the manuscript and the assistance of Mariëlle Duijndam. My sincere thanks as well to Donna Devine for her editing work on the book. In addition I thank my publisher Joris Bekkers of Boom legal publishing for the expedient publication of this dissertation and of course my respondents, who made this study possible.

I am grateful to many of my current colleagues at the UvA for taking the time to discuss whatever topic comes to mind. My colleagues at the Paul Scholten Center, Roland Pierik, Marc de Wilde, Bas Schotel and Nienke Doornbos among others have broadened my horizons and the members of the teaching team at PPLE College have shared with me insights from their own disciplines. I have learned a lot about US law from Alison Fischer and Robert Weaver, about psychology from Noah Millman, about politics from Peyman Jafari.

Finally, I would like to thank my family and a group of very good friends for standing by my side. My mother Elly, my father Paul and my brother Jelger have kept supporting me, especially when my spirits were low at times. René, Ed, Roel, Ruben, Snaus, Cleef, Thijs and all those others I could not mention, I wish to thank you too. It is impossible to conduct research without spending time with goods friends over a beer once in a while.

Table of Contents

1 THE AIR QUALITY CLASH: INTRODUCTION AND METHODOLOGY

INTRODUCTION: AN AIR QUALITY PROBLEM IN THE NETHERLANDS

When I relocated to my hometown of Amsterdam in 2007, I noticed something peculiar. People were debating which street in Amsterdam was the dirtiest. It had nothing to do with litter or garbage collection, but with air pollution. I quickly learned that the Netherlands is a highly polluted country, and that there were satellite images to prove it. During a meal in the city, I was told that air pollution was a health menace, and that it came mainly from car exhausts, although copying machines and printers also emitted the dreaded 'Particulate Matter' (PM10). Quite frankly, I had never before heard of Particulate Matter or the Dutch equivalent 'fijnstof'.

I was genuinely surprised, because I had just moved back from Istanbul, a city with over 12 million inhabitants and about 2.5 million cars in an area that is approximately ten times smaller than the Netherlands. It features heavily congested streets, coal-fired steamboats, and little in the way of air quality regulation. Nonetheless, air quality was more hotly debated in Amsterdam cafés than in the coffee houses of Istanbul. Apparently, in my four years of absence the Dutch had developed concerns about air pollution. How could this be?

As a socio-legal scholar my interest was piqued further when I learned that there were many court cases in which the issue of air pollution featured. Judges were apparently annulling or postponing infrastructure projects on account of noncompliance with the air quality standards for Particulate Matter, referred to as PM10 and nitrogen dioxide (NO_2). I wondered about the emergence of these rules. As far as I could recall, air pollution had not been considered a threat to health after perhaps the late 1980s.

At that time, I had no inkling that I would spend the next eight years researching the emergence of the air quality clash in the Netherlands. The opportunity to research this phenomenon meant that I had ample time to ask questions, investigate scientific reports, delve into court cases, and examine policy documents. I became especially interested in the social, political, and legal background of this problem, and the main question was still the one that had me puzzled originally: namely, how and why did air quality suddenly become an important social problem in the Netherlands?

In this book, the process by which bad air quality became a topic for concern will be reconstructed. This first chapter serves as an introduction to the problem, and presents the research questions and methodology. In the following section, a short overview of the air quality clash is given in order to acquaint the reader with the issue. Subsequently, the research questions, the research methods, the social constructivist perspectives and the use of ideal types are elucidated. The chapter ends with a short overview of the work.

1.1 The air quality clash,[1] delineation of the period under concern

With the term 'air quality clash', I denote a period between September 2004 and April 2010. The clash started when the highest Dutch administrative Court, the Council of State Administrative Jurisdiction Division, annulled the decision to expand the highway between the Dutch villages Vught and Ekkerswijer on account of non-compliance with the standards for air quality on the 15th of September 2004. This was the second road expansion to be halted by the administrative court and it became clear that the Government's ambitious road expansion programme was in jeopardy. At that time the environmental movement realised it had 'dynamite' in their hands (Interview Joris Wijnhoven)

During the period of the clash, air quality featured strongly on the Dutch political agenda. Air pollution due to PM10 and to a lesser extent NO_2 received a distinctive increase in media attention compared to the previous years. In addition, the subject featured more strongly in Parliamentary Documents and in the verdicts by the highest Dutch Administrative Court, as can be concluded from the table below.[2]

The period was marked by heightened anxiety regarding the health consequences of air pollution, most notably PM10, but also by concerns over the possibilities of developing infrastructural projects in the Netherlands. After the Vught Ekkerswijer decision, the highest Dutch administrative court annulled a large number of administrative decisions to realise infrastructural improvements.[3] Newspapers ran

1. The term 'air quality clash' is based on the report 'the Jellyfish clash' (Van Raak, Van der Brugge, Van der Krogt & Te Riele 2006).
2. See text box below. Figures presented in the text box come from various databases. Keyword is always 'air quality'. 1. Website NRC, search term = luchtkwaliteit, 2. website rechtspraak: search term = luchtkwaliteit and instantie = Raad van State, 3. Website Dutch policy documents, search term = luchtkwaliteit.
3. Between 2005 and the end of 2007 alone air quality was invoked as an argument in roughly 400 cases against the administration. Of these cases, the appeal was granted in roughly one fourth to one fifth of them (Janse 2008). Many cases however have started already in 2004 and these were not taken into account in the investigation.

headlines that indicated that infrastructure development in the Netherlands was 'blocked' due to the air quality regulations.

On November 1st 2007, a new air quality law (Stb. 2007, 414) was enacted, aimed at solving the situation by way of a new legislative instrument, the so-called programmatic approach. Afterwards air quality was still frequently discussed in the Dutch Parliament, but consensus was reached on a way forward and concerns shifted from air quality in general to specific measures included in the programmatic approach. The intensity of the clash declined, but two more milestones had to be reached before the clash was finally laid to rest. On the 7th of April 2009 the European Commission extended the deadline for compliance with the EUs' Air Quality Directives. This postponement granted the Dutch Government time to implement its novel approach. Finally, on the 31st of March 2010 the Council of State accepted the programmatic approach as a justification for projects with bad consequences for air quality. This acceptance took away the remaining fears among project developers and politicians that infrastructure development could be halted by legal procedures because of impact on air quality.

Box 1: Air quality in the media, in court cases, and in Parliamentary documents
In the period between 2004 and 2008, air quality featured significantly more often in court cases, newspaper reports, and Parliamentary documents.

A quick scan reveals that the keyword 'air quality' appeared almost 15 times more often in the newspaper NRC between 2004 and 2010 than in the period 1994-2004. In cases before the highest administrative court, air quality was mentioned almost 38 times more often during this period than during 1994-2004 In Parliamentary discussions, mentions of air quality tripled. Between 2004 and 2010, air quality was mentioned 3192 times, while between 1995 and 2004 it received 471 mentions. In a table, the figures look like this:

Source	Period	Mentionings
NRC (newspaper)	01-01-1994 – 31-12 2003	48
NRC (Newspaper)	01-01-2004 – 31-12-2010	689
Council of State (Administrative Court)	01-01-1994 – 31-12-2003	31
Council of State (Administrative Court)	01-01-2004 – 31-12-2010	1191
Parliamentary documents	01-01-1995 – 31-12-2003	471
Parliamentary documents	01-01-2004 – 31-12-2010	3192

(Staatsblad, Staatscourant, Tractatenblad, all Parliamentary documents)

1.2 Research question and sub questions

The sudden emergence of air quality as a social problem in the Netherlands is puzzling because air quality has improved in the Netherlands, at least over the last 50 years. On 5 December 1962, for instance, pollution concentrations in the Rotterdam Rijnmond area reached peaks of around 1500 mug/m^3 for Sulphur Dioxide and around 500 mug/m^3[4] for smoke, an earlier indicator for particulate air pollution (Buijsman 2003, p. 38). Currently, Sulphur Dioxide has almost disappeared from the air in Dutch cities. Likewise, concentrations of black smoke have fallen dramatically, and are now more or less one fifth of what they were in the 1960s (Buijsman 2009, p. 13).

Air pollution by particulate matter (PM10) is now regarded as the main health threat, but figures show that this threat is also decreasing in the Netherlands. In the 10 years preceding the air quality clash, PM10 concentrations declined (Buijsman 2007, p. 53; Hoogerbrugge et al. 2010, p. 13). In regard to particulate pollution, the air quality now in polluted Shanghai is roughly comparable to that of Rotterdam in 1970.[5] It is clear that the air in the Netherlands has improved spectacularly over the last 50 years (Buijsman 2009, p. 15).

This situation leads to the question how air pollution became a social problem from 2004 onward. It presents a conundrum for several reasons. First, after being hotly debated in the 1970s air pollution could hardly be registered on the attention scale of public and political debate since the late 1980s. Second, looked at from the comparative perspective I used in the introduction it seems unintelligible that air pollution is considered a problem in the Netherlands and not in Istanbul. Third, all indicators about air pollution show substantial improvements since the 1970s, and if we look back to the first decades of the twentieth century the improvements are even more impressive. How can we understand the rise to frantic prominence of the air quality condition in the Netherlands as a major and disruptive problem at the beginning of the twenty-first century?

As a socio-legal scholar I have used insights from both the legal and the social sciences disciplines. The sociology of social problems knows two traditions, one employing an objectivist and one employing a subjectivist perspective on social reality (Best 2008, Loseke 2015). In the objectivist tradition some sort of objective standard of harm is the reference point in the definition of social problems (Best 2004, p 14). With regard to air pollution we shall see that medical experts did present serious adverse health effects as grounds for demanding more stringent air quality regulation. However, this approach does not seem very useful considering the above mentioned reasons. In fact, these reasons lead to the rather obvious conclusion that the objective facts cannot be used as an explanation for the sudden eruption of concern. We have to look for an

4. Mug/m^3 refers to micrograms per cubic meter of air.
5. This assertion is based on a comparison between the table for particle pollution presented in Buijsman (2009, p. 14) and a ranking list compiled by Greenpeace (2012, p. 4).

approach that allows us to understand the seemingly contradictory development of growing concern and objective improvement.

The second, subjectivist, tradition in the sociology of social problems offers such an approach. This tradition is known as social constructivism. It must be noted that although social constructivism originated from dissatisfaction with the objectivist perspective (Jamrozik & Nocella 1998, p. 30), it has firmly established itself and many social constructivist ideas became widely accepted in the 1990s (Hajer 1995, p. 43). In its strictest form it holds that the way people agree to define situations determines whether a social problem exists or not, irrespective of any given objective set of facts. If for instance a 'star wars program' would be installed to fend of the danger of 'alien abduction' this strict constructivist approach would allow us to understand how and why this program came about, the improbability of alien invasion notwithstanding.

In less strict versions of constructivism the 'objective facts' are awarded some importance in the analysis while the focus would still be very much on the ways relevant social actors debated and defined those facts. This less strict version is known as contextual social constructivism (Loseke 2015, p. 8). Given the conundrum presented above, in which the objective state of our air quality is awarded some importance, I employ a contextual version of social constructivism to answer my research question:

How and why did air quality become an urgent social problem in the Netherlands in 2004-2010?

Scientific data on the 'facts' that medical experts present are considered in this study, but predominantly as claims that are being put forward in order to promote their goal of more stringent air quality regulation. In my research their status as 'objective truth' is to a large extent unimportant. However, related to the third sub question regarding the legality of precaution, the issue of 'scientific uncertainty' is crucial. This led me to analyse the degree of 'objective truth' of the negative health claims medical experts made concerning the impact of particulate matter to some extent. In chapter 2 I present a social construction of particulate matter expert health claims by describing them as qualified and contested by the participants in the relevant medical debate. By doing this I am able to offer some insight in the level of 'scientific uncertainty' that is present in this debate. However, contrary to the objectivist tradition, in my research scientific claims do not hold a privileged position when defining social problems.

The preliminary investigation of the literature undertaken in preparation for this research yielded three main aspects that could provide an explanation for the emergence of the air quality clash: the Europeanisation of Dutch environmental policy, the ecological modernistic turn in Dutch environmental policies made in the 1980s and a turn in the legal order to demand the prevention, even against very high costs, of environmental and health threats. These three aspects each form the basis of three sub questions.

Sub questions

This work is situated within the discipline of sociology of law. I am especially interested in how the regulatory context has influenced the emergence and eventual resolution of the air quality clash and vice versa. This concern comes to the fore in three sub questions, one question concerning the interplay between national and European regulation, a second question concerning the role of the then reigning policy discourse and a third question concerning transformations within the Dutch legal order.

Sub question 1: Europeanisation

It is self-evident that European integration leaves its mark on the policies of Member States. The increasing influence of European environmental policy on national policy is a well-studied phenomenon (Héritier, Knill & Mingers 1996, Börzel 2002, Jordan et al. 2003, Knill & Liefferink 2007). This process is known as Europeanisation. Europeanisation plays a role in the air quality clash too because the Dutch air quality standards cited by the Dutch courts as reasons for annulment of infrastructural projected were based upon European air quality regulation. Therefore it is necessary to consider the social construction of European air quality policy as well as the Dutch implementation of these policies.

This is all the more pressing because, 'Brussels' was frequently blamed by Dutch politicians for causing the block on infrastructure by promulgating too stringent and badly worded air quality standards. However, the Dutch Government is represented in the prime policy making institution of the European Union, the Council of Ministers. The Dutch Government, especially the environmental Minister, is a European policy maker. Therefore the question arises how the representatives of the Dutch Government acted in the European arena at the time the Air Quality Directives were being conceived. Moreover, a solution to the clash was also sought in Brussels, because the Dutch Government tried to persuade the European commission that the Netherlands needed more time to meet the standards. Its wish was finally granted based on a new Air Quality Directive promulgated in 2008.

Europeanisation has hitherto been a rather neglected field in the tradition of socio-legal scholarship, but the afore mentioned developments show that the interplay between European and national policy is of importance in understanding the air quality clash as well as the role of the Dutch Government within the European policy making process. These considerations lead to the first sub question:

How and to what extent has the interplay between Dutch and European regulation and policy contributed to the emergence of the Dutch air quality clash in 2004?

Sub question 2: ecological modernisation discourse

Secondly, Dutch national political choices and the peculiar characteristics of its environmental law have had an impact on the emergence of the air quality clash (SER 2006; Koeman 2010). In addition to documenting and analysing these choices, the study will focus on the political and discursive context within which these choices have been made in order to understand them.

I follow Hajer and various other authors in their opinion that the dominant way of conceptualising and institutionalising environmental problems at the beginning of the new millennium was a policy discourse referred to as 'ecological modernisation' (Weale 1992; Hajer 1995; Flynn & Bayliss 1996; Mol & Sonnenfeld 2000; Cohen 2000; Fisher & Freudenburg 2001). Ecological modernisation can be characterised as an optimistic view of the possibilities of solving environmental problems by innovation, market mechanisms, shared responsibility, and sound management. Policy choices that influenced the air quality clash are analysed in light of this environmental policy discourse.

The second sub question is the following:

How and to what extent did national political choices, in light of the then reigning policy discourse of ecological modernisation, influence the emergence and resolution of the air quality clash?

In order to investigate this question I will compare the events that took place in the context of the air quality clash to an ideal typical construction of two policy discourses, namely ecological modernisation and a policy discourse I termed 'limits to growth', after the report for the Club of Rome. To be able to increase the precision of my analysis, the policy discourse of ecological modernisation is divided in a strong and weak variation. The use of ideal types as an analytical tool goes back to Max Weber and is used among sociologists of law to analyse concrete phenomena against the backdrop of an idealised representation of a certain social condition. It is impossible to grasp the whole of social reality within the confines of an ideal type, but the abstraction from social reality that is obtained by its construction makes it possible to point out different features of a certain phenomenon the researcher considers important. Here the ideal type is used to situate the debates in the Netherlands and Europe concerning air quality within the environmental policy discourse over time and to investigate which similarities and discrepancies come to the fore.

Sub question 3: the legality of precaution

Various authors such as Frank Furedi (1997), Francois Ewald (1999), Adam Burgess, (2004), Roel Pieterman (2008) and Jaap Hanekamp (2015) describe a social transition towards a culture of precaution. In this culture, a 'paradigm of safety' (Ewald 1999, 2002) holds sway that demands a reduction or even the disappearance of all forms of technological and environmental risk. Environmental harm caused by human action should be avoided in order to achieve the ideal of a safe and sustainable society (Hanekamp 2015). Among a number of Dutch socio-legal scholars, the idea that a legal transition took place towards an approach based on precaution gained ground as well (Pieterman 2008, De Vries & Francot-Timmermans, 2013). I investigate whether the emergence of the air quality clash can be attributed to this legal transition.

For this purpose I operationalise this perceived legal transition by proposing a new type of legality, the legality of precaution. The term 'legality' is taken from Dutch sociologists of law André Hoekema and Niels van Manen (2000). It is defined as a coherent set of values, convictions, and attitudes that is dominant in the legal order at a certain time, and is constitutive of social behaviour. Hoekema and Van Manen consider that social arrangements are reflected in the legal order, and that the legal order is dominated throughout history by certain types of legality. Even though throughout history different types of legality dominate the characterisation of the legal order, this does not imply that when a certain type takes over, the earlier types completely disappear. Their characteristics live on within the legal order, but are not as prominently present anymore.

I consider the legality of precaution as a legal discourse in which the occurrence of damage is strongly morally condemned, prevention of damage is considered the prime imperative of the legal order and the damaged person is perceived as a victim of negligence by the relevant authorities. In order to characterise the legality of precaution I again take recourse to the construction of a Weberian ideal type. In this ideal type the characteristics of the legality of precaution that I consider relevant are listed. By comparing the emergence of the air quality clash with the ideal typical legality of precaution I answer the last sub question:

How and to what extent is the emergence of the air quality clash an indication that a shift towards a legality of precaution took place within the Dutch legal order?

In order to make a meaningful comparison possible the legality of precaution is contrasted with another ideal typical legality, the legality of risk and compensation. Whereas in the legality of precaution damage prevention was the prime imperative, in a legality of risk and compensation damage compensation is considered a paramount duty. Moreover the damaged person is not primarily seen as a victim of negligence but as a rights bearing member of a risk collective, established through private and social insurance. Damage is not viewed as a disgrace, but as an inevitable, though undesirable side effect of our modern way of life. The complete ideal types may be found in table 2 below.

1.3 Research methods and selection of research material

This research is essentially historical in character. Even though the period of most concern runs from September 2004 to April 2010, the roots if the air quality clash go much further back in time, to the early 1980s. Moreover some features of the regulatory context date from as far back as 1972. Therefore in this research almost 40 years of environmental policy and law are under scrutiny. The lion's share of relevant data has been obtained through desk research. Scientific literature has been studied regarding various disciplines and subjects such as law, sociology of law, policy analysis, epidemiology and particulate matter. Relevant court cases brought before the Dutch Council of State were reviewed as well. Moreover publically available policy documents were studied, both Dutch and European in origin.

Within the scope of this research, policy documents are defined as all documentation from Parliamentary sources in the Netherlands, such as acts of Parliament, minutes of Parliamentary and Parliamentary committee meetings, legislation, and policy plans and programmes. Such policy plans are referred to in the Netherlands as 'memorandums'. European policy documents reviewed include white papers, green papers, policy proposals from the Commission, reports from debates in the European Parliament, and where available, minutes from meetings of the committees and sub-committees that are involved in the preparation of relevant Community proposals and legislation.

It is impossible to scrutinise all relevant policy documents on 40 years of environmental policy making, therefore a selection had to be made. I zoomed in on the first three years of the air quality clash in the Netherlands, from September 2004 until November 2007, when the Air Quality Law was enacted. All Dutch policy documents containing the keyword 'luchtkwaliteit' were obtained from the website of the Dutch Government and reviewed.[6] By using the same key word policy documents from the period of 1999 until September 2004 and December 2008 until April 2010 were also reviewed, but only when they concerned key moments in the air quality clash, such as the air quality scare in Overschie in 1999 and the derogation obtained on the basis of the programmatic approach in April 2009. The study concerns an investigation of the origins of the general problem of air quality in the Netherlands, independent of its particular manifestation in specific cases. Therefore, I narrowed this set down by excluding documents that related to individual projects, such as the expansion of military airports,

6. Website Dutch policy documents, last accessed 27-09 2015.

Schiphol Airport, or the Rotterdam harbour, and anything related to aviation in general. The discussions relating to individual road expansions were also skipped, although the consequences for road expansions in general have been included. Also the scientific and legal controversies over the effectiveness and obligatory nature of soot filters have been considered too specific.

During the initial literature review the large influence of European rules on Dutch air quality regulation became clear. The Dutch air quality regulation of the time was based on two European Directives on urban air quality, 'Framework' Directive 96/62/EC and 'Daughter' Directive 99/30/EC.[7] Therefore European policy documents related to these two directives have been examined. They were obtained through the prelex system in which the political process of community proposals is detailed, and which contains links to the applicable documents. The prelex pages of these two Directives provided me with a starting point for searching documentation. If available, the documents listed on these two pages were reviewed. A broader perspective on the European environmental policy making process was obtained through review of articles of the UK based monthly journal ENDSreport and ENDSeurope.

European regulation contributed to the emergence of the air quality clash, but it also contributed to a solution. On the basis of a new Air Quality Directive promulgated in 2008 in the context of the CAFE (clean air for Europe) programme, the Netherlands obtained a postponement of the deadline to meet the air quality standards. The CAFE strategy and the resulting directive is the subject of a separate chapter in the book, and data for this chapter was gathered from the extensive EU website on CAFE (Website CAFE). The website contains the minutes of committees and subcommittees and provided a comprehensive resource of documentation on the period between 2003 and 2008. Documents relating to particulate matter and to the implementation of the air quality regulation by Member States have all been reviewed.

Since the EU Air Quality Directives were influenced by the UK clean air strategy promulgated in 1997, attention has been paid to the emergence of this strategy as well. I used UK Parliamentary documents and the debates of the House of Commons and Lords listed in Hansard to reconstruct its history. The data set consisted of the reports of the debates in the House of Commons along with the written questions posed and answered between 1988 and January the first 1996, in which the search term 'air pollution' appeared. Debates between 1996 and 1999 in which the search terms 'EU' and 'air pollution' appear together were also reviewed.

7. Directive 96/62/EC, Council Directive on Air Quality Assessment and Management (Framework Directive) and Directive 99/30/EC, Council Directive relating to limit values sulphur dioxide, oxides of nitrogen, particulate matter and lead in ambient air (Daughter directive).

In addition to desk research, open interviews have been conducted with scientists, civil servants, policy makers, and experts from the research institute RIVM, the Dutch Ministry of Housing, Spatial Planning and Environment (VROM), and the European Commission, along with a number of experts from the UK. These interviews functioned mostly as a reality check, but they were sometimes used to fill in the gaps where the data obtained contained hiatuses. I interviewed respondents who were in key positions when Dutch and European environmental policy was being developed and they are privy to exclusive knowledge. Therefore these interviews have at times been explorative in character as well, offering new avenues for research. Incidentally a certain interview proved to be one of only a few sources available. The interview with Kees Zoeteman for instance is one of only a small number of sources available regarding the way the World Health Organisation became involved in air quality policy in the early 1980s. Most interviews were conducted face to face, but the interviews with the UK respondents have been conducted by telephone. A list of participants is included as an annex at the end of this book.

As a sociologist of law I am interested in the history of the air quality clash, because it sheds light on regulatory and legal transformations in coping with environmental conflicts. This consideration informed my choice for policy documents and court cases as research material. These documents provide insight into the arguments used in debates between political actors representing various interests. New perspectives on coping with environmental conflict and regulatory proposals are laid down in memoranda and policy plans. For me as a socio legal researcher they are obvious places to look for answers to my questions. In my research I did not primarily focus on the decision making procedures within the bowels of bureaucracy, although civil servants are important actors when it comes to policy making. To answer my questions, regulatory changes in themselves are of interest and not primarily the question who exactly drafted a certain memorandum.

For the same reason the role of the media has not been paid the same attention as in other social constructivist studies. The role of the media as a vehicle for claims making is important to understand the social construction of the air quality clash, but my study did not primarily concern the role of public opinion. In addition the selection of material in which policy documents most strongly feature was not conducive to an in depth media analysis. The role of the media should not be neglected altogether though and therefore I have studied newspaper articles and television shows when they were mentioned in policy documents. Media attention for key moments in the air quality clash was investigated, such as the episode on political news show Den Haag Vandaag recounting the important debates in April 2005 and the visit by Minister Jan Pronk to air pollution stricken Overschie in 1999.

The study of policy documents has inherent limitations as well. The involvement of policy makers generally occurs later in the construction process. A focus on policy documents tends to gravitate towards the decisive stages in the problem construction process, and leaves out the activity of important actors mentioned above. Moreover, a subject in Parliament is discussed within a specifically political setting. In this setting, negotiations and conflicts are going on that outsiders may not be aware of when reviewing the documents. Politicians will discuss problems within the context of their own political agenda, and instead of an ideological description they may well vie for personal gain and glory. Personal scores may be settled and deals concluded in the hallways surrounding the Parliamentary arena, which are not laid down in the official documents.

The first step to address these biases when conducting this kind analysis is to be aware of them. A critical and distanced attitude towards the material is imperative. A second step involved analysing a significant amount of data from other sources, such as literature, grey literature and interviews in order to provide a counter weight to the political sources. I am confident that I have amassed a sufficient amount of data beside the policy documents that provide a counter weight to possible distortions. My choice of focus on the first years of the air quality clash may mean I missed a certain development occurring early or late. However, in this case too I am confident that this development will not have been major, because I would have learned it through other sources than the policy documents. Developments after March 2010 though have not been considered in this study.

1.4 A social constructivist perspective

This study fits within the social constructivist tradition of socio-legal research, and this perspective has a well-established track record in the study of social problems (Hajer 1995; Peper 1998; Best 2008; O'Byrne 2011). The social constructivist method was pioneered in the sociology of knowledge in the 1960s by Peter Berger and Thomas Luckmann (1966). Malcolm Spector and John Kitsuse (1977) applied this perspective to social problems. Traditionally the social constructivist is concerned with the activity of so called claims makers. Claims makers define a certain situation as problematic and when they are successful in convincing others of their view, the situation will be considered as such irrespective of objective facts, as also explained in section 1.2. The traditional version of social constructivism focusses on actors and the way they use their claims to further their own interests. Therefore I refer to it as actor-centred social constructivism.

In the 1990s a new social constructivist method gained ground, a perspective inspired by among others the work of French philosopher Michel Foucault. The discourse analytic method of social constructivism focusses on the type of arguments with which claims are brought forward. Discourse analyticians are not primarily concerned with actors making claims but with the question what arguments are successful in debates about social problems and what this success tells us about the beliefs and values embraced by the actors involved and the public at large (Weale 1992, pp. 58-60). In the field of environmental problems the works of Hajer (1995) and Dryzek (2005) are examples of this tradition.

I am interested in the actors who make claims and their problem definitions, because I am interested in the question how air quality became a pressing social issue in the Netherlands. Contrary to indicators that air quality is improving, it became a problem and I intend to find out which actors played a part in this shift. However I am also interested in the arguments employed within the Dutch and European debates, because they can tell me something about the changing political and legal context in which they take place. I am a sociologist of law and not a historian, my interest concerns the air quality clash as a possible result of, and catalyst for, legal transformation. Therefore I have chosen to combine concepts from both these social constructivist methods in my own research. The concepts I adopt from actor centred social constructivism are mostly taken from the work of Joel Best (2008). The discourse analytic concepts I use are mostly coined by Dutch scholar Maarten Hajer (1995).

In the following two sections I present my adaption of the terminology of these two authors. I make an eclectic use of their terminology and adapt them according to serve this study. Moreover I supplement them with a few concepts of my own or from other authors.

1.4.1 *Concepts taken from actor centred social constructivism*

Below I describe a number of concepts taken from actor centred social constructivism which will be used throughout the study. These concepts are the claims maker, the counter-claims maker, policy maker, interest and arena.

Claims makers and counter-claims maker

Claims makers are actors that define a certain situation as problematic, and they usually also provide a cause for this problem and a favoured solution. They act as ambassadors of the problem, and try to bring their definition to the attention of others. In order to make their claims heard claims makers often seek media attention. When a claims maker gains the attention of the media, he can be sure his definition is widely disseminated. However, not all claims makers target the media. 'Insider' claims makers such as scientists or captains of industry may try to influence policy directly, without taking recourse to the media (Best 2008, p. 66). Another subject of interest is the counter-claim lodged by a so called counter-claims maker. A claim that a certain situation is troubling is often opposed by groups that benefit from the status quo, or by a group with diametrically opposite views. For instance, the claim that abortion should be legal, inspired pro-life organisations to claim the exact opposite. Organisations are often considered the main candidates for the position of claims maker. An organisation has more resources than an individual so tends to be more successful at claims making than the individual. This is not necessarily so however, powerful individuals may also be claims makers. Especially the insider claims maker may well be an individual with special access to policy making circles.

Policy makers

Policy makers are a second category of actors of importance in this study. Policy makers regulate matters to address the situation considered as problematic by claims makers, and their role determines the eventual success of the claims. The Dutch Government and the European Commission are prime examples of policy makers taken into account in this study. Dutch political parties with seats in Parliament are policy makers as well, but their influence is mostly indirect. Most policy is proposed and drafted by the Dutch Cabinet of Ministers and is debated and amended in Parliament. In this study I consider political parties to be claims makers in the Parliamentary arena rather than policy makers. Civil servants within the Government bureaucracy should be considered as claims makers as well, but their activities are seldom in the spotlight. In this study too, their possibly substantial influence, is by and large left out because of reasons mentioned in the previous section.

Since this is a socio-legal study I do focus on the policy making activities of the Dutch Council of State, the Governments highest advisory body and the highest administrative court. One of the outcomes of this study is that the Council of State should be seen as a claims maker, albeit in a negative sense. By negatively commenting on policy proposals issued by the Government in 2005 it forced the Government to introduce a novel regulatory instrument to resolve the air quality clash. The Government feared that not heeding negative advice on policy proposals could lead to more annulments by the Council of State Administrative Jurisdiction Division.

Interests

Actors will use their position as claims maker or policy maker to further their own interests. I define an interest as a certain goal or benefit for which an actor is striving. In this study, I distinguish between ecological, economic, and political interests. Interests that focus on immaterial goods that increase wellbeing such as protecting public health, a clean environment, and nature conservation have been termed ecological interests in the context of the study. Interests such as infrastructure development, increased mobility, and technological innovation are considered economic interests. Political interests include gaining the favour of voters, gaining seats in Parliament, obtaining a favourable position for further negotiations, and so on. It is evident that these three clusters of interests are far from exhaustive, given the many different interests actors may have. In the context of this study, however, ecological, economic, and political interests play the most important part.

Arenas

Claims-making and policy making activities take place in a number of different venues. To distinguish the different venues in which the construction process of the air quality clash has taken place I use the term 'arena'.

Sociologists Hilgartner & Bosk (1988) use the term to emphasise that the features of certain venues in which claims are put forward determine their success. They apply the term to social problem research and describe how claims making activity occurs within certain institutions in which these claims have to compete with other problem definitions. The media is an arena of problem construction for instance, because various claims compete with each other for media attention. The courts of law are an important arena as well, because in court a certain claim may be considered and embraced by the judiciary. In the same vein national Parliaments and European policy making institutions are arenas in which claims may be accepted or rejected. During the air quality clash, some struggles took place in the Netherlands, some in Europe. Important struggles took place in the courthouse as well. In this dissertation, therefore, the arenas of note are the Dutch Parliament, the Dutch Courts, the European Committees and European political institutions like the European Commission, Council of Ministers and Parliament. In chapter 4, the UK Parliament features briefly. Attention is also paid to the emergence of the problem in the scientific arena especially in circles of epidemiologists and in Dutch research institutes. The media arena received relatively less attention, for reasons discussed above.

1.4.2 *Concepts taken from the discourse analytic approach*

For this study I find discourse analysis to be an attractive supplement to the actor centred approach because it builds a bridge between micro level interaction and macro level cultural change. According to discourse analysis, an actor does not choose his arguments freely, but is constrained in his argumentative action by the discourse he is drawing upon. This discourse is tinged with the beliefs and values prevalent in society. In fact, according to the discourse analytician they do not exist independently outside of discourse, but are sublimated within it. However, a discourse needs to be reproduced in concrete situations. Argumentative interaction is a key moment in discourse formation. Actors are actively engaged in the production and reproduction of discourse and may at times introduce a novel argument within an existing discourse. Actors are constantly engaged in discursive interaction and therefore also in discursive change. To paraphrase Karl Marx, the debates in which actors participate are both the product and the producer of discourse. By treating the history of the air quality clash as a series of discursive moves, we uncover the cultural context in which the air quality clash was able to play out. Discourse analysis provides me with a different set of concepts with which to analyse texts. I will introduce these concepts below.

Discourse

As the name indicates, discourse analysis revolves around the analysis of discourse. A discourse is defined as 'a specific ensemble of ideas, concepts and categorizations produced, reproduced and transformed into a certain set of practices through which meaning is given to social realities' (Hajer 1995, p. 44). I will use the same definition of discourse in this study. I favour this definition over other more language oriented definitions because it allows us to incorporate specific practices as belonging to certain discourses. I also feel this definition remains truer to Foucault's original use of the term.

Discourses are modes of speech and practices that belong to a whole field of activities. For instance, physics has its own discourse; radical environmentalism has its own discourse, and so does law.

The legal discourse for instance is characterised by a number of practices; for example, the ritual-like proceedings in court cases, or the Latin phrases used in discussing a legal case. The legal discourse is also characterised by the prevalence of the law as a central text, and by certain characteristic arguments that distinguish a legal procedure from any other, such as the reference to jurisprudence, to earlier court decisions, to the legal doctrine, along with the appeal to impartiality of judges and the independence of the judiciary from the administration.

Discourses cover different fields, and diverse discourses hold sway within different scientific disciplines and social groups. Just as law is guided by such routinized manners of speech and practice, so is medicine, environmental science, sports, left-wing activism, and so on. Discourses delineate the argumentative space within a certain field, and determine what kinds of arguments are and are not appropriate in discussions within it.

Storylines

A storyline is a narrative that gives meaning to specific social and natural phenomena by supplying a more or less coherent explanation for them. Actors use them in debates to characterise certain phenomena, provide a reason for their emergence and if applicable a solution as to how to get rid of them. They greatly reduce the complexities of phenomena under discussion, and make the exchange of viewpoints more convenient. All discourses contain storylines. The discourse on international safety and security for instance frequently contains the storyline of the 'war on terror'. Environmental discourse features many storylines, such as the fear of acid rain caused by industrial pollution or climate change caused by CO_2. During the air quality clash many storylines were iterated such as the storyline that ambient air pollution is mainly caused by transport and traffic, but also the storyline that the Netherlands became blocked to infrastructure development due to the stringent air quality standards and the important storyline that clean air is essentially a human right because these standards are laid down in law. The content of these storylines may well be disputed, but truth is not a quality applicable to storylines. Storylines may hold grains of truth or not, the point is that they are told, reiterated and used in debates to get the upper hand. If they consistently do so, they will take on an almost absolute quality, and people will treat them as the natural explanation for certain social and physical phenomena. In relation to social problems I consider storylines to contain the definition of a certain situation as problematic, the cause of it and the solution for it. When problems are constructed, different storylines compete with one another in order to determine the dominant conceptualisation of a certain problem.

Discourses evolve because manners of speech and practices evolve. Storylines are intermediate concepts through which a discourse might be modified by discoursing actors. A discourse may be defined as an interrelated web of storylines, some almost forgotten, some alive and prospering and some so dominant that they are not even noticed as being storylines. Changes in the storylines change the discourse itself.

Discourse coalitions

When different actors who normally discuss issues from diverse perspectives, start to employ the same storyline or set of storylines, one speaks of a discourse coalition. A discourse coalition refers to a group of actors that share the usage of a storyline or a particular set of storylines over a particular period of time (Website Maarten Hajer). Discourse coalitions are formed when previously unrelated actors become discursively related to one another (Hajer 1993, p. 45). Discourse coalitions can have considerable influence on actors to accept or reject storylines. When actors that come from different backgrounds become discursively related to one another, weight is added to a certain storyline. It is not only confined to certain actors, but becomes a shared way of thinking and speaking.

Discourse coalitions strengthen storylines, and frequently add new elements to it. A storyline is vigorous when it becomes more comprehensive, and when it manages to cement larger discourse coalitions. Likewise a storyline may become stale when no new elements are added to it, and when discourse coalitions fall apart that use it. When certain storylines about a given situation become dominant, they define the state of affairs in question among the actors participating in the debate.

Emblems

The last Hajerian notion I wish to illuminate is that of the emblem. Hajer discusses the emblem in the context of environmental problems, and defines an emblem as a case that dominates the perception of the whole environmental crisis in some period. In that sense, acid rain was the emblematic case in the 1980s, and air quality was the emblematic case in the Netherlands in 2005. I wish to propose a slightly different definition of emblem in the context of this study in order to relate it directly to the concept of storyline, Hajer himself does not do so. An emblem is a unique example that has particular force within a certain storyline. Emblems are the main discursive elements around which a storyline evolves. At some point in the air quality clash, for instance, the notion that living close to the highway was equal to the passive smoking of 17 cigarettes became current. The '17 cigarettes' became an emblem, because by mentioning it, one immediately invoked the whole storyline of bad air quality as a health problem caused by transport. The first area in the Netherlands in which problems became clear was the residential neighbourhood of Overschie. 'Overschie' is an emblem as well, and it too invokes the storyline of bad air quality as a health problem. In the same vein, the Dutch Delta Works is an emblematic example of Dutch ingenuity, invoking the storyline of the success of Dutch spatial planning.

Camps

One term that can neither be found in Joel Best, nor in Maarten Hajer's work, but used extensively in the later chapters of this study is the term 'camp'. In this study a camp indicates a set of actors bringing forth the same claims, using the same discursive means. The term connects the idea of claims makers with the concept of the discourse coalition, because the actors that comprise a camp use the same discursive means to bring forth their claims. Although the interests of the actors involved may be different, the crucial aspect of a camp is that the same claims are made. It differs from the term claims maker in that a camp always consists of more than one distinct organisation or individual claims maker. The terms discourse coalition and camp do overlap to a significant extent, in the sense that in the way the idea is used here, all camps are considered discourse coalitions. However, not all discourse coalitions make claims and therefore not all discourse coalitions are camps. If a certain issue has been settled, actors may still conceptualise a situation in a similar way and maintain their discourse coalition, even though they may not face opposition. A camp though is always pitted within one or more arenas against other claims makers or counter-claims makers.

1.5 Ideal types of ecological modernisation and the legality of precaution

The social constructivist research methodology is augmented by the use of ideal types. They are used to analyse the claims making and policy making activities and the storylines used against the backdrop of regulatory and legal change. Constructing ideal types of the discourses of ecological modernisation and the legality of precaution enable me to answer the last two sub questions raised above. An ideal type is formed by assembling elements from the given phenomena, but the ideal type is not intended to correspond to all characteristics of the phenomena in question totally. In fact ideal types are accentuations of phenomena and specific characteristics have been emphasised by the researcher. It is never found as such in reality, but it is a device with which reality may be ordered and categorised. An ideal type is an analytical construct and serves to ascertain similarities as well as deviations between concrete cases and the phenomena thus typified. In our case it serves to indicate discrepancies and correspondences between the events recounted in the context of the air quality clash, ecological modernisation and the legality of precaution. In sociology of law ideal types are a frequently used device to make comparisons possible, the already mentioned work by Hoekema and Van Manen on types of legality in the Weberian tradition being a prime example.

In this study, the social construction of the air quality clash will be related to two ideal types of ecological modernisation, a strong and a weak variation. Moreover the ideal typical constructions of ecological modernisation are contrasted with an ideal type of the earlier policy discourse, 'limits to growth'. In this section, the meaning and the background of these two concepts are explained.

1.5.1 *Ecological modernisation*

The term 'ecological modernisation' is taken from the discipline of policy analysis. I use it to denote the specific environmental policy discourse which according to Hajer (1995) became dominant in the Netherlands during the middle of the 1980s. Ecological modernisation has been conceptualised differently by various authors (Buttel 2000a). Firstly, it is considered an independent sociological school of thought. The works of Arthur Mol and Gert Spaargaren fit in that tradition as well as the classical eco-modernist literature of Jänicke and Huber. Secondly, ecological modernisation is used to describe processes in environmental management of the private sector and industrial ecology. Thirdly, ecological modernisation is considered a discourse, and Hajer is considered the main scholar to employ the concept in an analytic and social constructivist way (Buttel 2000a, p. 58).

The discourse of ecological modernisation enabled political scientists and politicians to deal with the challenge of environmental conflict in practice. The term was first coined in German social scientific debate about the future of environmental policy, and was initially expressed by environmental sociologists Joseph Huber and Martin Jänicke in the early 1980s. These early authors stressed in particular the need to address environmental problems by the use of technology and by *further* modernisation. This was a break away from mainstream environmental critique, which in general was suspicious of further industrialisation (Mol & Jänicke 2010, p. 20). Ecological modernisation has the theme of an increase in modernisation in common with Ulrich Beck's notion of reflexive modernisation (Buttel 2000b), but proponents are generally more optimistic about the role of the market and technology in this process than Beck was. If imaginatively managed, the market and technology may cause the social changes that authors such as Beck feel are needed to deal with the problems of the risk society. One core aspect of ecological modernisation is the conviction that the behavioural change necessary for an ecologically sustainable society may be realised by relying on the logic of the market.

The shift towards ecological modernisation started in the 1980s. It designated a new type of pollution-control politics that differed from those used in the 1970s.

Previously, environmental degradation had been an argument of the counter cultural movements to point out the negative effects of capitalism. In this view, environmental protection necessitated a thorough restructuring of the economy with the emphasis on de-industrialisation or 'counter productivity' (Mol & Spaargaren 2000, p. 19). Ecological modernisation challenged that assumption. In the logic of ecological modernisation, environmental degradation posed a significant challenge to the capitalist economic order, but it could be solved within the structures of capitalism. In fact, the answer to environmental degradation was considered to be found not in less modernisation but in more, albeit of a qualitatively different kind (Young 2000, p. 2). Other characteristic features are the promotion of shared responsibility between state and civil society, including non-state actors such as NGOs and the industry, for environmental protection, a reliance on preventative, long-term integrated policies, and the search for win-win scenarios if clashes erupt between economic and ecological interests.

Even though ecological modernisation is not anymore in the forefront of the theoretical debates concerning the relationship between economy and ecology, the air quality clash gives us the opportunity to investigate to what extent storylines that fit in the discourse of ecological modernisation were still used in 2010 to describe and discuss a problem in which environmental and economic demands are pitted against each other and whether the discourse was still able to reconcile these diverse interests. In order to undertake this task I construe ideal types with which I can typify this environmental policy discourse, based on the works of Winsemius (1986), Hajer (1995), Young (2000), and Dryzek (2005) among others.

1.5.2 *Ideal typical elements of ecological modernisation*

A number of characteristic features of ecological modernisation can be teased out from secondary literature and from the history of environmental policy. Ecological modernisation emerged as a reaction to the policies of the 1970s, which were generally perceived as sluggish and ineffective and presented its own set of solutions. The typical Dutch way of conducting ecological modernistic policy is described in chapter 3, and there the historical background of ecological modernisation is recounted in detail. This account is necessary because the typically Dutch characteristics of environmental policy making are of importance for understanding the air quality clash. I cannot undertake this full historical exploration in this introductory chapter, but I will give an account of the characteristic eco-modernist elements that have informed my ideal type of this discourse. The elements have been extracted from the secondary literature and from examples of Dutch political practice. For comparison, they will be positioned opposite the characteristic elements of the earlier environmental discourse, dominant in the 1970s, known as 'limits to growth.[8]

8. After the highly influential Report for the club of Rome, with the same title: *The Limits to Growth, a Report for the Club of Rome's Project of the Predicament of Mankind* (Meadows et al. 1972).

In what follows I consider characteristic features of eco-modernist discourse, these features pertain to: 1) the relationship between economy and ecology, 2) the type of science that dominates the presentation of environmental problems, 3) consensus and conflict, 4) the stance towards prevention and precaution, 5) responsibilisation, and 6) the stance towards long-term planning and comprehensive policy making. The following summation is presented to the reader in order to give an account for why I have chosen to focus on these specific features to construct my ideal type. The citations and examples of concrete policy are chosen as illustrations to make clear why I adopted that specific element in my typology.

1 Relationship between economy and ecology

Theorists agree that in the perspective of the ecological modernisation theory, ecology and economy are not mutually exclusive interests. Hajer mentions in this regard the 'positive sum game' format of ecological modernisation. Whereas earlier environmental theory presented a choice for either economy or ecology, ecological modernisation offers the 'beguiling vision' of a win-win scenario (Young 2000, p. 2). Adhering to ecological modernisation entails accepting the idea that economic progress and environmental wellbeing may go hand in hand. In its stronger forms, ecological modernisation discourse even assumes that environmental protection may be a very good investment and may benefit the economy. Not only will future costs of pollution clean-up be avoided if one embarks on a policy of preventing pollution in the first place, the technological innovations which pollution protection and prevention require will later turn out to yield a competitive advantage. Albert Weale provides a good example of such an argument when he quotes Laurens Jan Brinkhorst, Dutch politician for D66 and the former Director General of the European DG Environment, and a defender of ecological modernisation:

> *'Secondly, I have become very much concerned – and I think this concern is shared by other departments – that environment and technology, environment and competition have become brothers and sisters. It is not because of low prices that Japanese products are making inroads into all kinds of areas (whether we speak of cars or computers), but it is largely because of the quality of their products and in the field of cars at least, their very high emission standards'* (Weale et al. 2000, p. 78).

The first and arguably most important of ecological modernisation is the idea that ecological interests and economic interests are not mutually exclusive and may even reinforce each other.

2 The type of science that dominates the presentation of environmental problems

Within ecological modernisation, various frames of reference for the environmental problem are different from those used in the older environment-related discourse. An economic and managerial description of the environment gradually displaced an ecocentric, environmental scientific one. Ecological modernisation conceptualises the environmental problem by using managerial and economic language. For instance, in the Dutch 1989 Natural and Environmental Policy Plan (Natuur en Milieubeleidsplan, hereafter NMP) we may find the following section: *'We are presented with unexpected bills from the past for behaviour that may be typified as obtaining loans. Life cannot be lived on the tick, this goes for the budget deficit, as well as for the environment'* (VROM, V&W, Econ, L&V, 1989, p. 70). Hajer (1995) pays attention to this redefinition of the environmental problem, and Blühdorn and Welsh also consider that proponents of ecological modernisation reframe environmental problems in economic and management terms (Blühdorn & Welsh 2007).

3 *Consensus and conflict*

One of the most influential changes in ecological modernisation concerns the role of the state in environmental protection. In ecological modernisation theory, an environmentally ambitious Government intends to work together with market parties and not against them (Mol 1995, pp. 46/47, in Buttel 2000a, p. 61). This new strategy is rational if one considers that in ecological modernisation environmental interests and economic interests do not clash, but depend on each other. However, it is a definite shift from earlier times. Whereas traditionally market parties were seen as obstacles to environmental improvements, in ecological modernisation they are seen as allies. Industry and other important economic sectors should work together with the Government and with the environmental movement in order to solve the environmental predicament. As Stephen Young writes: *'The state appears to play a more limited role than in the 1970s, but it establishes supportive frameworks and promotes consensus based solutions'* (Young, 2000, p. 33). Also Pieter Winsemius, one of the prime Dutch ideologues of ecological modernisation emphasises consensus building: *'Essential for the process of internalisation of environmental values, is building consensus among those who are the primary targets of environmental policy, those whose action needs to be influenced and corrected most'* (Winsemius 1986, p. 62). It is therefore considered that in ecological modernisation, the state should establish the frameworks and conditions for environmental improvement by market-based means, and negotiate with interested parties to form a consensus. It should also facilitate and moderate negotiations between environmental groups and groups representing economic interests.

4 *Stance towards prevention*

Various authors (Von Prittwitz 1990; Hajer 1995) consider that environmental policy changed from a reactive and curative policy to a proactive and preventative one. Ecological modernisation fully subscribes to the need for proactive and preventative policy. It is assumed that a preventative approach is superior because clean-up costs that are incurred later will be higher than the costs of prevention now. The embrace of such a preventative policy is argued for in economic terms, as befits the eco-modernist discourse. Albert Weale calls the insight that *'the failure to address a pollution problem does not save costs, but merely displaces them elsewhere around the economy'*, one *'of the key propositions of ecological modernisation'* (Weale 1992, p. 83). To warrant a proactive preventative policy, the legal notion of the precautionary principle was proposed, which states that scientific uncertainty could not be an excuse not to take preventative measures this was an important indication of the shift in environmental policy and law (Flynn & Bayliss 1996; Hajer 1995; Andersen & Massa 2000).

5 *Responsibilisation*

Ecological modernisation has an affinity with philosophies of civic responsibility, self-regulation, and reducing the scope of the public sector. In this context Van Tatenhove and Leroy (2010) speak of the 'societisation' and 'marketisation' of environmental politics. In line with these commitments, environmental protection should no longer be the responsibility of some actors in the political arena, such as the environmental Ministry, but should become the responsibility of other Ministries, corporations, producers, the industry and – in the last instance – the public at large. Winsemius states this goal as follows: *'This is what I mean with internalisation of environmental responsibility, seeing it as a normal aspect of daily routines to act responsibly towards the environment'* (Winsemius 1986, p. 61). To this end, numerous awareness-raising campaigns were started in the Netherlands. In 2001, the EU explicitly incorporated the mobilisation of citizens into its strategy for sustainable development (EU Commission 2001a, 8). The state wishes to motivate the public in making the 'right' environmental choice. Moreover, the public is urged to participate in the debates on environmental policy and the environmental direction of society. Policies should be *'an opening up of the existing policy making practices and the creation of new participatory practices'* (Hajer 1995, 29).

This new role of the state as a moderator for consensus building meant a departure from the more state-centred days of the 1970s (Van Tatenhove 1993). In those days, the Ministry of Public Health and the Environment (Volksgezondheid en Milieu, henceforth known as VOMIL) saw environmental protection as a task entrusted to it, and fulfilled it. It did reach out to other actors in the field, but was not successful, partly because it presented a gloomy image of environmental protection (Hajer 1995).

6 *Adoption of strategic planning approaches and comprehensive policies*

The complexity of environmental problems reached a point where a strategic planning approach was considered necessary, and was preferred to ad hoc legislative measures. A preventative policy also made such a strategic approach more plausible. Integral policies were considered necessary to tackle complex environmental issues. Ecological modernisation entailed a long-term perspective on environmental problems (Andersen & Massa 2000). Pollution should be targeted in plans that transcend differences between forms of pollution. Piecemeal policies tended to displace pollution. A law against waste burning for instance in order to protect air quality could result in waste being dumped at sea instead, leading to marine pollution. Ideally, these comprehensive plans should transcend different policy domains as well. The most comprehensive policies and plans also aim at changing people's behaviour.

Such large-scale goals of social engineering certainly require long-term and broad strategies. The Dutch NMP is an example of that approach, and is considered an example of ecological modernisation (Weale 1992). According to John Dryzek:

'It (ecological modernisation TA) requires political commitment to the enlightened long term rather than the narrow minded short term and a holistic analysis of economic and environmental processes rather than piecemeal focus on particular environmental abuses' (Dryzek 2005, 168-169).

1.5.3 Ideal types of ecological modernisation discourse and 'limits to growth'

The ideal type of ecological modernisation presented above is not yet sufficiently fine-tuned. It presents a general outline, but says nothing about the level of commitment to more or less ambitious environmental policies. A proponent of minimal environmental policies may use this discourse as well as a highly ambitious environmental policy maker, intent on thoroughly reshaping production and consumption patterns. Therefore, a distinction between strong and weak versions of ecological modernisation is sometimes made in the literature (Spaargaren 2000; Christoff 2000, p. 228). I distinguish between strong and weak ecological modernisation as well, and the distinctions proposed are again ideal typical. In practice, mixes of these types are most likely to occur. Moreover, ecological modernisation must be distinguished from an alternative environmental policy discourse in order to facilitate comparison. Therefore it is contrasted with the discourse that reined over environmental policy in the 1970s, the discourse I named 'limits to growth', after the influential 1971 report of the same name. Even though this discourse reached the peak of its influence in the 1970s, elements of it can still be found in modern day discussions, and also in the air quality clash. Within this discourse, economic and environmental interests are sharply opposed. Economic growth is considered an imminent threat because of the danger of resource depletion, and technological progress is considered a hindrance more than a boon.

Within the scope of my research, the distinctions within eco-modernist discourse are important because they are used to assess the level of ambition of environmental policy. Weak ecological modernisation suits a technocratic endeavour to reduce environmental degradation by developing technological innovations and managerial administration. When actors apply arguments that conform to the weak form of ecological modernisation, they may well be playing down ambitious policies because they consider that technological innovation will solve the crisis in the end, and that robust economic growth is necessary to achieve these technological innovations. The weak variation accommodates the environmental movement into consensus-based structures, and manages to depoliticise environmental conflict without upsetting vested interests or having to change deep-seated structures of society.

Strong ecological modernisation, however, demands a far more ambitious set of policies, and adherence to it may require far-reaching shifts in society's approach to risk and blame. Strong ecological modernisation is characterised by the need for behaviour change among many different sections of society, including the industry, transport, producers and consumers. However, it must be kept in mind that ecological modernisation is a programme for evolutionary change, rather than a revolutionary upset of existing practices.

In contrast, a policy inspired by limits to growth is committed to a stabilisation of the economy and to a rigorous restructuring of society to avoid resource depletion. From this point of view, environmental considerations rule supreme, because our current reliance on industrialization, resource use, food production and the concomitant emissions lead to a situation of either collapse or decline (Meadows et al. 1972).

The difference between the diverse strands of eco-modernisation and the limits-to-growth approach is visible in all the aforementioned features. The positive sum game format may imply solely that economic growth is possible without increasing environmental degradation. In the weak form of ecological modernisation, environmental wellbeing is not necessarily threatened by economic growth. The weak type considers that environmental protection and economic growth can be adjusted in a way that they reinforce each other. The strong variation leads to more ambitious environmental policies based on the assumption that ecological policy can be a new motor for economic growth. In the 'limits to growth' discourse, the implication that economic growth may be beneficial to the environment is rejected.

Ecological modernisation presents ecological as well as managerial and economic problems. However, strong ecological modernisation would also introduce ecological terms to define economic conditions.[9] In strong ecological modernisation, economic and ecological discourses merge.

The consensus-making role of the Government can be analysed in a similar fashion. The weak form of ecological modernisation urges cooperation with market parties to combat environmental problems, but they should be restricted by regulation as little as possible. This may be achieved, for instance, through the use of self-regulation. In the strong version, participatory structures are developed that facilitate direct negotiation between industrial interests and the environmental movement. In the Netherlands, the 'Green Polder Model' (Weggeman 2003) is an example of such a practice. As regards 'limits to growth', opposition from market parties needs to be broken by a forceful state. Top down regulation and standard setting should force market parties in an eco-friendly direction.

A preventative approach is an integral part of ecological modernisation. This implies that policy that targets the source of pollution directly is preferred. Pollution that is

9. For instance, the intention to replace or supplement the Gross Domestic Product as an indicator of a nation's wealth with a Genuine Progress Indicator is an example of strong ecological modernisation, because not only is the language of ecology 'economised', but the language of economy is 'ecologised' as well.

not emitted does not need to be cleaned up afterwards. The weak variation focuses on the sources of pollution, such as industry, transport, and the like, and establishes rules and regulations to make these sources produce cleaner air. In contrast, strong ecological modernisation targets the people who own and operate these industrial installations, as well as the producers and owners of cars and lorries. Not only should they incorporate new technology to pollute less, but they should also adopt less polluting methods of production. This implies a change in behaviour rather than simply incorporating new technology. An example of strong ecological modernisation is the setting of environmental standards that are so tight that they force the creation of new technology, or force producers and consumers to change their patterns of production and consumption. The application of the precautionary principle befits strong ecological modernisation as well. In the European REACH regulation, for instance, producers and not public authorities are under the onus to provide safety information about the chemicals used in products.[10] The policy also calls for a gradual substitution of hazardous chemicals by safer ones.

Responsibilisation of the public and other actors can be achieved by stronger or weaker ways as well. The weak way involves targeting them with environmental campaigns to take their own environmental measures. In weak ecological modernisation, environmental pressure groups are enlisted as allies who need to influence other actors to internalise environmental concerns. A strong version of mobilisation involves designing curricula to raise environmental awareness and improve access to justice for citizens in environmental matters, and for instance to provide informational rights to pollution registries. In this model, the public and the environmental movement may become a true countervailing power to the industry. As regards limits to growth, more is required from the state; it needs to use its regulatory power to halt environmental degradation. This role of the state was more fitting in the 1970s than it is today, but voices can still be heard that argue for strong state intervention (Somsen 2011).

The distinctions between the different discourses apply to comprehensive policy making as well, because ecological modernisation transcends the sectorial approach to pollution that divides pollution into separate categories. Such a division is artificial, because if a pollutant is burned for instance it becomes air pollution, but if it is emitted in the rivers directly it returns as soil and water pollution. In the 1980s and 1990s, policies were devised that proposed to tackle all these forms of pollution at the same time. The European Directive on Integrative Pollution Prevention and Control is an example of such a rule. However, this form of integration still belongs to the weak category of ecological modernisation. In its strong variation comprehensive polices target the

10. Website REACH, last accessed 14-12 2012. REACH stands for Registration, Evaluation, Authorisation and Restriction of chemicals and aims to protect public health by improving the identification and evaluation of the properties of chemical substances used in products.

behaviour of polluters themselves, instead of merely focusing on an integration of types of pollution. The Dutch NMP and the EU's Fifth Environmental Action Programmeare good examples of such holistic plans. A holistic approach was recommended in the report 'limits to growth' as well (Meadows et al. 1973). During the time that this discourse reigned, however, the Dutch state did intervene but only through piecemeal sector based policy (Van Tatenhove 1993, p. 17). Notably, other influential texts in the 1970s environmental discourse did not recommend the holistic state centred approach, for instance Schumacher (1973) and Goldsmith & Prescott-Allen (1972). Realising their visions though would involve an even more thorough rearranging of society according to ecological needs.

The difference between strong ecological modernisation and the weak version is not only a difference in the ambitions of ecological modernistic policies. There is a difference in the intensity of institutional change that they demand. It is certainly true that the weak forms of ecological modernisation present an affirmative answer to Ulrich Beck's rhetorical question as to whether the problems of the risk society could be tackled in the same way as the problems in industrial society (Beck 1994, pp. 11/12). Weak ecological modernisation relies on the same offerings as did industrial society, market, technology and Government. However, that is not the case for the strong version of ecological modernisation. Strong ecological modernisation as conceptualised above is more than a technocratic commitment to solving the environmental crisis through market friendly regulation and technological innovation. The technocratic dimension is supplemented with a 'sociocratic' dimension (Jänicke 2000, p. 3; Mol & Sonnenfeld 2000). The limits-to-growth approach takes this sociocratic dimension to the extreme. According to this discourse, society and especially the economy should be rearranged along ecological lines.[11]

Regarding ecological modernisation, the strong, sociocratic, and reflexive form implies much more than technological innovation and market-based solutions. Its analysis of the ecological crisis and the answers it provides do not simply conform to the existing practices of industrial society. Strong ecological modernisation implies that ecological considerations will lead to institutional change. These new arrangements should overcome the boundaries between ecology and economy, and '*correct the prevailing bias towards economization and scientification* ... []' (Hajer 1995, p. 281). Strong ecological modernisation involves behaviour change, social learning, and new democratic institutions. It means internalisation of environmental norms and values, changing patterns of consumption and production, life style policies, and changing consumer preferences. It resembles limits to growth, but it is far more optimistic. The boundaries between economic wellbeing and environmental degradation should be transcended, and this makes possible a new type of development that caters

11. An example of this approach may be found in Schumacher's collection of essays bundled in 'Small is Beautiful' (Schumacher 1973) and in Goldmith & Prescott-Allen (1972).

Table 1

Policy discourse	Limits to growth	Weak ecological modernisation	Strong ecological modernisation
Relationship between economy and ecology	Inimical: economic growth threatens the environment.	Peaceful coexistence: Economy and ecology may progress side by side.	Positively intertwined: Ecological progress will lead to economic growth.
Role of science	Natural science and ecological science are used to assess the state of environmental degradation. Natural scientific considerations should steer policy.	Science provides the data to discuss ecological problems in economic, natural scientific and managerial terms.	Science is applied to take stock of environmental threats, and economic problems are discussed by including indicators of environmental performance. Resource use, depletion, and pollution become an important concern in economic management.
Consensus building	Market parties should be forced to commit to the goals of environmental policy through top-down regulation.	Negotiations between the Government, environmental pressure groups and industry on the topic of environmental regulation.	Broad mechanisms of participation for industry and the environmental movement in policy making.
Preventative approach	Ecological disturbance should be reduced to a minimum. This implies rigid application of the precautionary principle and an antagonistic view towards economic growth.	Pollution should be prevented by cleaner processes of production and if necessary application of the precautionary principle.	Targeting the polluting individual, enterprise and consumer him or herself. Broad application of the precautionary principle and internalization of it by addressees.
Responsibilisation	Education and legislation should be combined to raise the right environmental mentality.	Awareness raising through media campaigns, supplying information and subsidising environmental pressure groups.	Active involvement of citizens through education, granting rights to review corporate registries and access to justice, alliance between environmentalists and Government.
Comprehensive policies	The earth is considered one interdependent ecosystem. Ideally, policy is integrated and formulated holistically. In practice though, only piecemeal regulation could be realised in the 1970s.	Targeting multiple forms of pollution in multiple environmental media with the same policy.	Targeting multiple forms of pollution as well as establishing behaviour change with regard to polluters within the same policy.

to both. From a limits to growth perspective, such a vision is naïve. Instead, we should sacrifice our need for economic growth and make do with a 'steady state economy' in which population growth, consumption, and production remain stable.

After this clarification, the typology of ecological modernisation can be completed. It is presented in table 1 above, and is contrasted with the limits-to-growth discourse that is characterised on the left-hand side. This table will be used when policies and storylines are judged for their adherence to the discourse of ecological modernisation and in establishing whether weak or strong ecological modernisation is being advocated.

1.5.4 Ecological modernisation: its critics and its possibilities

Research interest in ecological modernisation peaked in the late 1990s and early 2000s, but it has since lost ground to other approaches. Ecological modernisation has a great deal of affinity with the notion of sustainable development; however, the two should not be conflated. Sustainable development contains elements of international solidarity that are absent from ecological modernisation. The concept of sustainable development though managed to overtake ecological modernisation in social science and policy research. Currently new conceptualisations of environmental policy have caught the spotlight such as social ecological resilience thinking, for an application of this perspective in socio-legal studies see O'Mally (2013).

Moreover, like every theory, ecological modernisation is not without its problems and critics. Critical comments about the usefulness of the theory have been made by Buttel (2000a), Blühdorn (2000), and Andersen & Massa (2000), among others. Much of the criticism concerns the different uses of the term as well as the imprecise nature of the concept. It is argued that ecological modernisation comes in many guises and means different things, dependent on the author. This deficiency is acknowledged, and in this chapter an attempt has been made to forge conceptual coherency by constructing a typology of ecological modernisation and discriminating clearly between a strong and a weak variation. Moreover, I treat this concept exclusively as a policy discourse in order to avoid conceptual confusion.

In addition ecological modernisation has drawn sharp ideological criticism. Some authors regard it as legitimating unsustainable trends in production and consumption (Blühdorn & Welsh 2007, p. 187). According to Blühdorn, ecological modernisation serves to keep the myth of modernity alive in the face of the ecological challenge (Blühdorn 2000). Foster et al. (2010) argue that it is merely a thinly veiled attempt to protect capitalist structures from the challenge they face in terms of environmental degradation.

Despite these criticisms I have chosen to use an ideal typical rendition of ecological modernisation discourse as one of the two analytical lenses through which the construction of the air quality clash is analysed. I have done so especially because ecological modernisation was an important policy discourse at the time of the clash and especially at the time the policies and plans were drafted that precipitated the clash. Blühdorn, who is otherwise critical of the notion, noted in 2000 that the practice of ecological modernisation was the dominant response of contemporary European societies to environmental challenge (Blühdorn 2000). Moreover, as both Cohen

(2000) and Hajer (1995) have stated, ecological modernisation seems remarkably congruent with how policy making in the Netherlands is often characterised, namely as a consensus democracy with policies that contain a peculiar mix of moralism and eye for economic opportunity (Andeweg & Irwin 2005). These considerations warrant the question whether ecological modernisation in the Netherlands still managed to bridge competing economic and environmental/health interests in environmental conflicts such as the air quality clash. In my study I argue that storylines taken from weak ecological modernisation resolved the conflict over air quality, although the discourse did lose the glory it had in the 1980s.

1.5.5 *The legality of precaution and the precautionary principle*

The rise of environmental problems is often cited as an indication that our society, sometimes referred to as 'risk society' (Beck 1986), or 'post traditional society' (Giddens 1994), is in a phase of transition. Ulrich Beck is probably the most well-known proponent of the thesis that newly emerging technological risks force society to adapt to these new challenges. However, various authors in the field of sociology of risk elaborate on his work, and question the nature of this social transition. They point out that society is changing in its tolerance of risk (Ewald 1999; Furedi 1997; Burgess 2004; Pieterman 2008; Hanekamp 2015). Summarised roughly, their proposal is that the point of the risk society is not that the risks themselves have increased dramatically, but that our attitude towards risk has changed. During most of the 20th century, faith in science, in technology and in progress, optimism concerning human ingenuity, and a focus on the control of nature reigned in industrial society. This shifted in the 1960s In Rachel Carson's book Silent Spring a pessimistic account of technological progress is given and a pessimistic account of the danger presented by reliance on growth for environmental health is given in the already mentioned report for the Club of Rome (Meadows et al. 1972).

The point of the authors mentioned above is that our relation to risk and uncertainty always changes and I relate the air quality clash to transformations regarding risk and uncertainty as well; however I take an explicitly socio-legal approach. I am interested in changes in the legal order, meaning the law and its institutions such as the judiciary, legislation, and legal doctrine. Therefore I place myself in the Weber-inspired tradition within sociology of law in which we also find Dutch sociologists of law, André Hoekema and Niels van Manen. Inspired by their book 'Types of Legality' (2000), I will construct a new ideal type of legality that I see emerging on the horizon, a legality of precaution. In the legality of precaution the prevailing attitude of risk management by compensation for the victims is gradually becoming eclipsed by an attitude of intolerance towards risk per se. The emphasis is on extreme prevention in order to minimise risk as much as possible, irrespective of costs. This new emphasis on precaution is linked to changes in the perception of human nature, society, and the threats deserving of most concern. Societal and legal concerns have shifted from a focus on short- and medium-term damage that is calculable and insurable to catastrophic, uncertain, and incalculable damage (Pieterman 2008). Humankind is seen as an environmental threat, and from this perspective the relationship with science and technology is ambivalent. Though they provide the tools for damage prevention, the increasing hold of science and technology on our lives is seen as one of the main causes of potentially catastrophic threats.

The most conspicuous legal pendant of this view as to the relationship between humankind and its environment is the precautionary principle. There are many versions of the principle, but in one influential formulation it is stated as such: '*Where there are threats of serious or irreversible damage, lack of full scientific certainty shall not be used as a reason for postponing cost-effective measures to prevent environmental degradation*' (1992 Rio Declaration at the UNCED Conference, Principle 15, in: Freestone & Hey 1996, p. 3). From a legal point of view, the most important aspect of this principle is that it may be required to take positive environmental measures before scientific proof of harm has been provided (Freestone & Hey, 1996, p. 13). Stronger variations of the precautionary principle may leave out the requirement of cost-effectiveness, and/or shift the burden of proof of harm.[12] Dutch scholar Arie Trouwborst describes the essence of the principle with the Latin phrase 'in dubio pro natura' (Trouwborst 2009, p. 108). The precautionary principle has since found its way into many environmental treaties and in European and international law. Article 191 paragraph 2 of the Lisbon treaty on the Functioning of the European Union stipulates that European environmental policy should be based on the precautionary principle.

I will investigate whether the air quality clash and its legal implications are consistent with the type of change from risk and compensation to precaution.[13] To that end, I construct ideal types of the legality of precaution and the legality of risk and compensation.

1.5.6 *Ideal types of the legality of risk and compensation and the legality of precaution*

Below, I will present the ideal typical reconstruction of the legality of risk and compensation and the legality of precaution. I discuss eight features of these types of

12. Such a strong version may be found in article 5 paragraphs a and b of the Earth Charter in which the precautionary approach is described as: a. Take action to avoid the possibility of serious or irreversible environmental harm even when scientific knowledge is incomplete or inconclusive.
b. Place the burden of proof on those who argue that a proposed activity will not cause significant harm, and make the responsible parties liable for environmental harm. (Website Earth Charter, last accessed 12-10 2015).
13. The legality of precaution will not replace the legality of compensation. When a type of legality becomes dominant all other types are not immediately eradicated. The older type of legality will still remain persistent in some aspects of the legal order. It will only cease to be the dominant type visible in the legal order.

legality I consider to be characteristic. These features are: the core imperative of the legal order, category of damage of most concern, moral reaction to damage, perception of the victim, level of proof and safety guarantees required, view on public participation, stance towards the possibility of social engineering by law and the value of most protection. They are indebted to a large degree to Pieterman 2008.

1 Core imperative of the legal order

Within the legality of risk and compensation, the necessity of compensation and the spreading of risk by insurance mechanisms are central features of the legal order. The misfortune of the victims has to be redressed because victimhood is the product of the complex organisation of modern industrial society. The arrangements that form and sustain our social order necessarily produce victims; think for instance of our reliance on mobility. The ubiquitous presence of motorised traffic takes lives causes injuries in the form of accidents. These injuries are considered bad luck, the inevitable by-product of our arrangement of society and the victim is entitled to compensation. This type of legality appeals strongly to social solidarity between victims and non-victims. Prevention is important in this type of legality, but it extends to calculable and certain threats and only applies where the costs of prevention are lower than the cost of compensation.

Within the legality of precaution, prevention of damage is the core imperative. The occurrence of damage is considered to be the product of mismanagement. Social arrangements such as mobility, production, housing and the environment should be managed in a way to provide maximum safety, and applying the precautionary principle is necessary to obtain a safe society. Uncertain threats should be accounted for and prevented if possible, regardless of costs, because if they are not prevented at an early stage, it may be too late. 'First do no harm' is an expression that typifies this type of legality (Pieterman 2008, p. 64). It appeals less to solidarity than compensation does, because ideally damage is prevented before it can occur. It does appeal to responsibility however, but now to the collective responsibility of producers, consumers, and the state to keep us safe. The precautionary principle plays an important part here, because it invites the relevant authorities to anticipate on potential threats. In such a situation, scientific institutes become more heavily occupied with the signalling of threats before they occur than with finding a solution to problems after they have materialised. Legally, this means that an administration is considered to be at fault, and should be overruled by the court if it has taken insufficiently into account what could *possibly* happen rather than what would *probably* happen (Furedi 2009). The necessity of prevention is extended from certain and calculable threats to uncertain, but possible threats. Even if the cost of prevention is predicted to be higher than the cost of compensation, prevention should be chosen for two reasons. Firstly, the prevention of harm is a moral duty and secondly, (excessive) prevention protects us against the possibility of the realisation of worst case scenarios, it is better to be safe than sorry.

2 *Category of damage considered of most concern*

The legality of risk and compensation is both forward and backward looking (Ewald 1999). It is forward looking in the sense that people anticipate risks and collectivise them in order to be insured when damage occurs. It is backward looking in the sense that if damage occurs compensation is provided for, based on the rationale of insurance. This rationale implies that short- and medium-term damage is considered to be of greatest concern. Damage that is calculable and probable can be insured, but long-term damage cannot. However, people had an optimistic view of science and progress (Pieterman 2008, p. 65), and for that reason long-term damage was of little concern to the public. Conflicts occurred over compensation after damage occurred, but relatively little over the need to take preventative measures.

The legality of precaution is forward looking. Potential damage must be prevented, even if it may only occur in the long term. The focus is on the typical problems of the risk society, such as environmental damage of the potentially catastrophic kind. This kind of damage is not insurable. Therefore ideally long-term damage is identified, and preventative measures are taken early in order to ward off the threat. Conflicts arise over when measures should be taken to prevent a future threat and how invasive these measures are allowed to be.

3 *Moral reaction to damage*

The legality of risk and compensation is not only a legality of compensation but also of prevention (Ewald 2002, p. 281). Ideally, damage is prevented, but there is a limit to prevention. Damage arises from otherwise productive and beneficial social arrangements such as mass production or mobility, and it is accepted that safety can only be provided to a certain extent. If damage occurs anyway, it is due to otherwise healthy social arrangements that we cannot do without. Even if we know that a certain activity, mobility for instance, will cause a certain amount of damage, no one will argue for banning all road traffic.

This is not the case in the legality of precaution. Damage should be prevented, and if it occurs, it means that the authorities – whoever they may be – have been negligent (Pieterman 2008, p. 71). The moral dimension that is not or is only dimly present in the legality of risk and compensation is conspicuous in the legality of precaution. It becomes incumbent on the authorities to demonstrate that they have exercised an effective amount of precaution.

4 *Perception of the victim*

Victims are considered to be people who have had bad luck in the legality of risk and compensation. They have been the hapless casualties of our social arrangements. To make sure that they can cope, they are caught within a tightly woven web of private and public insurance schemes against which they have rights. They can claim these rights, and the unfortunate situation can be redressed (Hoekema & Van Manen 2000, Pieterman 2008).

This consideration of the victim changes in the legality of precaution. Damage should not have occurred, and consequently the victim is a citizen that has been hurt in his legitimate expectations of care. He is not just unlucky, but has suffered the injustice of negligence, and therefore he can claim those rights against the people who are the responsible authorities. This is not explicitly concluded by Pieterman, but follows from the logic of his scheme.

5 *Level of proof and safety guarantees required in decision making*
Rules on evidence and justification are important for all institutions within a legal order. These rules determine the outcome of conflicts and limit the competency of important institutions, such as policy makers and judges to a large extent. In the legality of risk and compensation, the calculation and prioritisation of risks play a large role. Exact calculation is important for the correct and optimal spread of risk, as well as to determine an optimal ratio between costs and benefits. In fact, the development of cost-benefit analysis and risk assessment made possible the rise of this legality (Pieterman 2008). Weighing of interests is considered crucial, and cost-benefit assessment is considered an important tool to achieve an optimal solution. Administrative and legal decision making can point towards the optimal weighing of interests to motivate decisions.[14]

In the legality of precaution, scientific knowledge and calculation are important as well, but the emphasis on prevention leads to a reversal of the burden of proof. Governments and companies should present proof that their activities do not cause harm (Pieterman 2008, p. 72). Because this is logically impossible, the next best thing is demanded: namely, proof that the threat of harm is negligible. This means more research needs to be done as to the potential harm certain activities may cause, including research on less likely but possible worst-case scenarios.

6 *View on public participation*
Public participation in the legality of risk and solidarity is considered of secondary importance. The emphasis is on expert knowledge, and these experts are considered crucial in obtaining the knowledge of optimal outcomes of decision making. In the legality of precaution, faith in scientific expertise has declined, and the opinions and views of laypersons are considered more important. Ideally, all views should be taken into account, and the 'subjective' risk assessment of laypersons is considered important because they are 'experiential experts', experiencing the outcomes of policies and decisions at street level.

If it is impossible to let lay persons participate directly, interest groups and social movements that represent their interests must be taken on board, even if they have no official scientific credentials. Legitimacy is thought to be enhanced when different

14. Important in tort law is the 'Learned hand formula' in US jurisprudence (Posner 1982). A Dutch classic in this regard is the 'Kelderluik' judgment of 1965 HR 05-11-1965, NJ 1966, 136, LJN AB7079.

groups representing the interests of laypersons have broad participatory rights, including access to justice.

7 Stance towards the possibility of social engineering by law

Legal reformers in the early 20th century pointed to the possibility of realising desirable social goals through an instrumental use of the law (Schwitters 2008, pp. 53-58). This idea was contrary to formal legality that contained the ideal of legal neutrality. The state and therefore the law should not decree what is or is not a desirable social situation, because that is for the autonomous, rational individual to decide. It should only realise a general, stable, and predictable legal order within which economically active, free, and autonomous individuals are able to realise their own ends (Hoekema & Van Manen 2000, pp. 50-51). In subsequent types of legality, this consideration was abandoned, and an instrumental use of the law became pervasive. Through regulation, the state tried to achieve desirable end states. The environmental laws in the Netherlands from the 1970s are a good example of the type of regulation used: top-down, command and control type regulation that was piecemeal in character. Each environmental sector, such as air and water, was covered by a separate law. Such laws gradually governed all sorts of safety and health issues including accidents in the work place, child labour, and food safety.

In a legality of precaution, the instrumental use of regulation is expanded upon and piecemeal command and control regulation gives way to law based on programmatic long-term planning with a comprehensive and integrated character. The assumption is that law can and should stop environmental degradation. The prescription of a minimum standard is no longer the main aim, but long-term incremental improvement is desired. These long-term plans and programmes do not only warrant certain environmental quality standards, they also encourage behavioural change and innovation in the fields of safety and protection.

8 Value most worthy of protection

The legality of risk and solidarity is rooted squarely in the progress of modern industrial society. The discovery of the concept of risk and its usefulness in insurance schemes meant that it was possible to spread the risks associated with modern living. Moreover, the advent of socialism and communism was conducive to the view that damage was a social problem, caused by otherwise useful societal arrangement rather than the victim's private problem (Pieterman 2008, pp. 55-56).

The permeating optimistic notion of societal progress legitimised the further development of these systems to overcome problems of scarcity. However, according to Ulrich Beck, at the end of the 20th century we had overcome scarcity and social conflicts were no longer primarily about scarcity but about risk positions. In industrial society, struggles occurred over the distribution of goods, while struggles these days occur over distribution of the bads – notably pollution risks and other industrial hazards.

The typologies of the legality of risk and solidarity and the legality of precaution reflect these changes. In the legality of risk and solidarity, innovation and expansion of industry and mobility are stimulated, and the delivery of the goods of mobility and industry to the people is considered crucial to ease the burden of scarcity. Economic expansion is considered of primary importance, and the expansion of roads and establishment of a free and competitive market are prime political goals. State interference in the economic sphere is present, firstly to secure an optimal function of the market, and secondly to ensure that the quest for growth and development does not lead to victims suffering from it without compensation. Otherwise however, enterprises should be allowed to proceed with little regulatory and legal intervention. In the legality of precaution, this relationship is turned on its head. Ecological harmony is considered most desirable. The values in need of nurturing are immaterial ones, such as a clean living environment and public health. Business is allowed to proceed if it can show that it does not lead to victims. The use of the precautionary principle gives the state a mandate to interfere in the economic sphere in order to protect the environment or health, also if it is not clear that a certain activity will lead to any damage (Pieterman & Arnoldussen 2008).

Table 2

	Legality of risk and compensation	Legality of precaution
Core aspect of the legal order	Compensation of damage by spreading of risk is the prime imperative; hence, strict liability and public and private insurance schemes are essential. Prevention of damage if it is cost-effective to do so.	Prevention of damage is the prime imperative; hence, application of the Precautionary Principle is essential. Prevention extends to possible threats of an uncertain magnitude.
Category of damage considered of most concern	Short- to medium-term damage of a relatively certain magnitude and probability.	Long-term damage of the potentially catastrophic kind but of an uncertain probability.
Moral reaction to damage	Damage cannot be prevented totally, but compensation is in order when it occurs.	Damage is disgrace and should have been prevented.
Perception of the victim	Rights-bearing member of a risk collective.	Civilian damaged by negligence of responsible authority, agency, or enterprise.
Level of proof and safety guarantees required in decision making	Decision making procedures should rely on scientific knowledge and cost benefit assessment to determine optimal solutions.	Decision making procedures should be based on proof that the chance of harm is negligible, and proof is required from the actor undertaking the risky activity.
Stance on the possibility of social engineering by law	Society may be improved by piecemeal policies and an instrumental use of legislation.	Society may be improved by comprehensive policy making (i.e. long-term holistic planning) and societal awareness raising.
View on public participation	Civilians are seen as laypersons with little knowledge of risks or expert systems. Policy needs to be determined by experts.	Imperative because the citizen is an 'experiential expert', and policy and law need to provide opportunities for citizens to present their views.
Value most worthy of protection	Economic development: growth and expansion of construction, mobility, and innovation.	Ecological harmony: balance of eco-systems, public health, and safety from possible threats.

The considerations above lead to the construction of the following ideal types of legality, the legality of risk and compensation and the legality of precaution, see table 2 above.

1.6 What this book is about

After this introduction of the research methodology and perspectives, I like to present summary overview of the main themes of this study. This book offers some answers to the puzzling fact that from the end of 2004 to the first months of 2010 we witnessed an 'air quality clash' in the Netherlands. My purpose is to explain why air quality is so hotly debated as a social problem during these years after a period of some fifteen years in which the lack of attention suggested 'the problem' had been 'solved'.

The answers I shall provide rest on the use of a social constructivist approach. This sociological perspective holds that not 'objective facts' but 'social definitions' determine whether a social problem exists, what the problem 'really is' and how it should be 'solved'.

The chapters in this book for the most part follow a chronological order with some overlap because of the need to shift back and forth from the Dutch national to the EU supranational level. In this first chapter I presented key elements of my social constructivist approach which is inspired by the work of Joel Best on social problems and Maarten Hajer on discourse analysis. Using this perspective I will investigate three specific questions in order to offer my answers to the riddle that drives this study. First, I shall look at the interplay between the Netherlands and the EU. This involves investigating the relevant Dutch and EU regulatory developments. The Dutch developments are presented in chapters 3, 5, 6 and 7, while chapters 4 and 8 are concerned with EU regulation. Where appropriate, explicit attention is paid to the interaction between Dutch and European developments. For example, chapter 4 portrays the influence the Dutch have exerted on EU regulation by involving the WHO in order to arrive at scientifically supported air quality norms. Chapter 5 then analyses the way the EU Air Quality Directives were incorporated in the Dutch legal system.

The second specific question I investigate concerns the environmental policy discourse known as 'ecological modernisation'. I concluded my presentation of this discourse in this first chapter by constructing two Weberian style ideal types of a weak and a strong version, which I contrasted with the discourse known as 'limits to growth'. The development of the Dutch, weak type of ecological modernisation is traced in chapter 3. Here Dutch environmental policy will be characterised as a consensual way of developing and implementing environmental policies, with a keen eye for costs and benefits. The relevance of this weak type of ecological modernisation for the question why the air quality clash came about and was finally solved, is discussed in chapters 5, 6 and 7.

The third specific question I investigate concerns the rise of the legality of precaution. In addition to the types of legality developed by the sociologists of law Hoekema and van Manen I have developed this new Ideal type. Whereas the legality of risk and compensation is more geared towards preventing and compensating well documented and understood risks, the legality of precaution extends that attitude to uncertain threats.

Chapter 2 discusses the scientific debate about the medical expertise on the health threats of particulate matter in order to find out whether they concern well documented and understood risks or are perhaps better understood as uncertain threats. This issue is picked up again in chapter 5 when discussing the way the Dutch deal with EU directives before the air quality clash actually erupts. Scientific uncertainty also figures prominently in chapter 8 when the evaluation of Air Quality Directive 99/30 by the European Commission is discussed.

As sketched above, the three specific questions that need to be answered to solve the riddle behind this study, are all dealt with in several chapters. By tracing these questions in the research material I focussed on two arenas where the contestants in the Dutch air quality clash confronted each other. The first arena is the Dutch administrative court. As recounted in chapter 6, the highest administrative court, The Council of State Administrative Jurisdiction Division instigated the clash. In 2004 it accepted the claim of the contestants that prioritised the interests of health and environment that projects to further develop Dutch infrastructure should not be allowed as the air quality was below EU standards. It also closed the clash, as discussed in chapter 7, by accepting in 2010 the position of the Dutch Government that the programmatic approach to air quality improvement was a legally valid option to continue infrastructure development in such a way that Dutch air quality would be improved.

The second arena where the contestants clashed is the Dutch Parliament, especially the Second Chamber, which is the more political of the two chambers. How the contestants divided in two camps –the pro-health camp and the pro-infrastructure camp – confronted each other in Parliament is analysed in chapters 6 and 7.

Because of the extensive amount of material analysed and presented and the complexity of the study both at the conceptual and the empirical level, I have opted to recapitulate in the concluding chapter 9 the most salient elements of this research. This chapter first provides answers to the three specific questions and then wraps up the study with a summarising answer to the riddle of the air quality clash in the Netherlands from September 2004 to the 31st of March 2010.

2 PUBLIC HEALTH CLAIMS ABOUT PARTICULATE MATTER IN THE EARLY 1990S

INTRODUCTION

The main pollutant of note in the air quality clash was a pollutant known as PM10, or particulate matter. This pollutant was considered the gravest threat to public health and moreover the standards for particulate matter were exceeded in many parts of the country when the clash over air quality erupted in 2004.

Before delving into the development of Dutch and European environmental and air quality policy, I will discuss the way Particulate Matter became considered a dominant health threat by epidemiologists and how this claim was disseminated to European and Dutch policy makers. Because epidemiological scientists acted as the dominant claims makers in the early period of the social construction of air quality as a serious health problem, a lot of attention in this chapter is devoted to the question what the implications of these epidemiological claims are. Apart from presenting an account of how PM10 came to dominate the environmental agenda on air pollution, this chapter raises the question whether bad air quality by itself may be considered the most prominent threat or whether bad air quality should be considered as one in a complex of causes that play a part in the higher premature mortality rates among people living in impoverished neighbourhoods.

In section 2.1 the main characteristics of particulate matter are described. Section 2.2 is devoted to the discovery of PM10 as a health threat by epidemiologists and the main studies that showed that there was a correlation between air pollution and premature mortality. In section 2.3 the possibilities and limitations of epidemiological studies are highlighted and section 2.4 focusses on the uncertainties present in the two most important studies about PM10. Section 2.5 covers epidemiological studies in the Netherlands on PM10 and claims making activities by agencies responsible for public health. In the sixth and final section remarks are made in regard to the claims about air quality and to epidemiology as a science particularly conducive to a precautionary approach. Finally the question is raised whether air pollution should be considered a typical risk society conflict in the sense that it is primarily a question of who should bear the risks engendered by our current way of life and how these risks should be dealt with politically.

2.1 PARTICULATE AIR POLLUTION

Particulate Matter is a type of particle pollution. In common language, it is a form of dust, albeit a very small one, although claims that particulate air pollution in the form of smoke causes health damage go back as far as the 16th century (Brimblecombe 1987). The most well-known incident of high pollution levels is the great London fog of 1952, in which it is estimated that thousands of people died owing to the exceedingly high levels of air pollution (Buijsman 2007, p. 11).

In the early 1990s, the topic of particulate air pollution started to climb on the political agenda. During the air-quality clash of the 2000s, public concerns over PM10 rose sharply in the Netherlands, and worrying reports indicated that thousands would die prematurely because of it. We will discuss the medical and social construction of this threat, but first we will look at PM10, and determine what kind of pollution is under consideration.

2.1.1 What kind of pollutant is Particulate Matter?

In simple terms, Particulate Matter is a certain type of what we commonly refer to as dust. It is actually dust of a certain size, tiny particles. To be considered PM10, the particle must have a diameter smaller than 10 micron. Dust was one of the first pollutants to receive attention, but the division according to particle size is more recent. Up until the 1980s, dust was measured in different ways. In the United States, the most common procedure was the gravimetric method, by which dust was caught in a filter and landed on a highly sensitive weighing mechanism. By weighing the particles, researchers could assess the total number of particles per cubic meter of air. Initially, no difference was made regarding particle size, and what was measured was the total number of particulates suspended in the air (TSP, Total Suspended Particulates).

In Europe, particles have generally been measured differently. They were caught in a filter, and by measuring the device's blackness, researchers can evaluate how much soot, smoke, and other contaminants the air contains. This is the black-smoke method devised in the 1960s in the UK, and this type of measurement emphasised different characteristics of the pollutant in question. Particulate Matter or PM10 is generally measured by leading air through a filter, which strains out particles larger than 10 micron. Particulate Matter is measured by weight per cubic meter of air, and the weight per meter typically runs into the micrograms. The European yearly standard, for example, allows for an annual average of 40 microgram per cubic meter of air: 40mug/m^3.

It is important to note that this categorisation only tells us something about the size of the particles. It says nothing about their chemical composition, or whether they are naturally occurring particles or of human origin, caused by combustion, for instance.

In this sense, Particulate Matter is a catch-all term for a complex and always changing mixture of different particles (RIVM 1998; Buijsman 2007, p. 18).

The mixture contains in varying quantities:

- Inorganic secondary substances such as sulphates, nitrates, and ammonium particles. These are formed in the atmosphere by gasses such as SO_2, NO_x, and ammonium (NH_3). These particles fall within the 90% range in the category of PM2.5, and are usually of human origin.
- Carbonaceous substances: These particles are usually emitted directly into the atmosphere. Elementary carbon that can be found in soot, for instance, is of human origin. This category also contains a small number of polycyclic aromatic hydrocarbons (PAHs), which are of human origin as well and presumably relevant to health.
- Salts: PM contains salts, and these particles occur naturally. In the Netherlands, on average, 4 to 5 mug/m^3 of all PM10 consists of sea salt, which is generally considered to be benign. Along the coast, the amount of sea salt is higher than in the east of the country. Salt particles can be very small indeed, and are easily dispersed in the lungs, but also quickly broken down.
- Oxidised metals and silicium: These particles mostly occur naturally, but may be dispersed by human activity, such as driving or digging.
- Water: Inorganic secondary substances in particular can bind with water to form droplets. It is considered that 10 to 50% of PM10 concentrations consist of water.[1]

Nevertheless, not all these different particles are equally relevant for health. Elementary carbon and PAHs are considered to be harmful, but even that depends on the exact chemical composition. Salt, however, is not considered harmful to human health. Since PM10 concentrations all consist of a mixture of these particles, and since the consistency of this mixture differs depending on the natural composition of the region and the human activities taking place, it is hard to compare PM10 concentrations in one location with those in another. Close to the sea or in the Sahara desert, one will find a great deal of naturally occurring PM and less PM of human origin, while the reverse is true beside a busy inner-city thoroughfare.

2.1.2 *Different categories of PM*

The first type of particulate air pollution under consideration was smoke, measured either as black smoke or as total suspended particulates. PM10 is a further classification of dust of a certain size. The diameter of 10 micron was chosen because it was thought that these smaller particles could enter the lungs more easily than bigger particles. However, further diversifications were made. Currently, a smaller fraction of PM is measured as well, with PM2.5 consisting of particles having a diameter smaller

1. For a more elaborate version, see Buijsman 2007, p. 18.

than 2.5 micron. Even PM1 and PM0.1 are under scrutiny. These particles have a diameter of 1 micron and 01 micron, respectively. The reason for focusing on increasingly smaller particles is that they can penetrate the lungs more deeply. At present, some experts and interested parties suggest that these smaller fractions of PM are more relevant to health (Buijsman 2007, p. 78; TNO 2012, p. 6). PM1 and PM0.1 are known to be ultra-fine Particulate Matter.

PM10 was first regulated in the US in 1987. The UK followed in 1997, and, the first European air-quality standards for PM10 were established in 1999.[2] However, it was already known that it was not the most health relevant fraction.[3]

The reason PM10 was regulated by the US, UK and EU respectively, was because it was considered a dangerous health threat after epidemiological research established a correlation between premature mortality and high levels of air pollution. This association was found by comparing the mortality rates in areas with high levels of air pollution with mortality rates in areas with low level of air pollution. In order to understand the construction process of PM10 as a health threat we will now turn to an examination of the most important epidemiological studies.

2.2 The discovery of PM10 as a health threat

In the 1980s, particulate air pollution was not high on the political agenda. Instead, acid rain was receiving widespread attention, and scientists were concentrating on acidifying pollutants such as sulphur dioxide (SO_2) and later nitrous oxides (NOx). Urban air pollution – the kind that one encounters mostly in cities and that has harmful consequences for health – was given less notice. This situation changed in the 1990s, however, when Particulate Matter emerged suddenly as a grave threat to public health. In this section, the new focus on particle pollution is under examination, and we will see how research undertaken by epidemiologists in the 1990s was instrumental in establishing of a link between health damage and air pollution resulting from road traffic.

The situation changed mainly as a result of two epidemiological studies that originated in the US, and that had a significant impact on all other research involving PM10. These studies on the effects of PM on health were the Harvard Six Cities Study conducted by Dockery et al. (1993) and the ACS Study conducted by Pope et al. (1995).

2. The construction process of these European standards is the subject of chapter 4.
3. This is also acknowledged in the most important US studies: the ACS and Six Cities Studies. These concerned mostly PM2.5, and not PM10. In 2006, the US Environmental Protection Agency EPA dropped the annual standard for PM10 that had originally been set at an average concentration 50 mug/m^3 per year.

We will turn first to the Harvard Six Cities Study and the ACS Study, the results of which became available in the early 1990s.

2.2.1 *Two significant studies: Harvard Six Cities and ACS*

The first and possibly most well-known piece of research regarding the effects of particulate matter on public health is known as the Harvard Six Cities Study. From 1974 to 1988, pollution levels were recorded in six US cities: Watertown Massachusetts; Portage Wisconsin; Topeka Kansas; Harriman Tennessee; St. Louis Missouri; and Steubenville Ohio. Death rates in these cities were compared and related to concentrations of PM and other pollutants. Of these cities, Portage recorded the lowest levels and Steubenville recorded the highest concentrations of air pollution for various pollutants.[4] The air in Steubenville contained roughly two-and-a-half times as much PM10 and PM2.5 than Portage, and almost three times as much sulphate, a type of sulphur-based pollutant. Researchers found that people in Steubenville ran a 26% chance of dying prematurely. This means that a quarter of all the people in Steubenville died before their expected date of death, compared to people living in Portage. In epidemiological terms, this meant that a relative risk of 1.26 was found between people living in these two cities (Krewski et al. 2000b, p.ii).

A second epidemiological study that had a significant impact on the US debate on the regulation of PM10 as well as of PM2.5 was the American Cancer Society Study (ACS) (Greenbaum 2003, p. 1495; Krewski 2003, p. 1514). The ACS Study is larger in scope than the Six Cities Study, and was conducted by using a large cohort of 1.2 million people living across all 50 States of the US and in Puerto Rico.[5] Residents were provided with questionnaires regarding their health, and the death certificates involving this cohort were compared for the metropolitan area in which they lived. The principal results of these analyses showed that fine particle concentrations were associated with increased mortality from all causes, and involved cardiopulmonary disease in both men and women; however, an association between fine particles and lung cancer was not apparent (Krewski 2003, p. 1513).

The studies provoked controversy when they first appeared, and especially when the US Environmental Protection Agency (EPA) used them to argue for a revision of the standards for air pollution (HEI 2000). Controversial issues surrounding the research

4. The ranges of concentration between these two cities were 18.2 mg/m^3 for Portage and 46.5 mg/m^3 for Steubenville for PM10, 11.0-29.6 mg/m^3 for fine particles, (PM2.5) and 4.8-12.8 mg/m^3 for sulphate (Krewski 2003, p. 1512).
5. The investigations included 552,138 adult subjects who resided in 151 U.S. metropolitan areas in which sulphate data had been collected regularly in 1980 and 1981, and 295,223 adult subjects who lived in the 50 metropolitan areas for which fine particle (PM2.5) data were available (collected from 1979 through to 1983).

included the charge of hiding data (website Harvard School of Public Health),[6] various methodological issues (Moolgavkar 2005), and the measurements used to determine the types of pollutant involved and their concentration levels (Phalen 2002, p. 11).

The results of these two studies were reviewed by the HEI, which concluded that the studies withstood critical review. A re-analysis team led by Daniel Krewski at the University of Ottawa found that the data and the results of those two studies were generally robust. HEI released the following statement: '*Overall, the re-analyses assured the quality of the original data, the original results, and tested those results against alternative risk models and analytic approaches without substantively altering the original findings of an association between indicators of particulate matter air pollution and mortality*' (HEI website). According to HEI, the replication of the Six Cities Study indicated that in a city where there was 18.6 mug/m^3 more PM2.5 in the air, premature mortality from all causes was 28% higher. The ACS Study indicated that people living in the most polluted area ran an 18% higher risk of dying prematurely than people in the least polluted area, with the difference in air quality being almost 25 mug/m^3 (HEI 2000, p. 1).

The belief that air pollution by Particulate Matter had serious health effects became widely accepted among epidemiologists (Pope 2003), and was one of the most common arguments for air-quality policy as well. However, the amount of epidemiological research regarding long-term health effects of PM was small at the beginning of this century. At the time, there existed only five cohort studies, and these had supplied the epidemiological data (Buringh & Opperhuizen 2002a, p. 34). The two most prominent pieces of research remained the aforementioned Six Cities Study and the ACS Study.

2.3 Prospective epidemiology, possibilities and limitations

Before we can assess the relevance of these findings in the construction process, an introduction to prospective epidemiology – the branch of medicine that spawned them – is required. The implications of the studies cannot be assessed without some knowledge of the field, and of the difference between epidemiology and another branch of medicine, toxicology. Moreover, the strengths and weaknesses of an epidemiological study need to be mapped out before we can proceed with an assessment of the two pivotal studies.

2.3.1 *Epidemiology and toxicology*

Epidemiology is '*the study of the occurrence of health and disease in populations with a view to illuminating the causes of disease and, ultimately, its control or prevention*' (Kabat 2009, p. 19).

6. Website Harvard School of Public Health, last accessed July 2013.

A classic example that illustrates the epidemiological method is the work of physician John Snow, who analysed the spread of disease during a 19th-century London cholera epidemic. He noticed that cholera was more prevalent in some areas than in others, and began to recognise a pattern. By analysing the pattern, he concluded that a street pump in the Soho district had to be the source of the disease, and he was able to stop the epidemic by removing the pump's handle. This was one of the earliest successes involving epidemiological research (Kabat 2009, p. 21).

Essentially, epidemiology is based on making comparisons between groups, and then examining the differences. It compares rates of disease or mortality between groups that are differentiated according to various characteristics: for instance, sex, gender, age, socio-economic level, or nationality. By comparing these groups, epidemiologists try to construe possible hypotheses as to why a certain condition occurs predominantly in some groups rather than in others, and why certain groups display levels of disease that deviate significantly from the expected base rate. Epidemiology reveals associations between particular types of ailments and certain groups, with the classical example being that of finding highly elevated levels of lung cancer among smokers (Brandt 1993, p. 161). This discovery led epidemiologists to conclude that there was an association between smoking and lung cancer. Through establishing this strong correlation, epidemiologists were able to point with conviction to the risk of smokers developing lung cancer.

The notion of risk is central to epidemiology. While it can predict that smokers run a higher risk of developing lung cancer, it cannot predict whether a certain smoker will in fact develop lung cancer. By using statistical analysis, it can estimate how much greater the risk for a smoker would be in comparison with a non-smoker. A study undertaken by E. Cuyler Hammond in the 1950s showed that lung cancer deaths were 5 to 16 times higher among heavy smokers than among non-smokers (Brandt 1993, p. 162). In epidemiological terms, this means that heavy smokers ran a risk that was 5 to 16 times higher than non-smokers of developing lung cancer.

Even though the relative risk factor was high in this case, epidemiological studies still only uncovered a correlation between smoking and lung cancer. They did not establish causative mechanisms: namely, the way in which the body reacted to cigarette smoke and developed cancer. However, since the relative risk factor of heavy smokers was very high at 5 to 16, it was safe to assume that when such a high risk was found, the correlation between smoking and developing lung cancer was indeed causative: namely, indicating that smoking causes cancer, even though the biological mechanism by which smoking led to cancer was not known at the time.

Epidemiological studies may be conducted by using different research designs, and a familiar one involves the case control study. This is a retrospective method, because the subjects under scrutiny are patients who have developed a certain medical condition. In a case control study, subjects are questioned, and the answers to these questions may uncover a circumstance relative to that particular disease. The first investigations to test the hypothesis that smoking and lung cancer were related were conducted by questioning hospitalised sufferers of lung cancer about their smoking habits. This type of study is referred to as 'retrospective', because it looks back in time and tries to establish a possible reason that a certain disease developed.

Another type of research is the prospective study, such as a cohort study. Cohorts are groups that are formed according to certain characteristic features, and then followed over a prolonged period of time. In a prospective epidemiological study, a group is not selected according to the development of a certain disease, but because the members have all been exposed to a certain agent. The researcher investigates whether there is a relationship between the exposure and the development of a disease in group members. For instance, one can research the effects of exposure to traffic-related air pollution by comparing a group of people more exposed to such pollution with a group exposed to normal levels of air pollution, and determine whether certain diseases are more prevalent within the first group. In this way, for instance, Beelen et al. (2008) established that traffic intensity on a road situated closest to members of the first group was associated with premature mortality due to all causes: cardiovascular disease, respiratory disease, and lung cancer. This type of study is of special interest for the topic of our thesis, air pollution, because both the Six Cities Study and that of the ACS were prospective epidemiological studies.

It is crucial to realise that although epidemiological research may uncover correlations between pollution and adverse health effects, it cannot logically establish definitive causations, because it does not investigate what goes on in the body. The causative mechanisms through which pollutants harm the body are established by another branch of medicine: toxicology. In contrast to epidemiologists, toxicologists work in laboratory conditions, and test the effects of agents on bodies by way of exposing them to the agents in controlled conditions. Researchers often use test animals to observe the physical effects of a certain substance, and dissection may uncover pathways through which a particular agent enters the body, and reveal the effects it has. By means of such research, the toxicologist may uncover dose-response relationships and determine causal effects of substances. Whereas the epidemiologist may study long-term effects of a certain substance on a group of subjects, the toxicologist focuses on single bodies and generally on short-term effects, because it is difficult to observe long-term effects under laboratory conditions. Ideally, the two branches reinforce each other. The toxicologist establishes the plausibility of the epidemiological findings by uncovering the mechanism through which the body is harmed. If such a mechanism is indeed found, the chance that an association discovered by an epidemiologist is truly a causal one is strongly supported.

In this case, the differences between these branches are relevant because – as will be discussed later – toxicologists had a hard time accounting for the reports put forward by epidemiologists regarding the harmful effects of air pollution.

In regard to smoking, epidemiologists uncovered strong associations, including dose response relationships, whereas toxicologists did not yet find the mechanism by which smoking harmed the body. The association was so strong, however, that it was highly unlikely that such a mechanism was not present, even though toxicologists could not account for it at the time. This was a great triumph for epidemiology, because it illustrated that it was able to detect threats earlier than classical toxicological science could.[7]

2.3.2 *Possibilities and limitations of epidemiological studies*

Epidemiology can be applied to uncover risks that cannot be observed in toxicological studies, because either the bodily effects occur long after exposure has taken place or because toxicological experiments are not possible. For instance, one cannot endanger human test subjects. However, establishing causality is more complicated in epidemiology than it is in toxicology.

The strength of prospective epidemiology lies in its ability to detect trends in disease and to differentiate populations according to more or less sensitive groups. It uncovers potential threats and provides indications that certain groups run a higher risk than others of developing a certain disease. For policy makers, this feature is significant, because it allows for measures that target different groups. During episodes of high air pollution, for instance, older or asthmatic people may be advised to remain indoors, while healthy citizens receive no such warning. It is also very useful for insurance companies when they draft risk profiles. Moreover, agents that affect the body only after long-term exposure can be detected by following cohorts. This type of study has advantages over retrospective studies, since these researchers rely on cohort members providing personal reports with regard to exposure to certain agents. The account may be biased, however, because after becoming afflicted, the respondents themselves begin to mull over the causes. This is known as the information bias. Since prospective studies generally do not rely on reports from afflicted subjects, they avoid this kind of bias (Kabat 2009, p. 30).

Epidemiology scored a number of successes by relying on prospective studies that were used to bolster the findings of earlier retrospective research into the relationship between smoking and lung cancer (Brandt 1993). However, as with all scientific

7. The biological mechanisms by which smoking causes disease have now been identified, and can be found in US Department of Health and Human Services (2010).

research designs, prospective epidemiology has its limitations. One of the most vexing questions concerns when it is justified to make the leap from association to causation. As stated, epidemiology identifies statistical correlations but not causal biological mechanisms. Since laboratory conditions cannot be achieved in prospective epidemiological studies, it is difficult to exclude all kinds of interfering factors, ones that 'muddy' the association of interest because they are related to both the population and to the disease under scrutiny. Such interferences are called confounding factors, and may always account for the associations found, causing the researcher to infer causation from correlation in a misleading manner. If, for instance, one were to compare heart attack rates among the populations of Alaska and southern states in the US, one could mistakenly infer that the risk of having a heart attack is much higher in the southern states than in Alaska. However, the population of Alaska is younger than that of the southern states, and age is correlated with incidences of heart attack. The age variation explains the difference in heart attack rates among Alaskans and other US residents (Kabat 2009, p. 23).

Such a distortion of the data, or misinterpretation of it due to bias, is a persistent danger in epidemiological research, and for this reason the renowned epidemiologist Sir Austin Bradford Hill devised a set of considerations that helped to minimise the chance of wrongly inferring causation from correlation. In his classic 1965 essay, Hill identified nine such considerations.

1 *Strength*

According to Hill, the first criterion of note was the strength of the association found. The associations found in the studies in regard to smoking and lung cancer lay in the range of relative risks from 9 to 30 (Hill 1965, p. 296): that is, one ran a 9 to 30 times higher risk of dying from lung cancer as a smoker than as a non-smoker. Hill considered this a strong association, whereas he considered an association of 2 or more to be weak (Hill 1965, p. 296). Such a weak correlation did not rule out the possibility of causation, but made it more likely that some other confounding factor was the real underlying cause of the association. The strength of a correlation was an important point raised by other epidemiologists as well. Geoffrey Kabat argued that a relative risk below 2.0 was very small. Moreover, relative risk was usually given as the median within a certain margin, and this margin was due to the statistical margin of error. Kabat considered that if this margin included a relative risk of 1.0, meaning no extra risk at all, the statistical significance should be questioned (Kabat 2009, p. 33).

2 *Consistency*

It is more likely that an association is indeed causative when the findings are repeated over time in different studies and in different places. This makes it less likely that a confounding factor is in play, because the association remains persistent over time and within different studies.

3 *Specificity*

A cause may be more easily inferred when a specific type of disease is connected to a particular area or type of work. Working with asbestos and developing a specific form of lung cancer are examples of such distinct incidences of disease among a particular type of manual labourer.

4 *Temporality*

Epidemiology gives rise to questions regarding causality, and one of these is the age-old chicken and egg question: which came first? When an association is found between certain dietary habits and the onset of a disease, for instance, it is tempting to conclude that the diet caused the disease. However, a disease may also interfere with people's eating habits, and in that case the disease caused the change in diet.

5 *Biological gradient*

If a regular dose-response curve can be found – that is, if an increase in the dose of the agent suggests a similar increase in the incidence of disease – a causative association is more likely.

6 *Biological plausibility*

If there is a biological explanation present for the causation found, a causative relation is easier to accept. This criterion, however, should not be considered absolute, because there may be an as yet unknown biological explanation.

7 *Coherence*

However, a proposed causative relation should not conflict with what we already know and have scientifically established. If such a hypothesis is in conflict with other scientific facts, causation is unlikely.

8 *Experiment*

Ideally, some experimental discoveries are present that corroborate the epidemiological findings. For instance, if epidemiological research suggests that dust in the workplace leads to lung disease, one would want to conduct an experiment by reducing the dust and determining whether it leads to a decrease in lung disease cases.

9 *Analogy*

A causative relationship is more plausible if there are similar cases in which the same relationship can be established. For instance, when we know that a certain drug ad Ministered to pregnant women may cause harm to embryos, and we find a relationship between a similar drug and birth defects, it is more likely that the relationship is indeed causative.

These considerations safeguard the scientific rigour of conclusions drawn from the findings, and are still used by researchers as an evaluative framework (Kreis & Cromwell 2013, p. 145).

2.3.3 *Epidemiology and its critics*

After their success in discovering the dangers of smoking, researchers were tempted to uncover all sorts of other relations, especially regarding the incidence of cancer. In the 1970s, the Western diet came under increased scrutiny, and Richard Doll, a prominent epidemiologist well known from research involving the relationship between lung cancer and smoking, pronounced that the Western diet was a cause of many types of cancer (Le Fanu 1999, p. 353).

Other researchers tended to look elsewhere to uncover the cause of this disease, and environmental factors received increased attention, especially after publication of Rachel Carson's 'Silent Spring' in 1962. In this book, she warned of a bleak future resulting from man-made chemicals being released into the environment. Epidemiologists turned to exploring the relationship between all kinds of environmental factors and health problems, and uncovered a host of potential threats. Many pharmaceutical substances were labelled as threats, but non-medical substances had also at one time or another been related to conditions such as strokes, heart attacks, birth defects, and cancer. Epidemiologist Alvin Feinstein (1988a), for instance, noted coffee, water, sugar, saccharin, and alcohol.[8]

However, these associations rarely stood up to rigorous scientific scrutiny. According to Le Fanu (1999), almost all claims pertaining to health effects of environmental factors were false, because the quantities of pollutants to which we were exposed in Western society were too low to have effects on the body (Le Fanu 1999, p. 361). Le Fanu stated that the association between electromagnetic fields caused by electricity pylons and leukaemia was false, the danger of water pollution was overstated, and the putative causative relation between air pollution and asthma could not hold, because incidences of asthma were higher in less polluted cities than in polluted ones (Le Fanu 1999, p. 361). Kabat discredited the relationship between passive smoking and lung cancer, electromagnetic fields and leukaemia, DDT and breast cancer, and air pollution and lung cancer (Kabat 2009, p. 35). In all these cases, the relative risk factors were very small; moreover; it was difficult to assess the type and amount of exposure. It was difficult, for instance, to isolate a group that was subjected to a specific amount of passive smoking, and such exposure information was crucial in order to assess the data accurately and ascertain a significant association. Especially when exposure information is limited and only weak associations are found, it is difficult to infer whether the effects seen are indeed real or are a product of confounding factors (Kabat 2009, p. 36).

The tendency of epidemiologists to link environmental factors quickly to the incidence of disease, and to jump rapidly from correlation to causation, was criticised

8. For a long list, see Brignell (2004, p. 140).

by a number of practitioners in the field. I have already cited Geoffrey Kabat (2009) – who felt health hazards were hyped by epidemiologists – extensively, but one of the most scathing criticisms came from a celebrated epidemiologist. In 1988, Alvin Feinstein, then director of the clinical epidemiology unit at Yale University School of Medicine, published an article in the renowned journal *Science* (1988) in which he criticised the research method and biases of his fellow epidemiologists (Feinstein 1988a). Later that year, he also published a special commentary in the American Journal of Medicine, in which he formulated three succinct points of criticism (Feinstein 1988b).

1. The research methods applied lead to distortions. Groups are compared that seem equal but are not, due to some bias in the selection of the groups or in the agents that are considered responsible for the differences found between the groups. For instance, when one compares the survival rates of groups operated on in instances of cancer with those that did not have the operation, one compares biased groups. Patients that are termed to be operable at all will be healthier than patients that are termed inoperable, and therefore the mortality rates will differ, regardless of whether surgery indeed takes place. Feinstein was especially critical of observational studies. In these studies, groups are formed and are observed in real life situations, such as living close to a highway. Feinstein considered that such comparisons would often be biased by the many natural phenomena occurring in such situations, which could distort the comparison. If possible, such observational findings would need to be corroborated by other experimentally derived evidence (Feinstein 1988b, p. 476).
2. The interpretation of data is prone to 'delusion', because they are interpreted in the light of contemporary beliefs held by the epidemiologist in question. Feinstein cited two classic examples, in which unbiased data were interpreted in line with inaccurate beliefs held about pathogenesis at the time. Even though they were classic examples, Feinstein considered that such delusions were still commonplace (Feinstein 1988, p. 476).
3. 'The consensus syndrome' plays a role (Feinstein 1988, p. 477). Authoritative opinion is established by consensus mechanisms rather than by the application of rigorous scientific analysis. Conferences and panels are called, and authorities in those panels determine what the consensus is regarding certain medical problem in their field. These consensus mechanisms silence 'heretical' voices, and, according to Feinstein, 'hamper progress' (Feinstein 1988b, p. 477).

In addition to the above criticism, Feinstein was repeatedly critical of what he called the double standard in epidemiology (Feinstein & Horrowitz 1982; Feinstein 1988b). In order to accept that a certain agent has a therapeutic effect on a group, massive corroborating evidence is required, but when a certain threat to health is found, no such control checks are demanded. Accordingly, it is much easier to detect and publish findings relating to a possible threat than it is to confirm a possible benefit. This double standard steers epidemiology in the direction of detecting threats.

Criticisms similar to those presented by Feinstein were made by Dutch epidemiologist Paul Knipschild, who until his retirement in 2005 was professor of epidemiology at Maastricht University. In his farewell speech, Knipschild castigated the research done by his colleagues, and stated that he had become tired of explaining all the methodological shortcomings. He expressed a wish for a new type of critical review, such as that of Feinstein in *Science* (Knipschild 2005, p. 10).

According to Knipschild, Feinstein's article in *Science* made 'short work' of epidemiological research that looked for environmental factors as causes of disease (Knipschild 2005, p. 10). In his article, Knipschild levelled another set of criticisms at his colleagues: namely, the use of 'convenience cohorts'. These are groups of people whose habits and living conditions are analysed in order to uncover meaningful associations between lifestyle or environmental factors and incidence of disease. Cohort study is a well-known methodology. However, in the case of convenience cohorts, the groups are formed initially for a research purpose other than that for which the cohort is eventually used. For instance, a cohort formed in order to examine the effects of dietary habits might be used in air research on air pollution. This cohort is used for the sake of convenience, and is not customised in accordance with the research hypotheses under investigation. Feinstein also points towards 'data dredging' as a scientifically problematic practice. Data dredging occurs when *'a large number of statistical associations are explored in an automated manner for diverse individual groups and outcomes'* (Feinstein 1988a, p. 1259). In other words, a group is divided into many different sub-groups – for instance, according to age, gender, smoking habits, and so on – and is subsequently studied within the context of a long list of possible pollutants to see whether a statistically significant relationship can be found between a certain pollutant and the rate of mortality in one of these sub-groups. If this is done with a sufficiently long list of sub- groups, together with a long enough list of possible pollutants, at least one significant association is bound to show up, even if only by chance.

While they pointed out methodological flaws, double standards, and the tendency to interpret data according to preconceived beliefs, critics like Feinstein acknowledged that epidemiology had been successful at detecting threats. Before WWII, epidemiological studies had had *'a more profound impact on public health, infectious disease and individual longevity than any other branch of medicine'* (Feinstein 1988a, p. 1257). After WWII, rigorous epidemiological science contributed to the eradication of smallpox, to demonstrating that rubella causes birth defects, to the finding that fluoridated water protects the teeth, and to the finding that cigarette smoke leads to lung cancer (Feinstein 1988a, p. 1257). The criticism raised was against the practice of warning against all kinds of minor lifestyle and environmental risks.

Prospective epidemiology is a branch of medicine that uses complex statistical analyses to establish associations between lifestyle and environmental factors and disease. Whereas the classical studies on smoking produced strong associations, current research on lifestyle and environmental factors identify much smaller risk ratios. However, since many people are subjected to environmental factors like water or air pollution, even a small increase in the risk of developing a certain disease can lead to a large prognosis of premature mortality (Phalen 2004). The possibility of detecting minute threats makes it very useful to argue for preventative policies based on epidemiological findings. The tendency to focus on lifestyle and environmental threats is reinforced by the fact that threats are quickly identified and accepted as a possibility without thorough meticulous review. The 'double standard' identified by Feinstein holds that while therapeutic effects of substances are subject to rigorous control, screening and replication, epidemiological discoveries relating to threats are published quickly and subsequently eagerly picked up by the media. According to Kabat, epidemiology has become instrumental within a 'sociology of health hazards', functioning within the larger constraints of a society that emphasises prevention and precaution.

2.4 Remaining uncertainties in the Harvard Six Cities and the ACS studies

Now that we have a handle on the possibilities and limitations of epidemiological studies, we turn once more to the main epidemiological studies underpinning the claim that PM10 is a serious threat to health. The HEI review concluded that although the findings were robust, upon closer examination, a number of persistent questions still remained. In this section these uncertainties will be discussed and an alternative explanation is offered.

2.4.1 *Critical objections to the findings in the ACS and Harvard Six Cities studies*

When we review the two key studies and the HEI follow-up, we will see that some of the critics' objections pertain to these studies as well.

Exposure information
In order to establish meaningful associations, exposure information needs to be optimal. In many ways, exposure information in the two studies is not entirely satisfactory. PM10 and PM2.5 are complex mixtures of a differing constitution; for instance, we do not know whether PM from Portage can be compared to PM from Steubenville. To give a practical example, in the Netherlands an important fraction of PM10 will consist of sea salt, which is considered harmless. However, sea salt will not comprise a large fraction of PM10 found in the Czech Republic. Moreover, in the case of the Six Cities Study, air pollution levels from the 1970s were taken into account. Particles were only measured as TSP, Total Suspended Particulates, while PM10 and PM2.5 were measured later. The ACS Study took place later, and contains differentiated

measurements for PM10 and PM2.5 (Krewski et al. 2000, p. 7). Nevertheless, even current measurements of PM10 and PM2.5 are largely uncertain, and correction factors are applied to account for them (Buijsman 2007). The appropriate correction factor is subject to debate, and different countries use diverse factors; according to the RIVM, measurements display uncertainties ranging from 10 to 25%. In light of these findings, I doubt that measurements taken in the early 1980s offer a sufficient degree of certainty upon which to base accurate exposure information.

Information regarding illnesses and causes of mortality is also not the most desirable, as the studies use questionnaires and death certificates to relate exposure to pollution to cause of death. Death certificates, however, are notably untrustworthy sources of information. According to Feinstein, a certificate of death is '*so untrustworthy that very few thoughtful clinical scientists pay any attention to it*' (Feinstein 1988b, p. 477).

Small relative risk factors

The risk factors found in the Six Cities, ACS, and HEI studies are all quite small. In the table below, the relative-risk factors in the ACS and Six City studies for PM2.5 are displayed together with their confidence level, indicating a certain margin for error.

Table 3

Cause of death	Six Cities PM2.5	ACS PM2.5	ACS SO_4 (sulphates)
All causes	1.26 (1.08 – 1.47)	1.17 (1.09-1.26)	1.15 (1.09-1.22)
Cardiopulmonary	1.37 (1.11 – 1.68)	1.31 (1.17-1.46)	1.26. (1.16-137)
Lung cancer	1.37 0-81-2.31	1.03 (080-1.33_	1.36 (1.11-1.66)
All others	1.01 (0.79-1.30)	1.07 (0.92-1.24)	1.01 (0.92-1.11)

Table taken from Pope & Dockery 1999, p. 691

The HEI validation confirmed the findings and noted relative risks of 1.28 for the Six Cities Study and 1.18 for the ACS Study. These associations are still very small, and should be treated with caution, in accordance with Hill's previously mentioned considerations (Hill 1965). Scientists generally counter this finding by pointing out the consistency in the results over the course of different studies (for example, MNP 2005, p. 57). They apparently attach more value to the second of Hill's considerations than to the first. However, little epidemiological research has been carried out regarding the long-term effects of PM10, and at least one substantial study, involving US veterans, does not corroborate the Six Cities Study and the ACS Study. Lipfert, the epidemiologist who concluded that study, argued that epidemiological research regarding PM10 did not sufficiently meet the criteria laid down by Hill to be able to conclude that a causal relation had been uncovered (Lipfert 1994 in Phalen 2004). However, the MNP, for instance, consider that this study was less robust, and that the US studies taken into account were consistent (MNP 2005, p. 57).

Confounding factors

Confounding factors are ones that relate to both the group under consideration and to premature mortality. The presence of these factors makes it difficult to state definitely that a certain pollutant is the cause of premature mortality. HEI researcher Daniel Krewski himself cast doubts upon whether the premature mortality was in fact due to PM, or to PM in combination with something else. He demonstrated that mortality was associated significantly with sulphur dioxide, SO_2, in addition to PM2.5. This is ironic, because the combination of smoke and SO_2 was considered dangerous before the 1990s. Later on, science moved away from considering particles and SO_2 together. In an article submitted to the American Journal of Toxicology and Environmental Health, Krewski et al. (2003) concluded: '*Collectively, our re-analyses suggest that mortality may be attributed to more than one component of the complex mixture of ambient air pollutants in urban areas in the United States*' (Krewski et al. 2003, p. 1548). This sounds less convincing than the conclusion of the Six Cities Study and others that air pollution by itself was responsible for a large number of premature deaths.

In addition to the possible distorting presence of other pollutants, the review found a significant modifying effect of education on mortality attributed to PM. It appears that the relative risk of mortality due to PM exposure was less among people who had achieved higher levels of education (Krewski 2003, p. 1547). In other words, the higher the level of education, the lower the chance of succumbing to a pollution-related disease. This is odd, because why would a well-educated person be less prone to the negative health effects associated with Particulate Matter than a less educated one? This finding points to education as a potentially large confounding factor that leads to what Feinstein refers to as distortion.

Krewski and co-workers continued to monitor air pollution, using epidemiological cohort studies. In 2009, another analysis of the cohort appeared (Krewski et al. 2009), and the same association between a high level of education and mortality resulting from air pollution showed up (Krewski et al. 2009, p. 2). In their conclusion, the researchers attributed this to chance, but in section 4.3.3 I will construe a hypothesis in which the finding that air pollution mortality is related to education plays a part.

Experimental validation

The group used in the ACS Study was a convenience cohort, used originally to check incidences of cancer. Both studies and the follow-up study were observational ones in which cohorts were followed in real-life conditions. Those kinds of studies were criticised by Feinstein and Knipschild, who claimed the research should be corroborated by experimental evidence. One of Hill's considerations was the presence of a plausible biological mechanism as well.

However, at the beginning of the 21st century, numerous uncertainties still existed about the causal relation between air pollution and observed harmful effects in the human body (Hanekamp & Pieterman 2005; Van Bree & Cassee 2000, p. 33; MNP 2005, p. 56; EPA 2006, attachment A p. 5, Ayres et al. 2008, p. 76) Toxicological studies listed by the WHO (2003) found only mild or no effects, and effects were noted only at high concentrations (200 mug/m^3) (WHO 2003, p. 13).

Many different possible biological mechanisms were hypothesised, but none could lay claim to being the definitive explanation, though Buringh and Opperhuizen (2002) provided an ingenuous scheme with a number of feasible explanatory mechanisms and their possible complex interrelationships (Buringh and Opperhuizen 2002a, p. 38). Robert Phalen provided a list of the most commonly proposed biological mechanisms as well, and stated that all of them could explain the epidemiological findings. However, it was uncertain whether they were present in sufficient quantities in susceptible persons to produce illness or death (Phalen 2004). Currently, the most prominent hypothesis is that PM10 causes what is termed 'oxidative stress', which in turn causes inflammation of the lungs (MacNee & Donaldson 2003; Van Bree & Cassee 2000, p. 26; Ayres et al. 2008, p. 92). Oxidative stress occurs when the body cannot detoxify a reactive oxygen species (ROS), reactive molecules that contain oxygen. These molecules are produced naturally as a consequence of metabolism, but the hypothesis is that air pollution causes their number to increase and cause damage to cell structures. ROS and resulting oxidative stress have been implicated in a variety of adverse effects, such as aging and the development of cancer.

Oxidative stress as a biological explanation for the epidemiological results is in itself not significantly strong. The problem is that the ROS that cause oxidative stress occur naturally as well as through environmental factors; in addition, the occurrence of oxidative stress can be due to a multiplicity of causes. Air pollution is considered to be one of them (Yang & Omaye 2009), but diet is also implicated (Hou et al. 2013), as are sunlight (Wenk et al. 2001) and strenuous physical exercise (Powers & Jackson 2007)

It is also implicated in the occurrence of a wide array of disorders, from asthma and COPD to bipolar personality disorder and depression,[9] and therefore it is hard to pinpoint what exactly oxidative stress does. It has many causes and is implied in many symptoms, and as a result it is questionable as to whether air pollution and the symptoms it causes can be fully attributed to it.

2.4.2 *Assessing the claims of early air quality research*[10]

According to Kabat (2009), inaccurate exposure information coupled with weak associations is a sign that the claim of uncovering a causal relationship is questionable. In

9. Website oxidative stress, last accessed 12-06 2015.
10. Many more uncertainties are present in the case of air quality assessment and measurement (Buijsman 2007). Here I focus mostly on epidemiological science, because epidemiologists acted as the primary claims makers.

this case, experimental evidence is thin as well, although there are counter-arguments. The HEI study is updated frequently, a number of studies have found roughly similar estimates of relative risk, and we know air pollution can be harmful on the basis of situations in the past, such as the deadly London fog in 1952. The findings correspond to some of Hill's considerations but not to others and where they do not seem to correspond to the criteria, additional explanations may be provided. The weak strength of the association is accounted for by pointing towards the consistency of the findings; a weak level of specificity may be countered by arguing that a biological gradient is determined; and in answer to the lack of experimental validation, biological plausibility may be put forward. However, the uncertainties remain considerable and are persistent.

The conclusion of the HEI 2009 study, mentioning the mysterious inverted association between education and premature mortality due to PM may provide an angle for a further hypothesis: namely, one that considers air pollution to be an indicator for low socio-economic status, with the result that people living areas afflicted with different problems, will be found to die earlier. I will expand on this idea further in the next section.

2.4.3 *Air pollution as an indicator for low socio-economic status*

The epidemiological and toxicological findings regarding PM10 were highly uncertain, but they did lead to social unrest and to regulation at the EU level. The findings that linked air pollution to premature mortality were also persistent, even though the risks were small. However, it remained a concern due to the large number of people subjected to this small risk.

The consistent findings of cohort studies suggested that people who lived in polluted areas had more health problems than people who lived in non-polluted regions. The question is whether air pollution is solely responsible, or only one of many factors that explained those health problems? The association between air pollution and mortality was weak, and exposure-related information was not convincing. This means there is room for an alternative explanation that relates both to air pollution and to life expectancy. Socio-economic status, as indicated by the level of education, relates to both. Socio-economic status is related to air pollution, because people with a low income may well end up living in areas where the pollution is more prominent: alongside highways, for instance. A low socio-economic status is also related to premature mortality, as was admitted in the research: '*It is possible that educational attainment is a marker for socioeconomic status, which is known to be correlated with health status*' (Krewski 2003, p. 1547). This suggests that exposure to fine particles might be one of a number of threats that cause premature mortality in subjects with a lower social-economic status.

People in poor economic circumstances may well live in areas that are more highly exposed to PM10, such as busy roads, and may die earlier than people who are financially well off (Centraal Bureau voor de Statistiek 2010, p. 35; National Audit Office 2010, p. 4). In general, living in areas with bad air shortens life expectancy. Such areas tend to be poorer, noisier, and generally less wholesome than other places. Air pollution might be an indicator for whether living in a certain area is healthy altogether. Moreover, people with a lower socio economic status tend to be exposed to other unhealthy environmental factors such as poorer diet, less exercise and poorer housing.

According to Dutch scientist Rob Maas, it could not be ruled out that the PM problem is actually a socio-economic issue, as many associations can be found between cardiopulmonary disease and factors affecting less affluent groups. Strong associations can also be found between cardiopulmonary diseases and traffic noise. Moreover, cardiopulmonary disease is also related to the quality of housing and to the diet of low-income families (Maas 2007, p. 367).

Krewski and co-workers asserted that the association between education and premature mortality due to pollution could be due to chance. This is a possibility, but it is not a strong argument, because the findings themselves display weak associations that are minimally statistically relevant. The researchers themselves cast doubts on the statement by conceding that the same negative association between a high level of education and premature mortality due to lifestyle factors shows up in Dutch epidemiological research (Krewski 2009, p. 134). Therefore I consider it more probable that air pollution is indeed one of the many risks that people living in poor neighbourhoods are exposed to and that, taken together, indeed lead to a life expectancy that is lower than that of residents in affluent neighbourhoods.

2.5 Dutch epidemiological research and claims making by health agencies

The mentioning of Dutch epidemiological research by Krewski's team is not coincidental. Dutch epidemiologists have contributed significantly to our knowledge of air pollution and especially to uncovering associations between premature mortality and living close to highways. The results of these studies have led to claims making in 1999 by the Rotterdam Municipal health agency that the air pollution in the Dutch residential area of Overschie was problematic and a health threat to the residents of this area. In this section first the work of Dutch epidemiologists will be discussed and subsequently the claims about the health situation in Overschie will be recounted.

2.5.1 *Dutch epidemiological research*

In the Netherlands the research findings of US scientists were noted early on. Dutch health and environmental institutes turned their attention to PM10 early on and PM10 was measured in the Netherlands generally earlier than it was in other European countries (Hans Eerens, interview). This was partly the result of cooperation between Dutch and US health institutes in the 1980s and the modern measuring network present in the Netherlands (Clarenburg 1995; Buijsman 2003).

Dutch epidemiological scientists worked closely with their American colleagues as well. For the Netherlands, Dutch research conducted by epidemiologists Gerard Hoek and Bert Brunekreef is of importance. Brunekreef's work is of particular interest, because it played a part in the construction of PM10 as a serious problem for public health in general, not only in the Netherlands. He contributed to the first EPA standard setting for Particulate Matter, and to research supporting the EU air-quality Directives, according to the tribute paid to him in honour of his winning the Heineken Prize for Environmental Sciences in 2008. He won this prize for his ground breaking research on the health effects of air pollution.

Brunekreef was also present at WHO meetings in 1987 which resulted in the drafting of the WHO guidelines for air quality. As will be explained in the next chapter, these guidelines were a crucial factor in the construction of the EU air quality regulation. Additionally he was a member of the Dutch Health Council. He worked closely together with the epidemiologists who had conducted the Harvard Six Cities Study, for instance, resulting in a publication regarding the association between asthma and dampness in the home (Brunekreef et al. 1989), and later on regarding outdoor air pollution in 1991 (Brunekreef et al. 1991). All of this made him an authority on the topic of Particulate Matter.

In 1995 he received media coverage because of his research on the relationship between asthma in children and living close to highways. The notion that living close to highways is dangerous is a persistent emblem in the discussions in the Netherlands on PM10. One of the places where he conducted his research is in Overschie, a small residential area that would become the focus of a health scare over air pollution in the Netherlands.

2.5.2 *Claims made by the Dutch Municipal Health Agency*

The research of Dutch epidemiologists received increased attention in the late 1990s. In 1996, an investigation was conducted at Wageningen University on the health of children living near busy roads, and epidemiologists published findings that the children's health had been significantly affected (Van Vliet et al. 1996). The research was subsequently published in an international journal (Brunekreef et al. 1997).

One of the municipalities in which Brunekreef and his team conducted their research was the residential region of Overschie, which became an important area in the history of the air-quality clash, owing to a report from the municipal health service

(Gemeentelijke Gezondheidsdienst, henceforth GGD) published in 1999 (Hegger & Slob 1999). In this report, the researchers mentioned a problem with air quality in Overschie. The team proclaimed that living near the Overschie highway was damaging to health. They presented the research findings in the form of a trope by proclaiming that living in Overschie was like passively smoking 17 cigarettes a day (Hegger & Slob 1999, p. 4). This trope became a persistent emblem in the public's perception of the PM problem.

The use of this emblem was instrumental in attracting political attention from the Minister[11] Jan Pronk for Housing, Spatial Planning and the Environment, (Volkshuisvesting, Ruimtelijke Ordening en Milieu, henceforth VROM) (Ecorys-Kolpron 2002, p. 34). Minister Pronk of VROM paid visits to the area, and these visits generated media attention (Dutch newspapers: (NRC 11-02-1999; Trouw 11-02-1999). He quickly commissioned follow-up research conducted by the same team of epidemiologists who had conducted the research near Overschie, including Bert Brunekreef. This research was broader in scope, and confirmed the earlier findings (Aarts et al. 1999). Pronk received the results of the study in 1999, and sent it to Parliament.

In Parliament, he reiterated that living in Overschie equalled the passive smoking of 'a packet of cigarettes per day'.[12] By this time, he had become committed to the situation in Overschie, and he acknowledged the necessity of curbing the increase in traffic and transport. Partly on Pronk's instigation, a measure was considered for Overschie to push back air pollution in the area; to this end, the speed limit was reduced from 120 km/h to 80 km/h in 2002. This measure was based in the premise that vehicles that drive slower emit fewer pollutants. The scientific findings caused Pronk to oppose further expansion of roads, but before a clash between the Ministries of VROM and Transport and Water Management could erupt, the Cabinet resigned in 2002, owing to a wholly unrelated matter.

The issue of Overschie and the speed limits will be elaborated on further in chapter 6, but for now it is important to note that Dutch epidemiologists and public health agencies managed to mobilise public and political support in the Netherlands around the storyline that living close to highways was damaging to health. This storyline emerged from claims by Dutch and international epidemiologists that found a correlation between PM and premature mortality. The connection with highways remained

11. In the Netherlands the office of Minister refers to the highest officer of state appointed to head an executive or administrative department of Government. It is comparable to the title of Secretary of State in the UK. In the Netherlands on the other hand, the title Secretary of State denotes a second tier state official comparable to an Under Secretary of State in the UK. The Dutch political constellation will be further explained in chapter 3.
12. Handelingen I 1999-2000 nr. 21, p. 925. Because of their length, throughout this book references to Dutch policy documents and court cases will be made in foot notes rather than in text.

present throughout the debates in the Netherlands, and caused the air-quality clash to become a struggle about mobility and its possibly pernicious influence. Brunekreef remained a powerful figure in the PM10 discussions. In 2002, he published an article in the renowned medical journal *The Lancet*, in which he sounded the alarm bell over the health effects of transport. This article forced the Minister of Verkeer en Waterstaat (Transport and Watermanagement, hereafter V&W) to issue a letter detailing the adverse health effects.[13] These debates raised public awareness that transport could be more harmful than expected, a case made earlier by former Minister Jan Pronk.

In particular, the simplistic translations of complex chemical phenomena into a concrete threat by the GGD facilitated the creation of problems by attracting the attention of politicians and the media. The connection with passive smoking meant that the threat was immediately recognisable to the public. Epidemiologists themselves generally mentioned the uncertainties inherent in their findings, but these were overlooked in public and political discussions. The public recognition of the storyline that PM10 from cars and other vehicles damaged public health began for the Netherlands in Overschie.

2.6 Concluding remarks

In conclusion of this chapter 2 further remarks will be made on the important claims making work by epidemiological scientists and the characterization of the conflict over air pollution as a typical risk society problem. I want to highlight that the air quality clash is a clash over the perceptions of risk. The question is who should bear the risks generated by economically important arrangements such as transport and mobility.

2.6.1 *Experts as claims makers*

The claim that the emission of PM10 leads to severe health damage was mainly brought forward by scientists and health agencies. In the US the claims of scientists led to a revision of the standards for PM10 by the US EPA, in the UK they emerged together with a public health scare over asthma as will be recounted in the next chapter and in the EU they influenced policy making through a variety of ways. As will be explained, the EU based their regulation on WHO air quality guidelines that were in turn influenced by the main epidemiological findings discussed in this chapter and the studies were known by the EU expert committees assigned to investigate the desirability of setting air quality standards as well.

13. Kamerstukken II 2002/2003, 28 600 XI nr. 73.

In the Netherlands the storyline that bad air quality was bad for the health of children emerged through epidemiological research by Dutch epidemiologists, but especially through the claims making activities of the Rotterdam GGD. The emblem that the risks of air pollution in Overschie were comparable to passively smoking 17 cigarettes a day struck a nerve. This emblem bypassed difficult numerical risk estimates of which the implications were unclear to the public. Instead it offered an immediately comprehensible picture of the problem at hand. All through the Netherland air quality standards were exceeded in 2005 and the air quality in Overschie was not much worse than it was in the rest of the country, but this frame turned the area in an emblematic case of concern during the air quality clash.

In the case of the Netherlands and probably in the US too, experts functioned as the primary claims makers that bad air quality due to particulate matter was a significant cause for concern. In the Dutch situation the storyline that bad air quality was bad for health focused mainly on transport and traffic as an important culprit. It received an emblem in the situation in Overschie and the area would feature in many of the early political discussions on air quality. Also during the heydays of the clash, in 2005, 'Overschie' featured in a television documentary (Zembla) about the negative effects associated with traffic and transport (Website Zembla, last accessed 23-06 2015).

By itself the issue of Overschie cannot explain the emergence of the air quality problem, an investigation of larger political processes is necessary to give a complete account. The storyline that bad air quality caused by traffic, however, supplied proponents of environmental and health interests with ammunition in subsequent Parliamentary debates and extra Parliamentary social campaigns against bad air quality.

2.6.2 *Epidemiology as a precautionary science*

The social construction of air pollution as a health threat in the Netherlands began with warnings from epidemiological scientists that the effects of air pollution may be underestimated. A critical examination of this process reveals that epidemiology fits within a precautionary attitude of the population regarding health threats (Pieterman 2008). In this case epidemiology acted as a kind of 'precautionary science' in the social construction of problems regarding air pollution. With this characterisation I denote a science that is capable of detecting potential threats early on, but is not capable of determining the exact nature of these threats or their magnitude. Therefore the exact nature of the threat remains unknown and a solution is difficult to determine. This situation fuels the anxiety among the public about this elusive threat.

In the wake of Rachel Carson's bestseller 'Silent Spring', published in September 1962, epidemiologists looked for environmental factors that could explain the occurrence of all kinds of diseases. Meticulously, they began to identify possible environmental threats (Kabat 2009). In fact, epidemiology fits well within a society that is increasingly worried about all sorts of health and environmental threats, because it is a medical science that is exceedingly apt at detecting them. Researchers use a cohort, divide it into subgroups, detect their mortality rates over time, and then relate these to a wide range of different environmental variables in order to determine possible associations (Feinstein 1988a). When such associations are discovered, the alarm bells are sounded. For policy makers, it is easier to err on the side of caution, since overlooking a potentially grave health threat is sure to lead to political unrest.

The way in which research on air pollution was used to attract attention to the plight of residents in Overschie involved both the aptitude of epidemiology to detect threats and the possibility of attracting media attention by presenting findings in a certain ominous light. The risks of dying from lung cancer caused by air pollution in Overschie were minimal, and of a similar small magnitude as the risk associated with passive smoking. The chance of dying from lung cancer due to passive smoking was minimal as well. However, in the public eye, passive smoking was a potential threat, due to its association with active smoking. By linking the two, the threat of air pollution was made immediately tangible. It led to the first political measures of the 2000s with regard to air pollution, and provided the breeding ground for a new social problem to emerge.

2.6.3 *Conflicts over air quality as 'risk society conflicts'*

The scientific findings regarding air quality and its impact on health are rife with uncertainties. One of the oddities in the research is that the effects of air pollution are negatively correlated with the level of educational attainment. Moreover, the problems with air quality appear to be mainly associated with living close to high ways and other less wholesome residential areas. This raises the question whether the air quality problem should be considered in isolation or whether it should be considered part of a set of bigger social economic problems that are faced by poorer classes in society.

I consider that this social component would make the conflict about air pollution a typical risk-society issue. Air pollution is one of a number of dangers created by our current way of life, together with other risks relating to lifestyle and socio-economic class. In the present risk society, people compete with one another over the risks they should be expected to bear and how to reduce the ones imposed (Beck 1986). The conflict regarding air pollution is essentially, an environmental spatial conflict (De Roo 2003), and not a medical one. The conflict over air pollution is a conflict about the use of space and the risks associated with it. It is part of a larger conflict about who should bear the risks connected with everyday activities. A road through a residential area may be a profitable use of space for car owners, but it creates risks for the residents of the area. By and large, these risks are uncertain, and therefore policy makers cannot avoid them. The conflicts of risk distribution have to do with cultural values, such as the freedom of mobility versus the right not to be subjected to the polluting activities of others. I contend that the Dutch air quality clash which would unfold in 2005 is a clash over values such as the use of space for economic or residential purposes, the desirability of the expansion of roads and the perceived right of citizens to be protected from pollution by precautionary measures. These are mainly political questions and therefore we will turn to investigate Dutch and European environmental and air quality policy in the remainder of this book.

Timeline Epidemiological research and claims

1993	Harvard Six cities study published by Dockery et al.
1995	ACS Study published by Pope et al.
1996	Van Vliet et al. publish research about respiratory health of children living near highways
1997	Brunekreef et al. publish finding about respiratory health of children in journal Epidemiology
Jan. 1999	Aarts et al publish VROM commissioned follow up research about health of children in report
Feb. 1999	Minister Pronk visits Overschie
June 1999	Hegger and Slob (GGD) publish report on health in Overschie in which they compare air pollution to passive smoking
2000	Review of ACS and Harvard Six Cities Study by Krewski et al.

3 Dutch ecological modernisation and environmental regulation: 1972 – 2002

Introduction

After the investigation into the initial construction of PM as a health threat by scientific experts, we now turn to investigate the discursive political context within which the air quality emerged. We will look at the emergence of a policy discourse that shaped the way environmental politics has been conducted in the Netherlands up to this day, ecological modernisation. The emergence itself will be examined and the continuities and discontinuities in Dutch environmental policy will be analysed. Moreover, we will discuss whether ecological modernisation could live up to its promise to reconcile environmental and economic interests.

Considerable ground on the topic of Dutch environmental policy and its history has already been covered by various authors (Van Tatenhove 1993; Arentsen et al. 1993; Hajer 1995; Van Tatenhove & Goverde 2007). I wish to emphasise certain aspects that are important to understand the context in which the clash over air quality appeared in the Netherlands, and that also explain the Dutch position in the European negotiations on air quality.

The examination starts with policy making in the 1970s and early 1980s. The review of this period sheds light on several important ingrained characteristics of Dutch environmental policy, and understanding these characteristics helps us to comprehend the latter policies on air quality.

Subsequently, the turn to ecological modernisation will be described, with special attention being paid to how modern managerial discourse entered environmental policy, and infused it with the storyline that environmental policy creates economic opportunities. This storyline corresponds with one of the most important postulates of ecological modernisation: namely, that a positive sum game between environmental protection and economic development is possible.

In the third section, the aftermath of the turn to eco-modernism is reviewed, and a mixed picture of success and failure is presented. Ecological modernisation managed to alleviate environmental conflicts in some areas, but was less successful in others, notably transport.

I will conclude by arguing that the strong emphasis on consensus and economic opportunity led to a high degree of ambition on the level of principles and plans, but also to a pragmatic implementation of concrete policies. Following an old saying, Dutch policy is characterised in this chapter as an example of the tension between an idealistic 'reverend', who preaches that ecological interests need to be taken seriously, and a pennywise 'merchant', who keeps an eye on his wallet.

3.1 The early days of environmental policy in the Netherlands

This study comprises a time span of roughly 30 years, from the onset ecological modernisation to the eventual resolution of the air quality clash. However, to understand the emergence of ecological modernisation in the Netherlands it is necessary to go back even further, because ecological modernisation was the discursive answer to the political problems environmental policy encountered in the 1970s. The nature of Dutch ecological modernisation can only be understood by assessing the continuities and discontinuities of environmental policy as it was shaped in that decade.

Environmental policy making during its first years had four main characteristics: the tactic of mobilising allies behind the environmental cause; the adoption of a programmatic approach to environmental policy; the setting of environmental quality standards; and the entwinement of spatial planning and environmental protection. In addition, I will focus on the discrepancy of a highly idealistically charged also play a part in the air quality clash but it is of interest to note that they were central features of Dutch environmental policy from its very inception.

The emergence of the first Dutch Ministry dealing with environmental matters, the Ministry of Public Health and the Environment (Volksgezondheid en Milieu, hereafter VOMIL), will be outlined in section 3.1.2. In this context, the strategy of mobilisation will be discussed in section 3.1.3. Subsequently the long term programmatic nature of Dutch environmental policy is under discussion by reviewing the first environmental policy plan, the Urgent Memorandum on Environmental Hygiene. The use of environmental standards as policy instruments is discussed in section 3.1.4 and in section 3.1.5; the peculiar Dutch entwinement of the domains of environmental policy and spatial planning is under scrutiny. In the final sub section an overview is presented of Dutch environmental policy in the context of the limits to growth discourse and the discrepancy between broad environmental rhetoric and modest policy initiatives in practice. However, before we review the history and nature of Dutch environmental policy, some explanation of the Dutch policy process is in order for a sound understanding of the following section, but also in order to shed light on the event recounted in the next chapters.

3.1.1 The policy process in the Netherlands

The Kingdom of the Netherlands is a constitutional monarchy. The king is formally the head of state. Moreover the king is part of the Government, but the fact that Ministers are held responsible for acts of the king in the Netherlands ensures his position is symbolic. The term 'Government' in its narrow sense refers to the executive branch of the state formed by the Dutch Cabinet of Ministers. In this study the term is applied in a broad sense to include the whole machinery of civil servants and representatives and to refer to the Dutch Government without specifically referring to a certain Cabinet. When the term Cabinet is used, specific reference is made to a certain Cabinet of Ministers in power at the time. Usually Cabinets are referred to by referring to the prime minister chairing a Cabinet, such as the Balkenende Cabinets from the 2000s or the political colour of the parties forming the Cabinet such as the so called 'purple' Cabinets of the 1990s.[1]

The legislative branch of the state is formed by the Government and two chambers of Parliament, the second and First Chamber. The main function of the Second Chamber of Parliament is to control the Government and to draft legislation, together with the Government. Both the Government and Members of the Second Chamber of Parliament may propose legislation and legislative proposals have to be accepted by the Second Chamber in order to become law. The Second Chamber holds the right to amendment meaning it can propose amendments to legislative proposals.

The function of the First Chamber of Parliament is to study and check the legislative proposals accepted by the Second Chamber in Parliament. The First Chamber does not hold the right to amendment, or the right to propose legislation.

When a proposal is sent to the Second Chamber, it is first discussed by the commission of Parliamentarians that deal with the relevant field of policy. The Minister responds to the questions and remarks made by the commission. Subsequently, the proposal is discussed in a plenary session in the Second Chamber of Parliament, and Parliamentarians have the possibility of introducing amendments to the proposal. After the plenary discussion, there are votes about the proposal and the amendments introduced. If the proposal passes, it is sent to the First Chamber of Parliament. The procedure is generally quicker here because the senators do not have the right to amend proposals, and the First Chamber has a less political profile. If the proposal passes, it is then sent to be signed by the reigning monarch and the responsible Minister. The law is considered to be in force once it has been published in the 'Staatsblad', the official Dutch gazette (Van Deth & Vis, 2006, p. 87).

1. After the mixture of blue for the liberal VVD and red for the socialist PvdA. The third party in this coalition was the left liberal D66. See for more explanation on Dutch political parties appendix 3.

Within the Dutch system lower administrative bodies such as municipalities and provinces have a significant degree of autonomy, owing to the goal of decentralisation, prominent in Dutch politics. These bodies take administrative decisions about infrastructural matters for instance. These decisions can be appealed before the administrative court. In this study, the term 'lower administrative bodies' is usually used to refer to municipalities.

A legislative proposal must be sent to the Council of State for advice before it is sent to the Second Chamber of Parliament. The Council of State is a reputable body of the State and it consists of two separate sections, the Advisory Division and the Administrative Jurisdiction Division. The Advisory Division advises the Government on legislative proposals and the Administrative Jurisdiction Division acts as the highest administrative court. More on the Council of State may be found in the text box on page 131.

The procedure discussed above must be followed when the highest legislative instrument is proposed, a so called 'formal law'. Various laws though delegate powers to the Government and in those cases the Government may enact regulatory instruments of a lesser rank based on the competencies granted in formal laws. The most important of those are the General Administrative Order and the Ministerial Decree. A general Administrative Order is a decision by the Government, and must be based on a formal law that delegates decision making powers to the Government. The Ministerial Decree may be issued by one Minister alone, on the basis of delegation by formal law or by General Administrative Order. A General Administrative Order must be discussed in the Cabinet of Ministers and the Advisory Division of the Council of State must be heard, whereas such is not necessary in case of a Ministerial decree.
Each Cabinet Minister is responsible for a different field of policy or for a number of different fields, organised around directorates general or departments. The Cabinet Minister, or Minister for short, is the highest ranking Government official responsible, but he may delegate certain aspects of policy making to a Secretary of State. The environment for instance is at times in the portfolio of a Minister and at times in that of a secretary of state. If a subject is delegated to a secretary of state, then usually it is considered of less political importance.

For a sound understanding of the Dutch political landscape it is necessary to consider that the Netherlands is always run by coalition Governments. The Netherlands has a multi-party system and election votes are counted based on the principle of proportionate representation without a threshold. In practice no political party establishes an absolute majority of 51% or Parliamentary seats. There are 150 seats in the Second Chamber of Parliament. The political parties that may establish a Parliamentary majority together negotiate after elections in order to explore whether they think they could establish a working coalition Government. This coalition system ensures that Governmental power tends to rotate among numerous parties and that political parties favour consensus style policies because alienation of other parties is strategically unwise. After the next election one may need to enter negotiations with that party in order to explore possibilities for a coalition.

This system also leads to a Government consisting of ministers with different political colours working together within one Cabinet. Formally Government policy is unitary and Ministers are required to uphold that fiction. Even though in practice friction between various ministries sometimes occurs, the Dutch democratic system is consensualistic (Lijphard 1968) and has a long tradition of plan based policy, especially in the realm of spatial planning (Hajer & Zonneveld 2000, p. 339). Both these aspects are important in light of this study and the following sections.

3.1.2 *The institution of the Ministry of VOMIL*

In the early days of environmental policy, starting at the end of the 1960s and the beginning of the 1970s, environmental concerns were tied closely to issues of public health. The creation of the first Dutch Ministry for the Environment illustrated this connection. The Directorate General for the Environment was combined with that of Public Health to form the Ministry of VOMIL.

The Ministry of VOMIL was created in 1971, at a time when many Dutch citizens were concerned about the poor quality of the environment. In the Rotterdam Rijnmond area, citizens had clashed with politicians over plans for further industrialisation near the Rotterdam harbour and over the resulting air pollution (Boender 1985). The Netherlands had undergone a rapid phase of industrialisation, and heavy industries such as steel and the petrochemical industry had become dominant economic factors, ensuring the speedy recovery of the Dutch economy after the war. However, this type of industry was also highly polluting. In addition, complaints about nuisance from air pollution increased in the 1960s.

In the 1960s, the fledgling environmental movement was strengthened by a pervasive mood of social change in Dutch society (Kennedy 1995). At the start of the 1970s, 45% of the population ranked environmental degradation among the top five societal problems. That number was well above unemployment (5.5%), inflation (8.1%), and economic problems (5.5%) (Van Der Heijden 2000, p. 58).

The first Minister of VOMIL was Louis Stuyt, a medical doctor who joined the Catholic political party, KVP, just before becoming Minister. Later, this party would merge with other Christian parties to form the Christian Democratic Appeal (CDA). From 1973 until 1977, the position of Minister was held by Irene Vorrink of the socialist PvdA.

The new Ministry was not short of societal goodwill, but it lacked means. Moreover, it did not have competency in all environment-related issues, but was competent solely in environmental matters insofar as they had public health aspects. Non-health-related environmental policy belonged to other Ministries, such as the Ministry of Spatial Planning, the Ministry of Economic Affairs, and the Ministry of V&W. This last Ministry was to a large extent competent in the field of air pollution caused by transport and traffic (Van Tatenhove 1993, p. 20).

To increase its influence, the new Ministry used a tactic of cooperation and accommodation. During the course of the 1970s, it created various advisory committees and scientific institutes that allowed the Minister to spread the message of environmental care. One of the most important was the Preliminary Central Council for Environmental Hygiene (VCRMH, later the CRMH) in 1974 (Van Tatenhove 1993, p. 36). The new Council for Air Pollution (Raad voor de Luchtverontreiniging) also acted as an advisory council for VOMIL, and the National Institute for Public Health (RIV) started to work closely with this Ministry. In 1984, the National Institute for Public Health became the National Institute for Public Health and the Environment (Rijksinstituut voor Volksgezondheid en Milieu, RIVM). The VCRMH acted as a think tank for future environmental policy, and the RIVM was its scientific institute.

In addition to these formal allies, the Ministry approached the more moderate section of the environmental movement (Hanf & Van de Gronden 1998, p. 158; Van der Heijden 2000, p. 59). A pivotal role was played by the Stichting Natuur en Milieu (Foundation for Nature and Environment, henceforth SNM). This foundation was instituted as a co-operative effort by a number of environmental associations, and became a lobby group for the environmentalist movement. In addition, the Ministry co-opted sections of the environmental movement into the CRMH, by inviting representatives of four environmental pressure groups into this Council (Van der Hoek 1996). In its early days, accommodation was already part of the strategy of the Ministry of VOMIL, and this consensual spirit has remained in environmental policy.

In 1972, the new Ministry set out to publish its views on environmental policy in the seminal paper 'Urgent Memorandum on Environmental Hygiene'. This document was analysed extensively in Hajer 1995 and also in Arentsen et al. 1993. I will not repeat their analyses, but Hajer mostly signalled the differences between the Urgent Memorandum and recent Dutch environmental policy, and I would like to pay particular attention to certain abiding characteristics of Dutch policy. The first item of note was the importance attached to public awareness of environmental problems and behaviour change. The document contained an ecological vision in which the earth was understood as a number of interrelating eco-systems. Our way of living at that time did not conform to this ecological vision; a harmony once present in agrarian societies had been lost. The only solution was to direct our thinking towards more ecologically appropriate behaviour (VOMIL 1972, p. 6, 11). This behaviour came down to respecting the limits placed upon us by our ecological situation. This view of the environment as a limit to expansion was taken from the famous report for the Club of Rome, The Limits to Growth, also from 1972.

To achieve this more appropriate behaviour, environmental education and nurturing an 'environmental mentality' via 'influencing consumers' were suggested. The importance of public participation behind environmental goals was highlighted. The Urgent Memorandum stated: '*...normative reasoning needs to be made clear on a bigger scale than today*' and '*contemporary environmental policy demands a high volume of information directed at the public*' (VOMIL 1972, p. 22). Moreover, the document emphasised the desirability of involving the organisations of enterprises, such as the agrarian sector (VOMIL 1972, p. 55) and the waste removal sector (VOMIL 1972, p. 41). In general, the Memorandum considered that industrial sectors should accept responsibility in resolving the environmental predicament (VOMIL 1972, p. 24).

Another noteworthy and abiding feature of Dutch environmental policy was its preference for international co-operation. Environmental policy should be co-ordinated by international bodies, and deals preferably be made internationally. The EU, the World Health Organisation, and other international bodies were seen as the appropriate place to exchange research and expertise. International co-ordination had the added advantage that international competition was not affected by environmental standards that were applicable in one country and not in another (VOMIL 1972, p. 24). The risk of disruptions of international competitiveness was a core concern for the Netherlands (Hanf & Van de Gronden 1998). Moreover, pollution did not respect national boundaries. Strict standards in one country were pointless if neighbouring countries did not have equally strict standards; this was especially applicable to a small country like the Netherlands.

In brief, from the beginning, behavioural change was on the agenda in the making of Dutch environmental policy. The Ministry was modest in terms of official power, but had ambitious aims to change society. The international dimension of environmental policy was also acknowledged from early on.

3.1.3 *A programmatic and long-term approach to environmental policy*

The Urgent Memorandum was an early example of an approach that would later become trendsetting in Dutch, EU, and even global environmental policy: namely, the strategic long-term approach to environmental management. Although the Urgent Memorandum was sometimes very pessimistic about our possibility to live in harmony with nature, and at other times overly optimistic about our ability to clean up pollution, it did contain a tentative pathway to ecological development.

The Urgent Memorandum divided environmental policy into different phases. The first was the correction phase, in which the trend to increase the speed of environmental degradation had to come to an end. The second was the clean-up phase, in which environmental degradation had to be diminished, and the emphasis was on cleaning up areas polluted in the past. After environmental degradation was brought to a halt and cleaned up, a new phase started. In this new phase, environmental hygiene needed to become an integral component of structural policy, and to determine which activities should be stimulated and which should be discouraged (VOMIL 1972, p. 19).

In accordance with the Memorandum, the clean-up phase restricted the possibility to take large-scale ambitious environmental initiatives (VOMIL 1972, p. 22). Efforts were concentrated on cleaning up the most polluted areas, and local measures were chosen, such as cleaning up the Rotterdam Rijnmond area. More encompassing measures would have to wait for a later stage of policy development. According to the Memorandum, Dutch society was in the phase of cleaning up the worst pollution, but was on its way to a phase in which our awareness of the finitude of production factors and other limits would define our further economic, technical, and societal developments.

Although the programme was modest compared to later plans, the three-step development marked the beginning of a long-term vision. In hindsight, the perspectives of the Memorandum were far too optimistic, but the inclusion of environmental harmony as a long- term objective sounded very modern. The 1972 Memorandum contained the goals of later policy in seminal form. The strategic long-term perspective related to the ambition of behavioural change discussed in the previous sub-section. From the beginning, the goal of an environmentally sound society was on the agenda: namely, that through behavioural change, people would learn to respect ecological limits. This change did not come about overnight, and in those early years the emphasis was still on local clean up, but that was not the overall aim.

3.1.4 *Setting standards and providing permits*

Environmental policy does not have teeth if it cannot use policy instruments to protect the environment. In the 1970s, a system of environmental protection was developed that remains influential up to this day. In the Urgent Memorandum, it was already decided that pollution would be combatted by the prescribing of permit requirements for industries polluting the environment. In later years, this system was elaborated upon.

The regulatory policies in the 1970s followed from the pervasive understanding that protecting the environment demanded the limitation of economic expansion and production, in line with commitment to the points of view expressed in The Limits to Growth. Permits set pollution limits, and they did so by requiring certain technological fixes regarding the emission of pollutants: these fixes were referred to as 'end of pipe' solutions.

The permit requirements were to be based on standards that should be met for different kinds of pollutants and products. Quality standards were under consideration first, and stipulated how much of a certain substance the air or water or soil could contain. The manner in which quality standards would influence concrete permit requirements was kept vague in the Urgent Memorandum, but the standards themselves would have to be set by scientific institutions like the Dutch Health Council (Gezondheidsraad).[2]

Box 2: Of standards and values
Dutch environmental policy relied on the concept of setting environmental standards. In this chapter and in the following ones, mention will be made of various environmental standards.
One of the earliest standards involved **product standards**. These prescribed an environmental characteristic that a product must have: for instance, the sulphur content of fuels, or the requirement that a car had to have a catalytic converter.
Quality standards represented the desired environmental quality of the soil, the air, the water, and so on. They prescribed maximum concentration levels for a certain pollutant in the air, for instance. An example was the yearly standard for Particulate Matter of 40 Mug/m3 (microgram per cubic metre). According to this standard, on average, air could not contain more than 40 Mug/m3 or particulate Matter per year.
Additionally, **emission standards** prescribed limits to the amount of pollution a source, such as an industrial site, was allowed to emit. As a result of the 2001 EU National Emission Ceilings Directive, for instance, all sources in the Netherlands in total are allowed to emit 50 Kilotons of Sulphur Dioxide.
A fourth category of **standards** set limits on the exposure of humans, animals, plants, or soil types to certain pollutants. An example was the standard for radiation. For instance, a person could be exposed to a maximum radiation level of 1mSv (millisievert) per year for a maximum of five years above the naturally occurring background radiation.
For our topic, **quality standards** are the most important. These may be set by using **limit values**, which is a maximum concentration standard that has to be met. It confers a result-oriented obligation on an administrative body to reach that standard. Quality standards may also be set by using a **target value**, which is a stricter standard than the limit value, but it does not confer a result-oriented obligation. It merely states the desired concentration level to which administrative bodies need to strive.

2. Later on, emission-based standards were preferred. Emission standards determined how much of a given pollutant may be emitted into the air, water or soil. The advantage of emission standards is that they combat pollution at the source at which it is created. If a certain pollutant is not emitted into the environment, it does not cause harm. With its preference for emission standards, The Dutch Government conformed with the German approach to environmental policy.

Another term of note is the '**no-effect level**', which is the scientifically determined concentration level at which no adverse effects on humans or eco-systems are to be expected. Ideally, concentration levels for pollutants fall below the no-effect level. In practice, this level is hard to reach, and limit values are often set above no-effect levels. However, the no-effect level is an important guide for setting Dutch quality standards.

In the Memorandum, it was acknowledged that on the level of ecosystems, the environment was to be considered as a large interrelated eco-system. However, for the purpose of managing environmental pollution, the environment was considered to consist of different sectors such air, water, soil, noise, and radiation. Pollution abatement was organised in this sectorial manner. This sectorial division differed from the holistic outlook of the Urgent Memorandum, but was a consequence of the limited task granted to the Ministry of VOMIL (Van Tatenhove 1993).

Each of the various laws and regulations considered only one of the numerous different sectors. During the 1970s, sectorial environmental laws were created that corresponded to the different environmental sectors mentioned in the Urgent Memorandum. Laws were drafted that dealt with water pollution, air pollution, noise nuisance, and so on. These laws all established different permit requirements and different inspectorates for various types of pollution.

The sectorial laws emerged in the 1970s and the early 1980s. Below the different sectorial laws from 1969 to 1982 may be found in a table.

Table 4

Sectorial law	Year of promulgation	Reference
Pollution of Surface Waters Act	1969	Stb. 1969, 536
Air Pollution Act	1970	Stb. 1970, 580
Marine Pollution Act	1975	Stb. 1975, 352
Chemical Waste Act	1976	Stb. 1976, 214
Waste Act	1977	Stb. 1977, 455
Noise abatement Act	1979	Stb. 1979, 99
Ground water Act	1981	Stb. 1981, 392
Soil Clean-up Interim Act	1982	Stb. 1982, 763

During this decade, the groundwork was laid for the Dutch system of environmental protection by way of permits and standards. Permits postulated the requirements to which industries had to adhere in order in terms of causing pollution, and standards represented policy goals as the desired level of environmental quality.

3.1.5 The entwinement of spatial planning and environmental policy

In the Urgent Memorandum, it was envisioned that environmental policy and spatial planning would have to work closely together. Environmental criteria had to become a structural feature of regional and provincial development plans, and the link between spatial planning was stressed on a number of pages. In regard to mobility and transport, references were found to this connection on page 18: '*Measures from town planning and the instruments available to spatial planning in general form an important means to influence traffic flows*'. Problems with noise pollution also needed to find a place in spatial development plans, and with regard to soil pollution, the relationship between spatial planning and environmental policy was mentioned as well (VOMIL 1972, p. 4, 14).

In legislation, this link between spatial planning and environmental policy was concluded with the promulgation of the Noise Abatement Act of 1979 (De Roo 2003). This law set quality standards for noise nuisance, and stated that in areas where the standards were exceeded no environmentally sensitive functions such as housing could be conducted. Effectively, this provision created zones in which no housing or other activities could be undertaken because the area was subjected to pollution in excess of the standards; in this case, pollution was in the form of noise. One such zone, for instance, was the Schiphol area; the presence of the airport was the cause of considerable noise, and in effect this made conducting other noisy activities impossible.

The philosophy behind the Noise Abatement Act influenced the regulatory consequences of environmental quality standards, up to and including the standards for PM set in 2005. In areas affected by soil pollution, for instance, building became impossible. Standards for odour determined circles around polluting factories or waste incineration plants, within which housing was prohibited. This had a double effect; namely, enterprises that contributed to pollution were not allowed to expand if they were located within a zone in which the standards for noise or odour were not met. Moreover, other activities within these zones became prohibited as well. This situation gave rise to environmental spatial conflicts (De Roo 2003), because often an area was claimed by competing activities. Through this system of zoning, spatial planning and environmental regulation became intimately related.

Environmental legislation was strict with regard to spatial planning. It left no room for a weighing of interests, but created boundary conditions that spatial planning needed to take into account. The centrally established standards were considered to overrule any local weighing of interests (VROMraad 2009, p. 29).

While the connection between spatial planning and environmental policy was made in the early days, the Ministry of VOMIL had little influence over spatial planning policy. Spatial planning resided under the Ministry of Spatial Planning, which had a strong position in the Government at the time, and it wielded considerable influence over environmental planning via the Council of Spatial Planning (Raad voor de Ruimtelijke Ordening, RARO) (Van Tatenhove & Goverde 2007, p. 56).

3.1.6 The discrepancy between idealistic rhetoric and pragmatic politics

The early days of Dutch environmental policy making are often maligned, especially in comparison to the more successful 1980s. However, it was in the 1970s that the subsequent course of Dutch environmental policy was determined. Environmental policy began in response to the social problem of continued environmental degradation due to rapid industrialisation after the Second World War. The state of the environment was considered as deeply problematic and the apocalyptic rhetoric of the Limits to Growth report infused environmental discourse in the Netherlands. A system of permits and environmental standards was designed to control environmental degradation. Strategically, environmental policy strived from the beginning to create a large public consensus on environmental matters, and tried to mobilise the public behind environmental ideals. It did so by accommodating and mobilising its allies, and by trying to gain new ones.

When comparing the discourse of the Ministry of VOMIL with its political acts, one cannot fail to notice a discrepancy in the way issues were discursively framed and the way they were dealt with in concrete policies. Maarten Hajer had already pointed out the tendency for stark apocalyptic rhetoric in Dutch environmental policy (Hajer 1995, p. 269), and this was certainly true in the Urgent Memorandum. However, the reality of policy making was still dominated by concerns of economic welfare and industrialisation. The concrete measures proposed in the Urgent Memorandum were therefore only modest and piecemeal. The competitive position of Dutch industry was a concern that the Ministry of VOMIL always had to take into account. In fact, we may find this eye for economic concerns in the Urgent Memorandum as well as in the Memorandum Ambient Environmental Standards (VOMIL 1976, p. 11). The Dutch concern regarding competitiveness was also one of the reasons to stress the need for international collaboration (Hanf and van der Gronden 1998).

The relationship between economy and ecology was regarded as deeply problematic in the early days of environmental policy. The main ecological storyline within the *limits to growth* discourse in those days was the disharmonious way in which modern humans interacted with nature. Economic concerns were considered a threat to the environment, and, ideally, humankind needed to restructure its economy to respect environmental limits; an emphasis on economic growth was considered detrimental.

The Ministry of VOMIL was never able to realise its ecological ambitions, and was forced to settle for piecemeal policy making. Because of its limited competency, it was not able to wield much power, and resorted to a strategy to convince other

Ministries by providing scientific arguments and environmental rhetoric. However, it found itself pitted against the Ministry of Economic Affairs, which used arguments and insights from the field of economics and management to counter the natural-scientifically informed language of VOMIL (Fürst 2004, p. 108). As Leroy & van Wiering noted, the economic growth discourse of other departments consistently dominated the *limits to growth* discourse of the environmental department (Leroy & van Wiering 2007, p. 82).

VOMIL's policies were not very effective. The sectorial system of the 1970s, with its different permit requirements for every environmental sector, was cumbersome. It was a patchy way of legislating with every department and authority asking for different permits. According to Bressers and Plettenburg, Sulphur Dioxide (SO_2) levels had indeed fallen in the 1970s, but that was in spite of the permitting requirements rather than because of them, since the permits generally allowed the emission of one and a half times as much air pollution as was produced in reality (Bressers & Plettenburg 1995, p. 30). This ineffectiveness made the system unpopular with the environmental movement (Hanf and van der Gronden 1998, p. 163).

The system was equally unpopular with the polluting industries, because it led to bureaucracy. The permits were granted by different agencies, each of which represented different environmental sectors, and the administrative staff often adopted a standoffish or hostile attitude towards representatives of economic interests (Bressers & Plettenburg 1995, p. 16; Fürst 2004, p. 107).

The burgeoning wave of environmental legislation and the isolated position of the Ministry of VOMIL with respect to more mainstream fields of policy making became a problem in itself, and led to the reorganisation of the Ministry. However, it needs to be stressed that many features of Dutch environmental policy that remain intact to this day can be traced to these times: for instance, the system of standards and permits, a long-term programmatic orientation, close ties with spatial planning, and in particular an emphasis on consensus, education, and the raising of awareness.

3.2 Embracing ecological modernisation

In response to the regulatory problems encountered in the 1970s, and the recognition of transboundary environmental problems in the 1980s, the Netherlands developed an environmental policy with which it became a European front-runner. It based these policies on ideas that were in line with the ecological modernist line of argumentation and policy making.

In this section, I examine the history of the Dutch Ministry of Housing, Spatial Planning, and the Environment (Volkshuisvesting Ruimtelijke Ordening en Milieu, hereafter VROM). The VROM Ministry succeeded VOMIL as the Ministry responsible for the environment, and during its expansion in the 1980s it embraced and added to the discourse of ecological modernisation. The policy discourse developed in this era remained the institutionalised environmental discourse at least up until the new millennium.

3.2.1 *The role of modern managers in the new VROM Ministry*

In the 1970s, it had already become clear that environmental policy had difficulties holding its own position against other fields of policy. The way it was applied was contentious and ineffective, and environmental policy makers intended to better integrate the environmental dimension into Dutch policy as a whole. The 'Memorandum on Selective Economic Growth' (EZ 1976), issued by the Ministry of Economic Affairs, (Economische Zaken in Dutch, henceforth EZ), was one of the first attempts to make a consistent case for the integration of environmental concerns in broader economically oriented policies. This Memorandum on Selective Economic Growth was written under the responsibility of the Minister of EZ, Ruud Lubbers, later a Dutch Prime Minister. It proposed thinking in terms of 'facet policy'. The idea was to see all aspects of policy making, such as energy politics, spatial planning, and environment as well as foreign aid, as aspects of the same whole, a coherent Dutch national policy. According to the memorandum, economic continuity would benefit in the future from keeping environmental concerns a priority.

This development was an important turning point, because it showed that integration of environmental considerations in other fields of policy was now also a concern in more economy-minded Ministries. It also demonstrated that the environment was gradually becoming a mainstream topic. This 'rapprochement' had two effects; firstly, environmental concerns became the concerns of former adversaries, the economically oriented policy fields; secondly, economic considerations took hold in the environmental discourse as well. However, even though a certain détente emerged and Lubbers' facet policy managed to integrate the environment and the economy in the same political discourse, we cannot as yet speak of a discourse coalition. It would take another ten years before environmental and economic storylines became intertwined within the discourse of ecological modernisation.

The way the environmental Ministry eventually became reorganised in the early 1980s displayed the increasing influence of managerial and economic science in the environmental field. The first Cabinet of the 1980s embraced the plan to integrate environmental policy with mainstream concerns. This Cabinet, led by Prime Minister Dries van Agt, was short-lived, but it began an investigation – called Project Integration Environment Policy (Project Intergatie Milieubeleid: PIM) – into the possibilities of better profiling environmental policy within mainstream Dutch policy. After the demise of this Van Agt Cabinet, a new Cabinet emerged with Ruud Lubbers at the helm. Ruud Lubbers was the original author of the Memorandum on Selective Economic Growth, and the PIM project began to bloom.

The project was set up in accordance with the Dutch tradition of consensual and accommodative policy making. It aimed to overhaul the Department of the Environment in order to *'accomplish a large degree of completeness and coherency in the preparation, adoption, implementation and evaluation of environmental policy making'* (Staatscourant 1982a, 149, p. 6). The project involved many of the stakeholders including industry, the environmental movement (SNM and even the Dutch branch of Friends of the Earth), and other Ministries. External advisors obtained leading positions in the implementation of the plan. R.H.P.W. Kottman of Schoenmaker BV and D.J. Schoenmaker of Twijnstra Gudde consultancy agency played leading roles (Staatscourant 1982, p. 29). These two consultants were organisational experts, and the agencies involved had a modern managerial outlook regarding organisational problems. A preventative environmental policy was a specific aim of the plan, and this type of policy required a *'more conscious integration of the environmental aspect in other aspects of policy making'* (Staatscourant 1982b, p. 29).

The PIM project had important consequences. It merged the intimately linked policy fields of spatial planning and the environment into one Ministry. The new Ministry of VROM was responsible for both spatial planning and environmental policy. Within the PIM project, environmental quality standards were considered to be the cement between the two different policy fields, and they gained in importance as instruments to integrate environmental concerns into the spatial development of the Netherlands (De Roo 2003, 175). On the whole, however, the fragmented institutionalisation of environmental policy was retained. VROM was by no means a 'super Ministry', as many areas of environmental interest, such as agriculture and energy policy, were outside of its competency. Nevertheless, within PIM, the other Ministries committed themselves to a joint effort in strengthening environmental policy (Van Tatenhove 1993, p. 22).

PIM's importance lay in the reorientation of environmental discourse that this reorganisation had made possible. Up until then, the Ministry of VOMIL had been populated by environmentalists who shared a pessimistic outlook regarding possibilities to impact the environment positively through regulation. PIM, however, was carried out by modern managers and consultants. The new VROM Minister, Pieter Winsemius (VVD), had a background in management as well, and he was a firm believer in positive management theories (Winsemius 1986; Hajer 1995, p. 187; Schenkel 1998 p. 135). This involvement of managers and consultants made it possible for the environmental department to shed its aura of gloom and doom, and to embrace a more upbeat discourse that was more palatable to the economically inclined Ministries and industrial sectors.

3.2.2 *Accommodation and responsibilisation revisited*

When Winsemius took the helm of the Ministry of the Environment, environmental policy was viewed as a problematic field. The new Lubbers Cabinet had promised to 'roll back the state' according to the neo-liberal principles in vogue at the time, with President Reagan in the US and Margaret Thatcher in the UK. Nationally, the environmental question polarised the left and the right in the debate on the use of nuclear energy. Internationally, the acid rain problem had appeared on the scene. Acid rain, nuclear energy, and climate change were different problems than those that had demanded attention in the previous decade. They were transboundary concerns that did not cause damage primarily to public health but to eco-systems. These problems developed primarily in the international arena, and they demanded more penetrating, long-term, and global answers than the local health-based issues with which VOMIL was used to dealing. Winsemius found himself in the middle of all these different demands, and set out to reorganise the Ministry of the Environment to meet them. The way the new Ministry responded to these diverse pressures determined environmental policy up to at least 2000.

Especially in the first years of its existence, VROM was short on staff and financial means, just like its predecessor had been. Winsemius was a Minister from the conservative liberal party VVD, traditionally pro-economy and mobility. He was trained as a natural scientist, but before he became Minister of VROM he had worked for the consultancy agency McKinsey. He was certainly not averse to the goals of environmental policy, but he approached the Ministry as a modern manager and not as an idealist. He had to work with little means, and the deregulation agenda prevented far-reaching legislation. As a result, Winsemius employed a cautious and strategic approach to environmental policy, key notions of which involved long-term planning, internalisation, and responsibilisation.

Long-term planning was important for environmental policy, since prevention was its cardinal goal, and because it facilitated the involvement of other actors in environmental protection. In the summary of the first Memorandum written in his period as a Minister, known under the title 'More than the sum of its parts', we read: *'This long-term perspective is indispensable for reaching a preventative environmental policy which emphasises the responsibility for environmental protection of each of us'* (VROM 1984a, p. 6). The involvement of other actors was crucial, because Winsemius was working under the same conditions as the old VOMIL Ministers. He inherited a fragmented policy field, and the Ministry of VROM tried to consolidate slowly and to expand its position rather than launch far-reaching proposals for further policies (Van Tatenhove 1993, p. 22).

Since top-down regulation was not considered, increasing environmental awareness would need to bring about bottom-up changes in the behaviour of polluters. To this end, Winsemius' Ministry introduced what was called the 'target group policy' in the middle of the 1980s. Instead of focusing on pollution, the VROM Ministry started to target polluters directly. Polluters were considered by sector, such as the chemical industry, the electricity sector, the transport sector, consumers, and so on. The Minister began to negotiate directly with the representatives of these different sectors and the big economic industrial powerhouses. He tried to conclude 'gentlemen's agreements' with them, and in exchange for their cooperation to meet certain environmental targets, these representatives were given a seat at the table. In addition, if possible, voluntary agreements on pollution reduction were concluded. For different target groups, specific policies were made to ensure that certain environmental quality standards were met. Inventories were drawn up concerning which target group contributed what to environmental problems, and what kind of measures they needed to take to ameliorate them.

The aim of this policy of negotiation and consensus was known as 'internalisation'. Environmental demands should no longer be felt as external obligations prescribed by the Ministry, but as intrinsic ethical commitments. Representatives of target groups should become convinced that they needed to change their own behaviour in order not to spoil the natural resources that they used. The goal of internalisation was not confined to industry. Other policy makers needed to internalise environmental values as well, and so did the general public through awareness-raising campaigns. The environmental movement could fulfil a useful function as a mediator between the Ministry and the target groups.

To this end, the Ministry responsible for the environment subsidised the activities of environmental groups. Through these grants it created networks of environmental allies that could approach target groups and engage them with a similar level of expertise and connections. This network could also supply the necessary environmental education and awareness-raising campaigns. According to Van Tatenhove and Goverde, the subsidising of environmental groups and the resulting mobilisation of public opinion had to be considered in connection with the drive to integrate the environment in other departments. The resulting network of civil servants in the environmental department, the CRMH, and the environmental movement became thoroughly professionalised. According to Pieter Leroy, Professor in Environmental policy, the VROM Ministry, the environmental movement, and environmental science formed an 'Iron Triangle' in the 1980s (Van Tatenhove & Goverde 2007, p. 61).

The strategy of internalisation was helped by the emergence of the problem of acid rain on the political agenda (Hajer 1995). Acid rain was the popular name for the problem of acidification, a transboundary environmental problem that threatened the balance of eco-systems, and could cause, among other things, a depletion of fish stocks and the disappearance of forests, as well as damage to cultural artefacts. In its environmental awareness-raising campaigns, the VROM Ministry used stark images of a statue of the Virgin Mary purportedly defaced by exposure to acid rain. The very resource base of industry appeared to be under threat, and the internalisation of environmental standards was considered a fair deal in response to this grave threat (Hajer 1995).

The aim of shared responsibility was transmitted primarily by advertising campaigns to target the public, and by concluding covenants with representatives of important economic sectors. These covenants fulfilled a double role. They would make sure certain environmental aims were met, but they also involved the industry in environmental protection. One of the first was the KWS 2000 covenant, concluded in 1989 to ban the use of harmful hydrocarbons in many different products (Infomil 2000).

3.2.3 *Expansion of the integrated long-term perspective*

VROM's environmental policy in the 1980s preferred long-term programmes to ad hoc intervention, as internalisation and responsibilisation required a long-term vision. Moreover, the type of problems on the political agenda, such as acid rain and other threats to the eco-system, demanded a more holistic approach. This gradual reconfiguration of environmental policy took place in long-term environmental plans referred to as 'Indicatief Meerjarenplan' (Indicative Multi Year Policy Plans, IMPs).

Initially, the IMPs were sector specific; for instance, they existed for air, soil, water, and noise. They covered a four-year time span, and were updated on a yearly basis. In the middle of the 1980s, an integral IMP appeared, known as IMP M.,[3] which concerned general environmental planning for the following four years, and was updated yearly as well. In total, three updates appeared: 1985-1989, 1986-1990, and 1987-1991.

The IMP M. transcended the traditional sectorial approach. The environment was no longer considered as being carved up into segments such as air, water, and soil, but was seen as a unified whole in which certain problems occurred. Policy focused on curbing environmental problems such as acidification, eutrophication, the spread of dangerous substances, disturbances of the living environment, and so on, rather than on cleaning up specific environmental sectors.

3. The M. stands for Milieubeheer, the Dutch term for environmental management.

These problems needed to be tackled by the Ministry of VROM, but also by society as a whole. In this process, the target groups, the various Ministries, and environmental pressure groups had their specific role to play. As described above, target groups had to internalise environmental responsibility, and other Ministries had to incorporate environmental considerations into their own policies. In their turn, the environmental organisations were to supply environment-related know-how, and to provide information to the target groups and the public at large in order for them to incorporate environmental considerations into their actions.

The IMPs introduced more influential innovations. Firstly, for air quality, the introduction of the two-track policy was important. Standards that determined environmental quality were employed in addition to emission standards, and, ideally, these standards ensured that quality standards were reached. This system would ensure that the various standards promulgated by the Ministry all contributed in unison. As we will see in chapter 4, this system was adopted by the European Union when it drafted its own policy on air quality. The setting of environmental quality standards became known as 'the effect-based policy'.

Secondly, the notion of region-specific policy was introduced. Not all regions in the Netherlands would benefit from the same type of policy, so tailor-made solutions were sometimes necessary, and region-specific policy would make this possible. The notion of region-specific policy further strengthened the co-operation between spatial planning and environmental concerns, but also created space for some flexibility, since different regions could have different economic and ecological needs.

Thirdly, an 'integrative risk management' approach to harmful substances was introduced. It made weighing the seriousness of environmental problems possible; by quantifying the risk they posed to health or to eco-systems, and made it easier to set priorities. The risk management approach was closely related to the environmental impact assessment, and large-scale projects had to perform an investigation regarding their impact on the environment in terms of the risks they caused. Winsemius considered this instrument a 'think before you act' law, and he claimed that – together with the United States – the Netherlands was an international forerunner in these areas (Winsemius 1986, p. 70). These new policies were used to instil a sense of responsibility in target groups and civilians. Such instruments sometimes replaced traditional legislation (VROM 1984b, p. 7).

3.2.4 *The possibility of a 'positive sum game'*

One of the most crucial innovations in the 1980s was discursive, and was important for the coherency of this new type of environmental policy. Instead of presenting the environment and the economy as mutually exclusive interests, Winsemius argued that environmental policy could entail opportunities for economic growth. He contended that people would change their behaviour in a pro-environmental way if they had a positive idea of the cost-benefit ratio of that change, and if they expected positive reinforcement from their social environment. Hence, it was important to present economically viable environment-friendly alternatives in a positive light. In fact, people could change their behaviour in a pro-environmental direction without having any pro-environmental considerations whatsoever, provided that the cost-benefit ratio pointed in that direction.

Environmental consciousness itself was not enough, as this would attract only a small number of pioneers. However, a positive vision of the economic benefits of this behaviour could change the behaviour of a broad range of actors. The IMP Air 1984-1988 concludes: '*All measures that this IMP aims at, such as promoting clean technology, energy saving techniques, sustainable sources of energy, traffic circulation plans and clean modes of transport, need to be seen from this perspective*' (VROM 1983, p. 109). The change of tone was aptly summarised in the first IMP M. A quote from the IMP M 1984-1988: '*Environment and economy may reinforce each other, when managed appropriately*' (VROM 1984b, p. 17).

I consider the embracing of this idea to be a Copernican turn for environmental policy. The intuition that environmental measures are a strain on the economy is strong, and this perception hinders the willingness to take environmental measures. In the 1970s, a clean environment was considered to necessitate economic sacrifice, and it was assumed that increasing environmental well-being would entail a decrease in production and services (VOMIL 1974, p. 53).

Winsemius highlighted the benefits that environmental policy could have for employment and technological development, and this perspective became a mainstay of Dutch policy. It was used to convince an international audience as well. During a 1984 conference organised by the OECD in Paris, Winsemius was the chairman. At this conference, the desirability of integration of environmental and economic policies was proclaimed, as well as the need for preventative and anticipatory policies. Winsemius offered help to countries that were developing an environmental policy, such as Greece and Portugal, and this aid bolstered the Dutch position as an international front-runner in new and innovative policies in the 1980s (Schenkel 1998, p. 99). Dutch policy maker and Director General of Environmental Affairs, Laurens Jan Brinkhorst pleaded in the EU to take product standards seriously because they could lead to better economic performance (Weale 1992).

The notion that the environment could provide opportunities for economic welfare, and the conviction that if well managed, economy and ecology could reinforce each other, is fundamental for ecological modernisation. The gap between ecology and economy was bridged by this notion, because, without it, all efforts at development would be suspicious from an environmental perspective, and environmental protection would only be considered a drain on resources from an economic perspective.

It is important to note that although the discursive form changed, the fundaments of Dutch environmental policy remained intact. The permit system was augmented by new instruments of environmental policy but not supplanted by it. Similarly, environmental standards were still considered crucial for protection of the environment (Winsemius 1986).

The type of ecological modernisation employed was still weak, and did not change fundamentally the relation between environmental and economic interests. Although it was ambitious on issues like awareness raising and internalisation, economic reasoning in practice still held considerable sway over policies of the VROM Ministry. The prime negotiating partners for policy were the target groups, and these were economically important sectors, as their concerns determined the type of regulation to a significant degree. The judicious tactics of Winsemius meant that environmental policy became a continuous balancing act between the diverging interests of various economic sectors, pressure groups, and Ministries. Measures could hardly ever be radical or truly environmentally ambitious, because they were based on negotiations with the target groups. Environmental policy had to live up to its promise that economic welfare would not be endangered. Winsemius' motto was: '*Nobody should be asked to do the impossible*' (Winsemius 1986, p. 50; Hajer 1995, pp. 236/237). From an ecological perspective, this motto boiled down to a defence of feasibility concerns over ecological concerns, because no one could be asked to do something that was unreasonable from an economic point of view.

3.2.5 Dutch ecological modernisation: The coming of age

The breakthrough towards a stronger form of ecological modernisation came about in 1989 with the release of the report 'Take Care of Tomorrow' and the National Environmental Policy Plan (NMP). The Dutch national environmental policy plan NMP was the successor to the earlier IMP M. and contained the strongest articulation of holistic environmental policy planning with a high level of ecological sophistication. At the same time it did not shun economic language and presented environmental degradation as a sincere economic threat. The plan welded, as it were, the economical discourse of the environmental managers to an ecological discourse obtained from environmental science.

The NMP was an important milestone, because it established the Netherlands as a significant environmental forerunner in Europe, perhaps even worldwide at the time, and because the direction of the NMP determined Dutch environmental policy for the future. NMP sequels were issued all through the 1990s.

At the time the NMP was drafted, the environment was a heated topic of debate among the Dutch public and among scientists. Shortly before the NMP's appearance, a report was published by the RIVM, which highlighted in strong terms the environmental predicament of the country. The report was titled 'Take Care of Tomorrow' (Zorgen voor Morgen), but the Dutch title contained a double entendre, as *Zorgen voor Morgen* may also mean 'Tomorrow's Worries'. According to one commentator, this report *'could have been written by environmental groups like Greenpeace or Friends of the Earth'* (Kronsell 2000, p. 96).

The report categorised environmental problems on a number of scales that ranged from the most local to the most global. It warned that environmental problems tended to be pushed from local scales to the continental and even the global scale, causing widespread damage over a large area. In the report, the environment was characterised as a set of streams that connected reservoirs. Water, air, and radiation were considered essentially to be streams, whilst fossil fuels and biological substances were reservoirs. Streams were sources of energy that could be regenerated, provided the streams did not dry up or become overly contaminated. Air pollution contaminated the stream of air, for example, and water pollution spoiled the fluvial streams. A non-sustainable use of resources meant overexploitation was taking place, and the reservoirs could dry up and the streams become polluted. Ecological cycles needed to be closed in order to prevent waste and over-exploitation (Langeweg 1988, p. 2). The report argued that the state of the environment was deteriorating on multiple fronts at the same time, and that the situation regarding each of these fronts had repercussions for the others. The novelty of the report lay in the holistic description it gave of the environment (K. Zoeteman, interview 05-09 2012). This description was powerful, because the interdependent relations between environmental problems pointed to only one solution: namely, a radical change in consumption and production rather than piecemeal improvement.

The report struck a nerve in Dutch society, to the extent that even the Queen (Beatrix) commented on the ostensibly bleak state of the environment. In her Christmas speech of 1988, she proclaimed, *'The earth is slowly dying'*. This proclamation was an embarrassment to the Government of Prime Minister Ruud Lubbers, and established the environment as a priority issue.[4] At the time that influential environmental scientists were drafting *Zorgen voor Morgen*, policy makers were in the process of creating the first truly integral plan to create an environmentally sustainable society. Minister Winsemius had already hinted at the plan back in 1986, but his successor Ed Nijpels took it further. Initially, the plan did not proceed well, hampered as it was by opposition from other Ministries, especially in regard to the long-term level of ambition (K. Zoeteman, interview 05-09 2012). After the Queen's Christmas speech and *Zorgen voor Morgen*, however, the political environment was much more conducive to the idea of an environmental strategy (K. Zoeteman, interview 05-09 2012). In any event, Prime Minister Lubbers now supported his Minister for the Environment, Ed Nijpels.

4. Contrary to the Dutch 'Troonrede', a kind of Dutch 'State of the Union' speech written by the Prime Minister, the reigning monarch is responsible for the Christmas speech.

The NMP was drafted by people who had a background in policy making as well as in environmental science. The same people had in some cases been involved both in *Zorgen voor Morgen* and the NMP, and the ecological tone of *Zorgen voor Morgen* could also be found in the NMP. Quotes taken from it provided some telling examples of an eco-modernist framing of environmental problems. Environmental degradation, for instance, was likened to the budget deficit, a highly influential economic storyline in the Netherlands:

> *'From the past, we get unexpected bills for behaviour that may be typified as obtaining loans. What goes for the budget deficit holds for the environment as well, life cannot be lived on the tick. Paying back environmental loans includes paying interest'* (VROM 1989, p. 70).

In this economical/ecological vein, the NMP urged us to look at the environment not as a 'free good' to be exploited by anyone but as a *commons* that had a limited carrying capacity. It argued that environmental problems resulted from various 'displacement mechanisms, and costs like pollution were 'externalised', and these externalisations created 'environmental loans'. These were costs that might not be felt at the time, but would have to be repaid with interest by coming generations.

The policy advocated in the NMP consisted of an expansion of the strategies used in the IMPs. The internalisation of environmental considerations by target groups was now considered on a much broader and deeper scale, involving internationally operating economic actors. Product and emission standards retained their function, but needed to be set as challenging as possible to stimulate new sustainable solutions and technologies. Research institutes like RIVM, TNO, and ECN, as well as universities should include sustainable development as a major element in their strategies, and the state should strongly encourage major enterprises to do the same.

Households were already considered an independent target group, but now citizens were targeted as well in their role of consumers and employees. The people's choice of transportation and their behaviour 'in the living environment' would be the subject of campaigns to raise awareness. Access to information would be improved in order to increase public knowledge of environmental matters. Consumers were also subjected to positive and negative financial reinforcement, and in education the attention paid to the environment would be substantially broadened. This would make behaviour change possible because, as the NMP urged: *'Each of us is considered to know his responsibility towards the environment and subjects his actions to this responsibility'* (VROM 1989, p. 13).

Consumers determined which products would be produced, so their sense of responsibility needed to be strengthened (VROM 1989, p. 89). To this end, for instance, advertising would be subject to regulation that would prohibit it from giving 'a negative impulse to sustainable development' (VROM 1989, p. 221) According to the plan, we were all 'environmental managers' (VROM 1989, p. 115).

The environmental movement was stimulated, since it was seen as a 'tool' to encourage others to internalise environmental responsibility. These movements would signal environmental problems early, disseminate information to the public, and stimulate internalisation in all other target groups (VROM 1989, p. 37).

The integration theme was clearly present in the NMP, and in a sense was already embodied by it. The NMP was not only drafted by VROM but involved the cooperation of the Ministry of Economic Affairs and the Ministry of Transport and Water Management. In addition to the Minister of VROM, the Minister of Transport, the Minister of Economic Affairs, and the Minister of Agriculture and Fishery were also responsible for the plan. The plan had close links to policies drafted by these Departments, such as the Structural Scheme for Traffic and Transport (Tweede Structuurschema Verkeer en Vervoer SVV2) and the Fourth Memorandum on Spatial Planning (VINO/VINEX).

The long-term and transboundary focus of the document implied that the integration ideal could not be confined to the Netherlands alone. The plan urged Dutch representatives to make sure that environmental matters became the concerns of foreign countries and organisations. To this end, international forums for environmental policy needed to be strengthened and EU environmental policy stimulated. In line with the internationalist outlook, priority was given to problems that concerned higher levels, such as continental and global. This focus on transboundary problems implied that an ambitious international environmental policy was necessary.

In summary, the new economy/ecology discourse involving ecological modernisation sought to address members of the public in their role as consumers, and because the worst problems were considered transboundary, the discourse pleaded for an ambitious international environmental policy. It buttressed long-term goals by political action, although we will see in the next section that the concrete measures were not as ambitious as the overall political vision and strategy.

3.2.6 *Assessing the NMP; the turn to strong ecological modernisation*

With the NMP, the Dutch Ministry of VROM shifted to a discourse that could be characterised as strong ecological modernisation. The languages of economy and ecology were fully integrated. The message was put across starkly that current economic and ecological trends would lead to mishap, but that a different kind of development was possible and desirable, and with which they would both be served. The precautionary principle was endorsed, and its need was underscored by the language of the NMP. Discursively, the NMP broke new ground for more ambition, and the plan was hailed as highly advanced in Europe and the rest of the world (Weale 1992). The Netherlands had been eager to showcase its state-of-the-art environmental plan, and civil servants from the VROM Ministry were sent abroad to spread the message (Arts, Dieperink & Liefferink 2002; Pettenger 2007, p. 58; K. Zoeteman, interview 05-09 2012).

With the NMP, a development that started in the first years of the 1980s with the PIM project came to fruition. The failure of the old sectorial system and the emergence of transboundary environmental problems necessitated a different approach to environmental policy. The infusion of managerial and economic discourse within the existing discursive structures of the Ministry of VROM had made policy much easier to sell to the representatives of industry and other powerful economic sectors than had the divided and bureaucratic approach of the 1970s.

During the period of the formation of ecological modernisation discourse, traditional features of Dutch environmental policy were deepened and became more pronounced, but did not change. Firstly, long-term planning became advanced and institutionalised with the NMP. Secondly, consensus seeking not only became the norm but it also secured the place and the role of the environmental movement in this process. Thirdly, the link between spatial planning and environmental policy remained intact; in fact, the cooperation of the Ministers of Transport, Agriculture, and Economic affairs ensured that the Ministry of the Environment exerted more influence over other fields of policy. Fourthly, the system of permits and standards did not change, but the choice of instruments for environmental protection was expanded. Many things that had been detrimental to the development of environmental policy were discarded, and the system of sectorial division was replaced with an integrated holistic view. The negative hostile attitude was succeeded by a stance more amenable to business, and the prevailing belief on both sides of the political spectrum that environmental and economic interests were mutually exclusive was broken.

Nevertheless, the NMP was criticised because of a lack of penetrating measures. For instance, the measures taken were predicted to preserve only 20% of Dutch forests (Hajer 1995, p. 195). Maarten Hajer spoke of a carefully co-ordinated anti-climax with which the Dutch Government inflicted a crisis of legitimacy on itself (Hajer 1995, p. 195), and Gerda Dinkelman mentioned that the environmental movement was critical when the plan appeared (Dinkelman 1995, p. 126). Dutch ecological modernisation still rested mainly on instilling a strong sense of urgency into the public, and on a penetrating perception of environmental problems. Policy refrained from taking top-down measures, and further developed a preference for self-regulation, target group policy, and covenants. The discourse moved towards strong ecological modernisation, but it still rested on consensus between different segments of society, including representatives of powerful economic interests. Radical environmental policy measures were ruled out. In the end, the targets set in the NMP fell short of the ecological rhetoric and the recommendations put forward in 'Zorgen voor Morgen'. A year later, the ambitions of the NMP were raised, and a more purposeful version was released: the NMP+. The rise in standards was good news from an environmental point of view, but the targets set in 'Take Care of Tomorrow' could still not be met. In a sense, the NMP was a case of history repeating itself; discursively, it was strong, but in terms of concrete measures it was more modest.

The VROM Ministry was at the apex of its power, however, and potentially it could influence policy fields like spatial planning and transport. The success of ecological modernisation depended on the struggles that were about to ensue in the 1990s.

3.3 The aftermath of the NMP: accommodation, and conflict in the 1990s

The discourse of ecological modernisation legitimised the strategic orientations of the VROM Ministry. However, the question was whether environmental policy would end up in the same situation as it did in the 1970s: namely, as part of the problem rather than the solution. To examine this question, we will investigate whether the integrative and consensual themes embodied in the NMP yielded results. Since it is impossible to investigate the reception of the NMP in every area of environmentally important policies, we will look specifically at environmental cooperation with the industry, the political participation of the environmental movement, the attempt to ecologically modernise the transport sector, and conflicts over infra-structural projects that emerged in the 1990s. We will pay particular attention to the process of juridification of these conflicts, as the courts also played a central role in the air quality clash.

3.3.1 Environmental cooperation with the industry

In the 1990s, this style of conducting environmental policy by focusing on target groups and regulated covenants continued, despite personal changes in the Ministries, and in different coalitions governing the country. In fact, environmental policy shows a great deal of continuity during this decade. After the initial NMP, four more versions appeared. In the second NMP, the term 'Self-regulation within boundaries' (Zelfregulering binnen kaders) was introduced (VROM 1993), and it denoted a further sophistication of the target group policy. The emphasis shifted further away from 'command and control' legislation to regulation based on consensus with the interested parties.

In the 1990s, this type of consensual environmental politics was considered innovative and successful (OECD 1995, p. 200), as was the covenant system as pioneered by the Netherlands (Hafkamp 1997, p. 260; Jordan et al. 2003, p. 116). These agreements were integrated within the comprehensive framework of environmental planning and management, and in contrast to Germany for instance, they were most often binding, and contained ambitious long-term targets (Delmas & Terlaak 2002, pp. 16/17).

During these years, covenants were concluded in a range of sectors and for a range of different products. In line with the overall aim of internalisation, the covenants did not necessarily have to lead to a reduction of emissions, but could, for example, also contain a requirement to make it obligatory for companies to draft yearly environmental reports.[5] At the end of the 1990s, more than one hundred environmental covenants were in force (Van der Jagt 2006, p. 19).

After 2000, criticism of the covenant approach was voiced in administrative circles. The fourth NMP, in force from 2001, promised to critically evaluate compliance with the voluntary agreements (Van der Jagt 2006, p. 21). The rise of the covenant illustrated how environmental policy was gradually shaped around consensual instruments and negotiation, but the criticisms voiced in 2000 indicated that there were doubts over its efficacy.

In the 1990s, the environmental consensus held reasonably well for relations between environmental policy and industry. According to the OECD in 2003, the industry had been responsive and proactive in environmental matters, and had integrated environmental concerns into their own industrial policies. The OECD stated, '*The characteristic policy mix of regulation/licensing plus economic instruments plus environmental agreements continues to be productive*' (OECD 2003, p 21).

However, not all industries adopted willingly the tasks thrust upon them. For instance, the chemical industry and the steel industries had to be threatened with stringent Government legislation before they gave in (Pettenger 2007, p. 59). The industry had been sensitive to the idea that if it negotiated with the Government, formal regulation could be avoided and environmental well-being and continuous production ensured (Hajer 1995, p. 189). Dutch industries ranked among the most energy efficient in the world in 2005, according to the Ministry of Economic Affairs (EZ 2005, p. 20), whose report cited an International Environmental Agency's review from 2004 as a source.

I conclude that with respect to the industry, the ecological modernistic approach of environmental management through consensus influenced industrial sectors not normally amenable to environmental protection. In this respect, Glasbergen and Driessen noted that although the changes in Dutch environmental policy did not come about without conflict, we witnessed a '*clear development taking place [...] of [...] Environmental politics increasingly taking the form of organising various learning processes in collaborations of stakeholders on environmental issues*' (Glasbergen & Driessen 2002, p. 23).

5. Kamerstukken II 1997-1998, 25 605. nr. 2

3.3.2 *Increased participation of the Environmental Movement*

Another integrationist aim of the NMP beside participation of the industry was the participation of environmental groups in policy making. Already in the 1970s and 1980s, the environmental pressure groups were considered allies of the environmental Ministry. They were able to fulfil a bridging function between the Ministry and other target groups such as industry and the public.

The environmental consensus forged around the NMP and its brand of ecological modernisation had indeed led to increased participation of the environmental movement and its concerns within policy making as well. In 1989, employer organisations and trade unions decided to make the environmental question a topic of regular consultancy. The Social Economic Council (SER), one of the most influential advisory councils of the Dutch Government, proclaimed that maintaining the ecological system was a goal of a higher order even than social economic targets (Dinkelman 1995, p. 259).

The process of accommodation continued in the 1990s. In this decade, the Dutch accommodative strategy to overcome social conflicts between employers and employees made headlines as the 'polder model'. In the later 1990s, this idea was extended to a 'green polder model' (Weggeman 2003). This was seen as a way to accommodate environmental interests more strongly in the typical consensus-minded institutions of Dutch policy making by involving environmental interest groups early on in decision making.

Environmental interests had already been represented since the 1970s in some institutions, but access had generally been restricted to the Foundation for Nature and Environment (Stichting Natuur en Milieu; SNM). In the environmental movement, this group was considered to be moderate, and more prone to negotiate than to protest. The green polder model also included more radical and newer environmental movements such as the Dutch version of Friends of the Earth, Milieudefensie. Milieudefensie joined the SER in 1999. The public by then was well aware of environmental degradation, so raising awareness was no longer an issue for the radical environmental groups. Milieudefensie started to focus less on action and more on bringing sustainable alternatives to the attention of the public.

The green polder model was an attempt to bring environmental groups closer to the policy making process, and to give them a seat at the table. In this respect, the NMP's integrationist policy was successful, because the environmental movement had become a serious negotiating partner. In section 2.4.5, however, we will see that many of the negotiations between the environmental movement and other interested parties in the field of infrastructure planning broke down. In 2000 and 2001 the enthusiasm for the green polder model already waned (Weggeman 2003).

3.3.3 Ecological modernisation of the transport sector

One of the main themes in Dutch environmental policy in general and in ecological modernisation in particular was the integration of environmental policy in other fields of policy making. While the OECD was positive about the integration of environmental concerns within the decision making practices of industrial sectors, it was not positive about integrating environmental concerns into transport policies and the agricultural sector. It concluded that the Netherlands was not yet moving in the direction of sustainable transport. Moreover, economic policies and ambitious economic targets for the transport sector presented grave environmental challenges (OECD 2003, p. 177). In the air quality clash the harmful emissions of cars and lorries attracted much attention and so a look at the attempted ecological modernisation of transport and mobility should not omitted.

The OECD verdict is interesting because a serious attempt had been made to bring transport policy in line with environmental policy. At the time of the NMP, the Dutch Ministry of V&W unveiled its own plans for the future. Transport policy was laid down in 'structural schemes', programmes that determined future road planning, expansion of public transport, and the relation of transport within the wider scope of spatial planning. In 1989, the Dutch Government unveiled the Second Structural Scheme for Traffic and Transport (Tweede Structuur Schema Verkeer en Vervoer, SVV2).

The first draft of the SVV2, (part A), dated from December 1988, and contains the policy proposal (V&W 1988). The final proposed version of the SVV2 (part D), dated from 1990 (V&W 1990). This plan was very much an adaptation of eco-modernist policy to the field of transport and mobility. The aim of the plan was twofold: namely, to relieve the Dutch roads of the heavy burden of congestion, but also to restructure transport in order to reach a sustainable transport system. Because congestion was a severe threat to the Dutch economy, reducing congestion was a mainstay of Dutch transport policies, as the transport sector was economically important (Brokking 2001). The focus on sustainable transport was new, and the SVV2 part D was a plan that focused more than any earlier one on public transport as the future of mass transit in the Netherlands.

Uncharacteristic for a transport and mobility programme, further expansion of the roads was portrayed as unsustainable, and was one of the 'nightmare scenarios' discussed on page 14 (V&W 1988, p. 14). In order to guarantee the high quality of the living environment in cities, the accessibility of cities by car would have to be limited (V&W 1990, p. 10). The alternative to road expansion was the further development of public transport. Moreover, various strategies were adopted to restrict economically non-viable transport.

To this end, a kind of target group approach was used in the SVV2. For instance, public transport and freight transport needed to gain access to preferential lanes, and commuting by car and using the car for recreational purposes was dissuaded. 'Unnecessary driving' was to be reduced, and, if possible, both consumers and freight transport should use rail instead of road transport to fulfil their mobility needs. The switch from the automobile to another, cleaner mode of transportation was referred to as 'modal shift'. In the SVV2, modal shift was an important policy target.

The public was to be targeted by awareness-raising campaigns, and to be educated on the deleterious effects of using a car. Communication efforts were to make sure that civilians were acutely aware of the '*nigh insurmountable problems*' that mobility posed to the living environment (V&W 1990, p. 89). A change in people's perception of the car was deemed inevitable in this context, with the car being treated not as a status symbol but as an ordinary object like any other (V&W 1990, p. 17). In a vein similar to what VROM had done earlier, the Ministry of V&W intended to mobilise NGOs and segments of society to disseminate these policy ideals:

> '*Societal organisation form the intermediary segment towards the public. Co-operation from these organisations is of essential importance to create a consensus in society for the often drastic measures which are necessary to realize a sustainable transport and traffic policy*' (V&W 1990, p. 125).

The intertwinement of spatial planning and the environment was well established, but in the SVV2, spatial planning policy was used to achieve the goal of reducing mobility growth. People would, for instance, be stimulated to live close to their work. In practice, this meant that some locations rather than others would also be more easily reached by car. To stimulate public transport, labour-intensive companies and facilities that attracted a large number of people should be easily reachable by public transport. These locations were dubbed A locations. Less labour-intensive companies, such as businesses in the service sector, should be reachable by car. These locations were henceforth known as C locations. B locations were mixed: that is, reasonably reachable both by public transport and by car. This ABC policy also had ramifications for the number of parking places allotted; for instance, 'A locations' in the city would have only 10 parking places per 100 employees.

A further measure to reduce the use of the car was the introduction of a pricing mechanism. Charging for use of the road was introduced in the SVV2 for the first time in the Netherlands. In the first draft, the instrument of choice was road charging per kilometre. These extra charges were justified by pointing out that drivers would pay for the extra costs of environmental clean-up that they caused. However, the internalisation of costs was not the only motive – extra charges would also help attain the target of limiting traffic growth (V&W 1990, p. 40).

Though the plan had a definite environmental slant, the economic importance of the Dutch position as a gateway to Europe was not forgotten. The transport sector would not be curtailed, for instance. The SVV2 emphasised the competitive possibilities of the Dutch transport sector. The distribution function of the Netherlands for the other countries in Europe should be strengthened, and this would mean an increase in mobility. A number of infrastructural improvements were foreseen, such as harbours, transport corridors to Germany and Belgium, and an expansion of Schiphol Airport. All these projects generated not only economic growth but potentially also extra mobility. All the measures that curbed mobility needed to compensate for this increase in order to reduce congestion for economically important transport. Some of these new infrastructural projects were very ambitious and controversial, such as the expansion of Schiphol Airport and the Port of Rotterdam, along with rail links with Belgium and Germany.

Within the SVV2, classical eco-modernist elements could be discerned. Firstly, there was the dual goal of economic expansion and environmental protection. Secondly, a policy of responsibilisation towards drivers started. Thirdly, spatial planning was used to influence the behaviour of motorists in an environmentally friendly way. Fourthly, advertising campaigns and environmental groups should forge a consensus on driving responsibly for the sake of the environment. However, in order to be able to stimulate the economically important transport sector without having to expand the roads, the plan restricted the freedom of the ordinary motorist. The rail links should in principle make modal shift an appealing option, but one of these trajectories ran right through a nature area.

Box 3: Policy innovations stemming from the SVV2

The SVV2 led to policy innovations, which were of interest from an ecological modernist point of view. Of note was the introduction of a more programmatic approach to infrastructure development through the 'MIT' (Multiyear programme on Infrastructure and Transport), in which projects had to be announced, forcing the department to work on a more programmatic basis. Another innovation originating from the SVV2 was the 'Route Law' (Tracéwet), laying down consultation procedures for adapting and constructing roads. These procedures granted the public significant rights of participation and access to the administrative court, but they also caused decision making to take a long time. This raised annoyance among political parties in favour of mobility legitimated large investments in public transport and in mega projects aimed to facilitate (rail) transport. Experiments were conducted with separate lanes for car-pooling, and car-pooling was stimulated by public awareness campaigns.

3.3.4 *The failed ecological modernisation of the transport sector (1989 – 2001)*

The SVV2 was an ambitious attempt to ecologically modernise the transport sector and mobility as a whole, but it was unsuccessful from the point of view of accommodation and consensus building. Car users tended to see the SVV2 as an anti-car policy and as pro-public transport (Vrijssen 2010). In particular, the issue of road pricing through the charge per kilometre created resentment among motorists, united in the Algemene Nederlandse Wielrijders Bond (Dutch Motorist Association, ANWB). Successive Cabinets were not able to introduce a form of road pricing. However, attempts to persuade the ordinary motorist not to use the car but to embrace public transport also failed (Raad voor Verkeer en Waterstaat 1998, p. 6). It was easy to see why the ordinary motorist was against the SVV2 measures, as drivers would have to pay for them in various ways. Motorists were targeted with all kinds of extra charges, but they reaped none of the benefits. Transport had preferential lanes, good rail links, and expanded corridors to the German hinterland, but the motorist was left with selective accessibility, limited access to cities, and few parking spaces in his or her work place.

The final version of the SVV2 was introduced by the third Lubbers Cabinet that had taken its seats in November 1989. After a campaign in which environmental issues played an unprecedented role, the VVD was ousted from power and replaced by the socialist PvdA. The Christian Democrat CDA remained the biggest party and Ruud Lubbers (CDA) took on a third term as Prime Minister. In the field of the environment this Cabinet continued the work of its predecessor, even strengthening the SVV2 and releasing a reinforced version of the NMP. Fuel prices were also raised quite steeply. The measure became popularly known as 'Kok's dime' (het kwartje van Kok), named after the Minister of Finance – and later Prime Minister – Wim Kok. Ordinary motorists though resented the plans in the SVV2, especially road pricing and the rise in fuel prices. They found allies in the motorists association ANWB and in the VVD, now in the opposition.

In the latter half of the 1990s, we witnessed a gradual relinquishment of environmental ambitions in transport policy as accessibility and combating congestion became more important throughout the decade. After the third and last Lubbers Cabinet, the PvdA formed a Cabinet with its notorious political adversary the VVD with the left liberal D66 as the glue that held this 'purple Cabinet'[6] together. The intention of these three parties was to finally oust the CDA of its traditional hold on power and the Christian Democrats found themselves in the opposition. Margreeth de Boer from the PvdA became the Minister of VROM and in that capacity would oversee the European negotiations over the Air Quality Directives. Annemarie Jorritsma of the VVD became Minister of V&W.

The motto of this new Dutch Cabinet in the 1990s was 'work, work, work' in order to combat unemployment. Economic issues gradually came to the fore. A storyline gained in ascendancy in which the Netherlands was positioned as a 'distribution country' and a 'gateway to Europe', and to fulfil this ambition the large-scale infrastructural projects of the expansion of Schiphol and Rotterdam, and the construction of a railroad through the nature area Betuwe and down to Belgium, needed to be realised. This pro-mobility storyline established itself firmly at the end of the decade. Congestion was seen as a major public nuisance during that time. The strong ecological

6. After the mix of the collors red, the colour of the PvdA and blue, the colour of the VVD.

modernisation of the transport sector faltered on the ever-present concern in the Netherlands for economic efficiency and its place in international competitive relations.

This gradual shift was outlined in a number of policy documents after the SVV2, for instance in the Memorandum 'Samen Werken Aan Bereikbaarheid' (SWAB, Co-operating for accessibility, V&W 1996), issued by Minister Jorritsma. In the 'Perspectives Memorandum on Transport' from 1999, (Perspectievennota Transport) efficient use of road capacity replaced the transition car to rail and from car to bicycle as a main instrument for an efficient transport system. Road pricing should ensure further efficiency and relieve congestion. In the long term, new technologies were to present solutions to both congestion and environmental degradation. The scarcity of infrastructure was the number one priority of the Ministry, and a behavioural change on the part of the driver was no longer considered. In fact, even the Advisory Council of the Ministry of VROM, the VROMraad, emphasised the social value of mobility: '*The positive1 contribution of mobility for the individual and for society can hardly be overstated*' (VROMraad 1999, p. 17).

The idealistic expectations of the SVV2 died down during the 1990s, and were replaced by a business-like approach and a less ambitious form of ecological modernisation. New technology and especially road pricing should lead to a more efficient use of the road, and to an internalisation of environmental costs by the polluter. The end-point in this development towards an ecological modernisation based solely on technology and road pricing was reached with the proposal of a new transport plan to replace the SVV2. One of the last actions of the purple Cabinet was the proposal of the National Traffic and Transport Plan (Nationaal Verkeer en Vervoersplan) in the year 2000. All the idealistic and normative overtones of the SVV2 disappeared and one of the new plan's core messages was '*Mobility is allowed and belongs to a modern society*' (V&W 2001, p. 3). However, the motorist would have to pay for his or her choices. In the NVVP, road pricing as a market-based regulatory instrument was the central mechanism to adjust individual interests of mobility and the public interest of the environment: '*Individual and collective interest find each other if all societal costs of mobility are included in the price one pays for mobility*' (NVVP 2001, p. 10).

The introduction of this form of payment for car use instead of simply car ownership became the central ecological modernistic element in the NVVP. It led to environmental gains, because road users were to pay compensation for the damage they inflicted upon the environment. All the idealism of behaviour change through other means was gone, however, and nothing remained of target group policy, modal shift, or the ABC policy of parking places.

In its turn, road pricing became the last vestige of ecological modernisation to be knocked down. Road pricing by way of a charge per kilometre was still highly controversial, and societal resistance was mobilised by the ANWB, the Dutch Association of Motorists (Website Wegenwiki NVVP). Its traditional ally in Parliament was the VVD. At the last moment, the coalition partner withdrew its support for the plan, just before the purple Cabinet was forced to resign in 2002. The 2002 elections resulted in a victory for pro economy and infrastructure development parties, which would form a new coalition, including the VVD. This new coalition led by Prime Minister Jan Peter Balkenende would almost do away with ecological modernisation all but in name. Road construction and expansion, without road pricing, would become a spearhead of this Cabinet.[7]

Upon evaluation in 1998, the SVV2 received a scathing review by the scientific bureau of the Ministry of V&W, the Council for Transport and Water Management (Raad voor Verkeer en Waterstaat). *'In no other societal sector does the effectiveness of the administrative system seem to be so low as in transport and mobility'* (Raad voor Verkeer en Waterstaat 1998, p. 5). This indicated that already in 1998 the strong ecological modernisation of transport and infrastructure was considered a pipe dream. In fact, the second half of this decade saw environmental conflict recur over the ambitious mega projects that were planned to facilitate the transport sector.

3.3.5 *Legal conflicts over the megaprojects*

The ecological modernisation of transport policy failed because of sustained opposition from motorists united in the ANWB and pro-mobility parties. However, the infrastructural projects foreseen in the SVV2 also suffered setbacks. In this case, the environmental movement was the party that broke the consensus.

Empirical research showed that Dutch pressure groups in the Netherlands took matters to court much more often than did their colleagues abroad. According to De Saedeleer, between 1997 and 2001 an estimate of 4000 cases had been brought before the courts. Neighbouring countries like Germany, Belgium, and the UK were estimated to have had only around 100 such cases, and only France was involved in more than 1000 cases (De Saedeleer et al. 2002, p. 3).

Table 5: Estimated absolute number of court cases brought by environmental associations

	Belgium	France	Netherlands	Portugal	Italy	Germany	UK	Denmark
1996-2001	146	1197	4000	57	117	115	102	4

Table taken from De Saedeleer et al. (2002), the number for the Netherlands was based on rough estimations.

In the 1990s, legal conflicts erupted over proposed megaprojects mentioned in the SVV2, and more extensively in two Memorandums on spatial planning, in the fourth Memorandum on Spatial Planning, known as VINO, and in the Fourth Memorandum on Spatial Planning Extra, known as VINEX. The VINEX partly reflected the

7. Balkenende would run four Cabinets in total. These Cabinets are in power during the years of the air quality clash, recounted in chapters 6 and 7.

environmental concerns of the age. The ABC policy, for instance, was developed most fully in this document, but it also shared the economic consideration of making the Netherlands a prime distribution country. Ambitious projects like the above-mentioned rail links and expansions at Schiphol and Rotterdam had to facilitate the flows of extra transport.

The construction of the Betuwe rail link and the expansion of Rotterdam Port and Schiphol Airport were especially controversial. The Betuwe route was a thorn in the side of the environmentalists, because it was planned through a nature conservation area. The expansion of Schiphol Airport was controversial, because of the increase in noise nuisance for the inhabitants of nearby living areas. Understanding this resistance was important for understanding the air quality clash, since this clash involved struggles over infrastructural projects as well.

In the context of the green polder model, negotiations between stakeholders about expansion of Schiphol Airport took place in the TOPS negotiation (TOPS: Tijdelijk Onderhandelingsplatform Schiphol, Temporary Negotiation Platform Schiphol). The environmental movement was present, including the radical Milieudefensie. The expansion of Rotterdam Port was discussed in the Project Main Port Development Rotterdam (PMR), but the TOPS negotiations between Schiphol and environmental pressure groups failed.

In the face of the failing negotiations, pressure groups and local groups of residents went to court to protest, among other things, the expansion of Schiphol Airport and the Betuwe rail route.[8] In the Netherlands, access to the administrative court was open, efficient, and cheap (VROMraad 2008; SER 2006), and environmental pressure groups had relatively easy access. In order for a complaint to be eligible, the claimant had to be a 'stakeholder', which meant its interests needed to be at stake. The concept was broadly defined in Dutch administrative law, and it included NGOs and lower administrative bodies who could initiate proceedings against administrative decisions (Marseille 2011, p. 239). Open access to the administrative courts fit the strategy to mobilise the public and to involve it in environmental matters, but it also made going to court an attractive option if negotiations failed. Conflict between environmental pressure groups and the administration was often adjudicated. In particular, conflicts surrounding the big spatial projects often ended up in administrative courtrooms (Weggeman 2003, p. 83).

This fight over the use of space and the juridification of these political clashes was a thorn in the side of the Government. In November 1997, a Commission chaired by Jos van Kemenade issued a Memorandum titled 'Met Recht Verantwoordelijk' (Responsible Law). In it the complaint was raised that environmental pressure groups used

8. In Parliament, conflicts over the megaprojects were researched by a Parliamentary Commission chaired by Adri Duijvestein. He presented his final report in 2004, and the Parliamentary documents that contain the reports can be found on this website (Website Duijvestein last accessed 22-02 2012).

the law as a mechanism to obtain a second chance if political negotiations failed. This situation was considered undesirable, and parties were urged to use the law and litigation in a responsible manner, preferably as little as possible.

From the point of view of the Government and other policy makers, this complaint was understandable. However, from the perspective of the environmental movement, the law was a legitimate mechanism to redress a situation they considered unjust. The environmental consensus of protecting the environment by further economic development was mostly between policy makers and little attention had been paid to whether it was legally workable. In the case of environmental spatial law, the way quality standards worked in the Netherlands was consistent with a limit to growth discourse in which limits needed to be set on economic expansion. Together with the broad mechanisms of participation and easy access to the courts, the law provided means to block the plans.

These legal opportunities had been put in place to make sure economic interests would not overpower ecological interests. However, also in the time of ecological modernisation they were still useful. This time not to set limits on growth, but to create a balance of power between environmental and economic interests. They were a relic of the past though and did not foster consensus. Environmental spatial conflicts would become a distinct feature of environmental politics in the 1990s and 2000s.

Apparently, consensus on the approach to environmental policy did not translate easily into consensus with regard to other conflicts, such as those between environmental interests and the interests of motorists, or mutually exclusive claims on the use of space. Clashes over the megaprojects could be seen as the prelude to the air quality problem, or conversely, the air quality problem could be considered as being borne out of clashes over the megaprojects. The construction of infrastructure created drastic interventions involving the landscape and other environmental concerns. These demands caused anxiety among people who lived close to the areas under consideration, and these people were a prestigious target for environmental groups who did not see the need for further economic expansion. In the Netherlands, space was a scarce commodity worth fighting for.

Legal battles over the megaprojects displayed an inherent strain in eco-modernist policy discourse. On the one hand, bringing infrastructural projects to fruition became almost impossible, because of the number of procedures that had to be undertaken, the number of hearings in which complaints could be voiced, and the strict limiting character of environmental standards. These impediments were the result of demands to involve the public in decision making that had an environmental impact, and to make sure that spatial development proceeded without damaging the environment. On the other hand, the positive sum game format of ecological modernisation demanded economic growth, and – as a means to this end – infrastructure development. In the long run, environmental well-being and economic progress could coincide on an abstract level, but residents who lived near Schiphol, for instance, did not benefit from these abstract improvements, nor did commuters who were stuck in traffic.

3.3.6 *The nature of environmental conflict and the position of the VROM Ministry*

The history of conflict in the 1990s highlighted the downsides of the optimistic eco-modernistic discourse used by VROM. Its ambition to merge ecological and economic language resulted in large-scale legal conflicts that this discourse was meant to avoid. The desire to keep the Netherlands competitive, while not encouraging road transport led to the expansion of rail links and other megaprojects that themselves engendered opposition. Ecological modernistic discourse propelled the Ministry to a prominent position at the end of the 1980s, but at the end of the 1990s economic thinking once again dominated VROM's environmental discourse. This becomes clear when one compares policy documents from the beginning of the 1990s with those at the end of the decade. At the beginning of that decade, the political programmes regarding spatial planning and transport were geared towards VROM's environmental discourse. Notions like the target group policy emerged, along with the use of spatial planning to serve the environment. At the end of the decade, an economy-dominated discourse was firmly in the driver's seat with regard to mobility and infrastructure development. Lip service was paid to sustainable mobility, but no more than that. This shift illustrated the declining power of the Ministry of VROM. It was no longer able to export its once innovative storylines. Gradually, ambitious environmental policy was viewed again as problematic, as it obstructed quick decision making with regard to the expansion of roads and other types of infrastructure development. The fact that the V&W Ministry strongly dominated discussions on the megaprojects was a sign that VROM had lost the power it held in the early 1990s.

3.4 Concluding remarks

In the preceding sections, I have provided a historical account of the rise to prominence of the VROM Ministry and the concomitant institutionalisation of a new direction for environmental policy. The setbacks have been discussed in the latter sections. I will use the final section to draw conclusions on the nature of Dutch ecological modernisation and on Dutch environmental policy in general.

3.4.1 *Continuity and discontinuity in Dutch environmental policy*

By accepting eco-modernist discourse, the Dutch Government let go of its earlier commitment to a discourse tinged with the idea of setting limits to growth. These two discourses were antithetical regarding their perception of the role of economic development and the regulatory instruments to be used. Ecological modernisation argued for economic growth and flexible policies to combat environmental degradation, while the discourse of limits to growth demanded limits on economic expansion through top-down regulation. It is remarkable, however, how much continuity there was otherwise in the policies of VROM. Many policy innovations that we consider modern and fitting with ecological modernisation had already been considered in the 1970s, and were envisioned in the Urgent Memorandum on Environmental Hygiene published in 1972. This phenomenon is little noticed by environmental scholars, but important in light of a proper understanding of the air quality clash. Many features of Dutch environmental policy are apparently ingrained in Dutch decision making practices and cannot easily be overturned.

The Urgent Memorandum contained a holistic vision of the environment as an integrated whole consisting of eco-systems. For practical purposes, legislation covered the environment sector by sector, but that did not correspond to its theoretical outlook. Moreover, the Urgent Memorandum urged for partnerships with the environmental movement, and for reaching out to industrial sectors. Though the Ministry of VOMIL, which had released the document, was not successful in finding partners among industrial sectors, the intention was there. Finally, the policies from the 1970s also called for a preventative approach, for the use of sound scientific knowledge and for reaching out internationally to stop environmental degradation. These were all considered novelties in the 1980s, but these intentions were already present in Dutch environmental policy from its inception onwards.

Discursively, these programmatic aspects were compatible with many of the characteristics contained in ideal typical ecological modernisation. We may therefore conclude that Dutch environmental policy was ecologically modern 'avant la lettre'. In practice, however, the Ministry could not realise these intentions. Antagonism between the Ministry of VOMIL on one side and more economically oriented Ministries and powerful economic sectors on the other thwarted many of its more ambitious intentions.

Interestingly, things started to look up for the Ministry in the 1980s, at a time when deregulation was a spearhead of policy, and the Government became committed to rolling back state interference in societal affairs. The department of the Environment was combined with the department of Spatial Planning and Housing, which made it possible for environmental policy to lose its focus on environmental hygiene and to focus on the environment as it related to the use of space. The new Ministry of VROM had neither the means nor the ideological commitment to start a regulatory campaign, so it devised instead a strategy of consolidation and consensus building.

A number of other continuities in policy are of special importance. Environmental spatial law was only augmented rather than being drastically changed. The permit system remained the cornerstone of Dutch practical environmental policy, and permits set limits regarding emissions of harmful substances. Moreover, environmental quality standards determined the amount of pollution permissible in a certain area, and permits could no longer be granted if these limits were transgressed. These policies were more reminiscent of ideal typical 'limits to growth' than of ecological modernisation. The internationalist outlook and the goal other reaching out to other actors beyond the Ministry's direct sphere of influence was also present in the Urgent Memorandum and not a particular novelty of ecological modernisation.

In terms of environmental radicalism and in terms of a precautionary approach to environmental problems the discourse of ecological modernisation embraced by VROM was actually a step back from the discourse of limits to growth. It is paradoxical that by taking a step back in ambition, the VROM Ministry by and large succeeded where VOMIL failed, namely in getting the environment high on the political agenda. In the next sub section I will argue that this success is due to the emergence of a new storyline on the economic efficacy of environmental protection.

3.4.2 *Environmental protection as necessary for business: The powerful storyline of win-win scenarios*

In the previous sections, the rise of a managerial discourse in environmental policy has been recounted. Discursively, the storyline that managed to forge new alliances between economic and environmental actor was the storyline that environmental protection could provide economic revenues. This storylines, which I refer to as the storyline of win-win scenarios was the most innovative line of the ecological modernistic approach of policy. Ecological modernisation relied on market parties, capitalist logic, and technological innovation for beneficial environmental results. The storyline was introduced into Dutch policy by Pieter Winsemius, who used his knowledge of modern management and positive reinforcement to make environmental policy appealing to the target groups. He assumed that people would accept the demands of environmental policy more easily if they could be convinced it was in their economic interest to do so. By appealing to a storyline that fitted well within traditional capitalist logic, he was able to argue for environmental policy and to circumvent ethical appeals for a more harmonious life or for other ecological/ethical arguments. The keyword became 'internalisation', the drive to persuade economically important sectors and other Ministries to incorporate environmental concerns into their own policies.

It is of interest that the VVD spearheaded the development of this innovative policy discourse. In further chapters of this study the VVD will often be portrayed as a political party that advocates infrastructure development and rolling back environmental regulation. However, during the late 1980's the VVD was certainly not merely a 'pro economy party'. Ecological modernisation managed to weld environmental protection, a traditionally leftist subject, with principles of industrial self-regulation and faith in the operation of the market traditionally embraced by the VVD. I conjecture that the steadfast reputation of the VVD as a defender of entrepreneurialism and market party autonomy, made it possible for the Minister to rally the industry behind his course. He spoke the language of the manager and was not frowned upon with suspicion. It is the irony of environmental history that the most noteworthy and ambitious environmental Ministers were delivered by the party that was least expected to do so.

In the early and the mid-1980s, Winsemius seduced the industry with this win-win storyline that portrayed environmental protection as necessary to protect economic interests and the environmental movement and environmental scientists created a sense of urgency among the population by presenting a worrisome account of environmental risk. In 1989, environmental scientists presented a bleak image of environmental degradation in a report called 'Take Care of Tomorrow'. In response to this report, the VROM Ministry came up with its first large-scale integrated National Environmental Policy Plan – the NMP. Other departments had to commit themselves to the policies of the VROM Ministry by co-signing the NMP.

In the NMP, the ecological aspects of policy were translated into economic language. Through the use of metaphors derived from economics and management, the case was made that protecting the environment now would be cheaper than having to clean it up in the future. The representatives of economic interests could be reached more easily through language and logic that was familiar to them. It seemed no coincidence that VVD Ministers who generally supported business interests introduced such logic into environmental policy. It made environmental protection appealing to their own rank and file, and promised that instead of economic stagnation it would lead to growth if correctly managed.

In 1989, the Ministry of VROM was at the peak of its discursive power. VROM managed to export its concepts to other Ministries such as the Ministry of V&W. In the later 1980s and early 1990s, we saw concepts developed by VROM appear in plans issued by this Ministry. Furthermore, the environmental movement was accommodated and it gained seats at the negotiation tables of various Dutch advisory councils and other consensus-building institutions.

By introducing the storyline that environmental protection created economic opportunities, Winsemius and his successor Ed Nijpels managed to form a discourse coalition between the industry, the VROM Ministry, and the moderate elements of the environmental movement. This storyline and the introduction of managerial language within the discourse of the VROM Ministry established it as a credible negotiation partner for these different actors. This discourse coalition accepted the positive sum game postulate of ecological modernisation, and subsequently this discourse was used in Dutch environmental policy. Initially, a weak form of ecological modernisation held sway which was supported by gentlemen's agreements with industrial sectors. Gradually opportunities arose to introduce a stronger version of ecological modernisation, especially after the publication of the worrying report 'take care of tomorrow'. In the NMP and also in the transport plan SVV2, a strong version of ecological modernisation is introduced with particular emphasis on behaviour change in an ecological direction.

3.4.3 *The Achilles heel of ecological modernisation; transport and infrastructure*

The most celebrated Dutch environmental policy, the NMP, introduced ecological themes into Dutch policy, such as a view towards production and consumption in terms of ecological cycles and streams of energy. It aimed at behavioural change, by urging everyone to assume responsibility for the environment. Even education and advertising, for instance, should be made subject to the message of sustainable development. It appeared that strong ecological modernisation was about to become a dominant policy discourse, but this study shows that it failed on two important terrains, transport and infrastructure development. These terrains would become battle grounds during the air quality clash.

The NMP had a counterpart in the Structural Scheme for Transport and Traffic: the SVV2. In the SVV2, behavioural change was envisioned with regard to the mode of transportation. People were urged to make far less use of the car, and to opt for a form of collective transportation. People had to be made aware of the detrimental effects that auto-mobility had on the environment. However, the SVV2 failed to bring about eco-modernistic changes in the transport sector. In the SVV2, green plans to modernise traffic and transport were unveiled, and many policies were designed to curb transport and traffic. However, the economically important transport sector was by and large exempt from the burdens, and its growth was even encouraged. The tab had to be paid by ordinary motorists, who resented this strategy and went on to block reforms, even when the sharp edges had already been filed off. They were represented by powerful a powerful organisation such as the ANWB and also the VVD, back in the opposition in the first half of the 1990s, represented their interests.

Moreover, the desire to construct megaprojects such as an expansion of Schiphol Airport, the Port of Rotterdam, and the realisation of a rail link to the German hinterland carried more weight. Gradually, successive coalition Cabinets started to lend its ear to more economically motivated storyline, such as the wish to position the Netherlands as a prime distribution country in Europe.

The megaprojects in particular were areas where the ecological modernistic consensus faltered. The Government's efforts to introduce the typical highly institutionalised Dutch consensus democracy to the environmental domain failed when the polder turned into a legal battlefield over important infrastructure projects such as the Betuwe rail link and the expansion of Schiphol airport. The environmental movement was empowered by Government support and the ecological mind-set of the public, but it became disillusioned when it could not make many gains at the negotiation table. The environmental movement took their claims to court when they were not heeded, leading to large scale juridification of environmental/spatial conflicts. The legal conflicts over megaprojects foreshadowed those over air quality ten years later. The introduction of ecological modernisation in these fields caused polarisation instead of consensus.

It is curious that environmental policy lost ground so quickly in the Netherlands. In 1989 the Ministry of V&W still took over concepts developed in environmental policy and utilised them to ecologically modernise transport policy, while in the second half of the 1990s economic issues firmly retook the driver's seat. In theory the Cabinets that ruled during the 1990s, first a Cabinet of CDA and PvdA, than Cabinets of PvdA, VVD and the environmentally conscious D66, were more susceptible to environmental concerns in theory than the CDA / VVD Cabinets of the 1980s. Changes in the political colour of the Cabinets in power cannot explain this loss of environmental ambition. Partly the relinquishment of environmental ambition had to do with an economic downturn and an emphasis on the importance of the Netherlands as a transport and distribution country.

On the other hand I also conclude that for an environmental cause to succeed it is important that it is championed by people and parties that enjoy a certain amount of trust with the industry and other vested economic interests. Ecological modernistic discourse can be more convincingly 'sold' to the sceptics by a VVD Minister. The party is traditionally a reliable partner for interests of industry and transport. Ecological modernisation is dependent on the success and consensus building skills of Government officials. The absence of the VVD during the first half of the 1990s in the Government may well explain why ecological modernisation did not manage to win ground among the motorists. It was resented by motorists who could easily rally against the socialist Minister of Finance requiring them to pay more for fuel and against the socialist Minister of VROM. The Minister of V&W was Hanja Maij Weggen of the CDA. She became resented among car owners for her road pricing proposals.

With the VVD in opposition it became much harder to reach out to the traffic and transport sectors and these sectors themselves toughened their opposition. The VVD committed itself to promises to end congestion and therefore, when it returned to power in the mid 1990s it had nothing to gain by supporting environmental policy. In the purple Cabinets it could profile itself as a defender of economic interests and distance itself from the Minister of VROM. By 2001 the VVD was adamant in its opposition to road pricing and even scuttled the successor to the old SVV2, a plan proposed by a Cabinet it participated in itself.

The conflicts over the megaprojects and the failed ecological modernisation of transport, display the serious drawbacks of Dutch eco-modernistic policy. It forces the

Government to balance carefully, because it needs to establish consensus between interests traditionally regarded as inimical. The Government is under constant pressure to deliver on its promise than environmental protection and economic growth can progress together. In its heyday, during the early 1990s, weak ecological modernisation was successful. Nationally polluting industries signed covenants to clean up their act. During the negations over the megaprojects and in the case of transport and traffic it was far less successful. Its propensity for consensus also fosters a mentality in which making hard choices is postponed. Winsemius' motto that 'No one should be asked to do the impossible' is telling in this regard. The eco-modernist style of policy making avoided difficult societal question when interests were not easily aligned. Therefore it gave rise to charges of dishonesty from social groups as diverse as car owners and environmentalists.

3.4.4 The merchant and the reverend

Dutch ecological modernisation succeeded partially in breaking up traditional animosities, but it did so at a cost. The discourse coalition between economic and environmental interests also meant that environmental policy discussions became influenced by the managerial and economically oriented dialogue of its coalition partners, and lost much of its earlier radical character. Environmental policy committed itself to the same logic of growth and development as traditional economic policy. Achieving economic growth became a target of environmental policy, as well as safeguarding the environment, nature, and public health. Even its language became tinged with economic justifications.

This strategy seemed to pay off in the early 1990s. Strong ecological modernisation seemed to be the course Dutch politics was heading for. However, after the release of the NMP, Dutch ecological modernisation gradually lost many of its strong characteristics. After release of the NMP in the early 1990s, and especially in the latter half of the 1990s economic considerations were on the rise again, and the Netherlands embarked on a path to become a prime distribution country, undoubtedly attracting environmentally harmful transport and traffic in the process. The Netherlands postured as an environmental leader abroad, but it was especially adept at exporting policy principles and policy discourse. It did so with a keen eye for economic interests.

We may conclude that where environmental policy was concerned – and to cite an old Dutch maxim – the Dutch Government acted as both merchant and reverend. This emblem describes a mentality of preaching idealistic rhetoric while making sure that profits are maintained. In environmental policy, the Government – like a reverend – preached high-minded environmental ideas, but – like a shrewd merchant – it also kept a close watch on costs and profits. This characteristic was not unique for environmental policy, however; Andeweg and Irwin (2005, p. 206) contended that in the area of foreign policy the same comparison could be made.

While both the 'merchant' and the 'reverend' preside over Dutch environmental policy, during some years the merchant is on top and at other times the reverend leads the way. VVD Minister Winsemius may be considered an environmentally wise merchant. He 'sold' environmental policy to a sceptical audience of industrialists. However, after 'take care for tomorrow' was published, the reverend took over in environmental policy. During the 1990s he had to tone down his preaching again because economic considerations came to the fore.

The policy discourse of ecological modernisation fits this merchant and reverend mentality. The discourse can be used to argue for an ambitious environmental policy with a strong emphasis on precaution and behaviour change, or for a cautious policy which demands minor economic adjustments and fits with free market capitalism. Pieter Winsemius had introduced a weak form of ecological modernisation in Dutch politics, a mercantile form. The discourse gradually acquired characteristics of strong ecological modernisation during its 1989-1992 heydays. After these 'golden years' of environmentalism, the discourse lost many of its strong features, especially in regard to transport and infrastructure, and assumed a weak form again. It gradually weakened until in 2002 new PM Jan Peter Balkenende almost does away with it altogether as we will see in chapter 6.

Dutch environmental policy displays a pendular movement and is caught between the merchant and the revered, with the merchant holding the upper hand most of the time. Weak ecological modernisation ruled throughout most of the 1980s and most of the 1990s with the late 1980s and the early 1990s being exceptions. During these years the strong ecological modernisation took over and the Dutch reverend preached successfully, not only at home, but also in Europe as the history of the social construction of European air quality policy displays. This realisation is of importance: The Netherlands exported policy principles abroad that were recommended by the reverend but became blocked by the merchant, at least in the Netherlands. Whereas in the Netherlands policy principles did not lead to equally tough measures, they did in the European Union, in any case with regard to air quality policy. In the Netherlands weak ecological modernisation generally held sway, but strong eco-modernistic principles were exported to the EU.

3.4.5 *Dutch environmental policy and the legality of precaution*

Early Dutch environmental policy was characterised by the discourse of limits to growth. Potentially this discourse may lead to the institution of a precautionary legality among legislators and the judiciary. The discourse also had some effects on the legal order pushing it in a precautionary direction. The link between environmental standards and spatial planning is a case in point. On the whole though, this discourse did not have far-reaching precautionary effects. The Government did apply laws and set standards but they were by and large ineffective and still based on what was considered economically reasonable. Pollution abatement of the era succeeded despite the permit system and not because of it.

The onset of ecological modernisation brought more opportunities for a precautionary legality to gain ground. In the early days of Winsemius' Ministry feasibility concerns still reigned but gradually eco-modernist discourse became stronger. The strong variation of ecological modernisation seeks to influence consumers and producers and urge them to seek prevention of environmental degradation out of an environmentalist but also an economic rationale. Because is considered to prevention pay off, resistance against more far-reaching environmental measures lessened. The strong eco-modernistic policy plans such as the NMP and the SVV2 leave more room for a legality of precaution. The NMP sought to utilise legal instruments such as criminal liability and broader product liability in the service of the environment and the SVV2 urged the prevention of mobility growth out of environmental reasons. Both plans tried to instil environmental awareness in the public through education and advertising campaigns.

However, the strong ecological modernistic period lasted relatively briefly. The ecological modernisation of the transport sector and mobility in general failed in the 1990s. Economic considerations started gaining the upper hand again and environmental interests receded to the background again. The environmental awareness raising campaigns and relatively open access to the administrative courts led to juridification of environmental conflicts but the Council of State did not posture itself as the guardian of environmental interests at the time. Moreover, the logic of ecological modernisation, even in the strong variation, forms a dam against notions of excessive prevention. Excessive prevention would harm the economy and for ecological modernisation discourse to succeed it should show that the economy would actually benefit from environmental measures. In order to do that it sought win-win solutions and that in turn required weighing costs and benefits, a core element of the legality of risk and compensation.

We may conclude that although ecological modernisation and the legality of precaution shared an affinity with a precautionary approach and strong ecological modernisation could even have fostered a legality of precaution, the variation dominant in the Netherlands was not conducive to the emergence of this type of legality. Economic consideration pulled too much weight during most of the years between 1972 and 2000 for a legality of precaution to take root.

Timeline The Merchant and the Reverend, Dutch ecological modernisation

1969-1971	Air pollution protests in Rotterdam Rijnmond area
1971	Ministry of VOMIL founded
1972	Report for the Club of Rome published
1972	Urgent Memorandum published
1976	Minister Lubbers released 'Perspectives Memorandum'
1982	Project PIM unveiled, restructuring Ministry of VOMIL
1982	Ruud Lubbers takes over as Prime Minister, Ministry of VROM founded, Pieter Winsemius first Minister of VROM
1984	First IMP. M. released
1986	Ed Nijpels takes over as Minister of VROM
1987	Release of Brundtland report "Our Common Future"
1988	Release of In Care of Tomorrow
1988	Presentation of first draft Tweede Structuurschema Verkeer en Vervoer, SVV2
1989	First Nationaal Milieubeleidsplan, NMP, released
1990	Release of SVV2 final version
1990	Release of NMP+, a fortified version of the NMP
1994	'Purple' Cabinet takes over, Wim Kok Prime Minister
1996	Minister Jorritsma (V&W) releases 'Samen Werken Aan Bereikbaarheid', anti-congestion plan.
1999	Perspectives Memorandum on Transport released
2002	New Transport Plan, NVVP, proposed and rejected in Parliament.
2002	Resignation of the purple Cabinet and formation of the first 'Balkenende Cabinet'.

4 | THE SOCIAL CONSTRUCTION OF EUROPEAN AIR QUALITY LEGISLATION: 1983 – 1999

INTRODUCTION

After the general investigation of Dutch environmental policy in the previous chapter, we turn our attention here to the policy making process in the European Union. It was in this political arena that the regulations that caused problems for the Netherlands were conceived. The onset of an ecological modernist discourse in the EU however, turned out to be an important influence on these regulations as well. The question how EU air quality regulation in the 1990s was shaped and the influence of various actors including the European Commission and the Dutch Government in this process is the topic of this chapter.

Two EU directives drafted in the latter half of the 1990s are of particular importance for understanding how the Dutch air quality clash came about. The first of these, referred to as the 'Framework Directive', contained the strategic elements of the new EU air quality policy. The second directive – referred to as the 'Daughter Directive' – contained the specific air quality standards for, among others, NO_2 and PM10. In this chapter four aspects most relevant to this study that shaped these two EU Air Quality Directives will be examined in particular.
These are:

1. The rise of ambitious eco-modernistic ideas in European environmental policy making around the beginning of the 1990s;
2. The influence exerted by the World Health Organisation on the Daughter Directive;
3. The economic evaluation supporting the Daughter Directive;
4. The lead taken by the UK during the proposal and negotiation of the Daughter Directive.

These aspects were of course not the only influences that shaped the directive, but were the most prominent ones found, based on the Explanatory Memorandums, the secondary literature, and respondents' answers. The influence of eco-modernist ideas became conspicuous when the explanatory memorandum and the main ideas found in the framework and the Daughter Directive were analysed. The two directives both cite the EU's eco modernistic 5th Environmental Action Programme as a basis for these policies. These directives as well as secondary literature also displayed the prominence of the WHO. The economic evaluation came to the fore in secondary literature and moreover, such an economic evaluation became a mainstay of more recent European environmental policy. The economic evaluation was also mentioned in Dutch policy documents. The role of the UK was mentioned mainly by various respondents in face to face interviews and became a subject of study later on. Secondary literature, UK policy documents and Dutch grey literature support the notion that the UK played a pivotal role in the negotiations on the Air Quality Directives.

The rise to prominence of ecologically modernistic ideas in the EU is discussed in section 4.2; the involvement of the WHO in air quality politics is considered in section 4.3; the economic evaluation is examined in section 4.4; and the role of the UK is explored in section 4.5. In section 4.6, the actual adoption of the Air Quality Directives is mentioned, and section 4.7 contains concluding remarks. In the following section, I provide information about the decision to update the EU air quality policy in the 1990s, and I present an analysis of the proposals made by the European Commission.

4.1 Update of the EU air quality policy in the 1990s

In the 1990s, two Directives concerning air quality were promulgated: the 'Directive on Ambient Air Quality Assessment and Management' (96/62/EC, the Framework Directive) and the 'Directive relating to limit values for sulphur dioxide, nitrogen dioxide and oxides of nitrogen, particulate matter and lead in ambient air' (99/30/EC, the Daughter Directive). They were adopted in 1996 and 1999, respectively. The Framework Directive contained the overall strategy and provided for the possibility to set air quality standards that would subsequently be set in 'Daughter Directives' based on the framework of Directive 96/62/EC. Directive 99/30/EC was the first Daughter Directive. Three more followed in the early 2000s, but were of no relevance for the air quality clash, and so will not be discussed.

These Air Quality Directives included more polluting substances than the previous ones, and the standards they set were far stricter. The directives had to be implemented in national legislation, and the European standards formed the basis of the Dutch legislation that eventually led the highest Dutch administrative court to cancel all sorts of infrastructure projects.

The first preliminary proposals for these directives were under discussion between representatives of the national Government and European Commission officials in the spring of 1993 (Wettestad & Farmer 2003, p. 6). In October 1993, EU's DG Environment representative Kathleen Cameron outlined the new European plans for air quality legislation. Cameron warned that the standards themselves would be *'quite stringent'* because the goal was *'to provide a high degree of protection for man and the environment'* (ENDS report Oct. 1993).

As Cameron's words indicated, the new Air Quality Directives embodied the EU's wish to tighten its air quality policy in order to harmonise different strategies and to ensure good air quality in the various Member States.

The new policy was a response to the perceived failure of older directives. The first air quality rules set by the European Union dated from the 1980s, during which four substances had been regulated by way of Air Quality Directives. These substances were sulphur dioxide, known to be harmful to human health and a major cause of acidification, smoke, lead, and NO_2 (Bennett 1991). These early standards were lenient and lacked force, and respondents from the RIVM referred to them as 'almost merely reporting obligations' (Rob Maas & Hans Eerens, interview). According to a report released in 1995, even the modest obligations imposed by these directives were not taken seriously by most Member States (European Commission 1995). In the Netherlands as well, compliance with the directives was lacking (VROM 1992), and at the end of the 1980s and the beginning of the 1990s, ambient air quality lacked urgency in many Member States.

The lack of urgency was a thorn in the side of the European Commission and the DG Environment. Various respondents mentioned that Prudencio Perera, Director General of the DG Environment at the time, came up with the idea of revamping EU air quality regulation. However, the information I gathered on this issue is contradictory. Whatever the case, there are more structural explanations for the drafting of new Air Quality Directives and their contents than the influence of one high-ranking Commission official.

In the proposals themselves, and in the attached Explanatory Memorandums, the aims, the context, and the justification for the proposals are revealed. They offer significant information regarding the inquiry into the social construction of the EU air quality policy from the late 1990s.

Box 4: The EU policy making process

In order to understand the social construction of the air quality policy in the EU, some basic knowledge of EU policy making is necessary. Policy in the EU is made by three organs in unison, each of which has its own separate function. The **European Commission** submits a proposal for legislation. In this case, it is handled under the cooperation procedure (formally known as the Article 252 procedure), and sent to the **European Parliament** and to the **Council of Ministers**. The EP discusses the proposal in a first and second reading. The EP may propose amendments, and sends its opinion to the **Council of Ministers**. The Council of Ministers discusses the proposal as well, and when the Ministers agree among themselves, the Council produces a new proposal, a situation known as **'common position'**. Generally, the Council of Ministers consists of Ministers who hold the portfolio of the subject under consideration within their Member States. For instance, the **Environmental Council** consists of Ministers responsible for the environment.

The **common position** is sent to the European Parliament, and the Parliament may propose amendments to the Commission. If the Commission accepts all or some amendments, it sends the proposal to the Council of Ministers. The Council can either accept the amended proposal in a second reading, or reject it and adopt a proposal that differs from that proposed by Parliament. If it adopts the amendments proposed by the EP, is can adopt the proposal by qualified majority. If it decides to accept a proposal that differs from that proposed in a second reading by the EP, it needs unanimity to adopt the proposal.

The Council of Ministers and the European Parliament are considered the EU's main legislative bodies, but the initiative for legislation lays with the Commission making it a powerful body in practice. The Commission may issue **Decisions**, **Communications**, propose **Directives** and propose **regulations**, among other for this study less relevant regulatory instrument. A Decision pertains to an issue relevant to a single Member State. Commission Decisions are directly binding. In Communications the Commission announces policy plans. Directives are pieces of European legislation that need to be implemented in the national law of a Member State. Directives generally impose the achievement of certain results on a Member State, but the Member State itself is free in choosing the means of achievement. Regulations on the other hand are pieces of Community legislation that directly bind Member States, no national legislation to implement them is required.

An important player in the actual negotiations in the Council of Ministers is the **President**. The **Presidency** of the Council rotates among the Member States every six months. The presidency is an office that allows representatives of the country holding the presidency to decide on the agenda of Council Meetings, and to decide which legislative proposals will be handled with priority (Elgstrom 2003). The success of a presidency is generally measured by the progress it has made on the various legislative proposals under consideration (Würzel 2005). Presidents are expected to act as honest brokers in the negotiation phase, but they also engage in shaping agendas to further national interests.

As the Air Quality Directives demonstrate, it takes many years from the time a proposal is introduced to its eventual adoption. It is common to divide into three phases the process during which a European Directive is proposed, negotiated, and eventually implemented in national legislation. At the time the proposal is drafted, experts deliberate about the scientific and economic rationale of the Directive under consideration. It is common to refer to this as the **expert phase**. When the proposal is drafted and negotiations begin, the Council of Ministers and the European Parliament become involved. This is referred to as the **negotiation phase**. After the proposal is adopted and the Directive is finalised, the Directive needs to be implemented in the national law of the Member States. This is known as the **implementation phase**

The **European Commission** is the EU's highest administrative body, although not the highest legislative body. The **Council of Ministers** together with the **European Parliament** officially pass European legislation. The most important pieces of legislation are directives and regulations. While not the principal legislative body, the **Commission** has considerable power to shape the EU regulatory agenda, because the **Commission** proposes the directives and regulations. These proposals are subsequently discussed and amended by the **Council of Ministers** and the **European Parliament**. The Commission can also revoke its proposal or force the Council to adopt a proposal by anonymity if the Commission feels that the proposal has been diluted too severely in the subsequent negotiations.

What is referred to as 'The European Commission' is in fact a multifaceted actor comprised of a large number of **Directorates General**, each of which administers its own field of policy. In practice, environmental matters are dealt with mostly by the **DG Environment.** The **DG Environment** caters to the interests of environmental protection, and therefore favours a clean Europe. One of its core interests is the harmonisation of the environmental regulation of Member States, and bringing it into line with European standards. It is at times opposed by other DGs such as the DG Econ, when they consider the proposals of the DG Env. to impose too much of a burden on Member States.

4.1.1 Proposal for the Framework Directive COM(94)104

Proposal COM(94)104 (European Commission 1994), was a Framework Directive on ambient air quality assessment and management. It included stipulations about how to assess and manage air quality, and it prepared the ground for more directives in which standards for air quality would be promulgated. After the Framework Directive was adopted, four more Daughter Directives followed in which these standards were laid down. These air quality standards were mandatory throughout the entire territory of the EU, except in the workplace. The reason for this exception was that other directives covered air quality near workplaces and industrial installations.

The Framework Directive was presented initially as a harmonisation effort, and it contained the principles for effective and uniform air quality measurement as well as rules for the provision of information by the Member States to the Commission and other agencies. It set '[c]*ommunity wide, minimum harmonization criteria*' to enable '*a community wide assessment of ambient air quality*' (European Commission 1994, p. 12).

However, the Framework Directive also contained significant strategic ideas. The text indicated that in addition to air quality standards, information about air quality had to be made publicly available and presented to relevant organisations. Moreover, measurements and methods of assessment needed to be harmonised in order to be able to compare developments in different Member States. This made accurate assessment of progress possible.

The text specifically mentioned that the best way to safeguard air quality was the parallel development of ambient air quality standards and emission/product standards (European Commission 1994, p. 2). This mention of the parallel development of emission and quality standards was interesting because it resembled the Dutch idea of the 'two-track policy', referred to in the previous chapter. In the 1980s, The Dutch Government intended to base environmental protection on a harmonious mix of emission standards and quality standards. The explanatory Memorandum praised the Netherlands for basing its environmental policy on the development of both these instruments European Commission 1994, p. 7-8).

The proposal promised that long-term targets for the protection of health and eco systems would be set. A booklet published by the Commission stated that the Directive would force Member States to introduce additional far-reaching environmental policies, such as *'innovative methods of city planning'* and/or *'environmentally targeted public transport policies'* (European Commission 1998, p. 12). We may conclude that the air quality standards functioned in the context of a broader strategy to 'green' Member State policies. This is an important aspect, because it indicates that there was a policy philosophy behind these directives that went beyond the mere promulgation of standards for air quality.

Europe's 5th Environmental Action Programme (5th EAP), in force since 1993, provided the rationale for these new strategic ideas (European Commission 1994, p. 1). Environmental Action Programmes gave an indication of forthcoming EU policies and the strategic direction that environmental policy was taking. According to the Explanatory Memorandum to proposal COM(94)104, the 5th EAP demanded that fundamental levels of environmental care and protection be established by community legislation, that policy be optimised by using statistics and indicators, and that long-term objectives be set. The standards should be based on the recommendations of scientific bodies, among which were the World Health Organisation (WHO) and the United Nations Economic Commission for Europe (UNECE) (European Commission 1994, p. 16).

The document mentioned the number and types of pollutants to be regulated, but it did not give specific numerical values. To that end, Daughter Directives would be proposed. These directives contained the concrete standards but otherwise fit within the framework and relied on the same strategy.

4.1.2 *Proposal for the Daughter Directive COM(97)500*

The first Daughter Directive, 99/30/EC, stipulated standards for the concentration of, among others, PM10 and NO_2 in the air. The entry date was 2005 for the PM10 standards and 2010 for NO_2. Both substances gave rise to political trouble in the Netherlands.

The Daughter Directive was proposed in 1997, after the formal adoption of the Framework Directive. Proposal COM(97)500 (European Commission 1997) highlighted continuity with the Framework Directive. Its objectives were to lay down standards for four pollutants: sulphur dioxide (SO_2), NO_2, PM10, and lead. Moreover, it provided regulations for the provision of information to the public, and filled in some technical details regarding assessment and management of those pollutants.

The wording of the Explanatory Memorandum stressed that the standards were based on scientific evidence. In the context of the directive, scientific evidence meant knowledge obtained from health and environmental science. Scientific insights were compiled in working documents by a working group consisting of representatives from Member States, NGOs, and the Industry, as well as scientific institutes such as the European Environmental Agency (EEA) and the World Health Organisation (WHO). However, cooperation with the WHO was mentioned in particular. The proposal stated that air quality standards would be based on the WHO Air Quality Guidelines for Europe. A common agreement to work on air quality standards together was signed with the WHO (European Commission 1997, p. 5), and the WHO 1987 guidelines were to be updated in 1997.

Not only environmental and health sciences were involved; managerial and economic science also played a crucial role. The directive was supported by a cost-benefit analysis undertaken by a team of researchers from several European Universities. The conclusion of the assessment was that the proposal would incur costs but yield benefits that far outweighed the costs in terms of health gains. The costs and benefits, including health gains, were evaluated in monetary terms (European Commission 1997, p. 4). Especially for PM10, the policy was considered to be highly cost-effective. The Commission advised caution while interpreting the results of the cost-benefit assessment, but stated: '*...In so far as benefits are quantifiable and bearing in mind the limitations inherent in the estimation, benefits (ECU 408 to 5 900 million) are expected to exceed the costs (ECU 5 to 285 million)* (European Commission 1997, p. 25). In summary, the Commission expected costs to run into the millions EU wide, but benefits were scheduled to run into the billions.

4.2 The 5th EAP and the rise of eco-modernistic EU policy

The 5th European Action Programme contained the target that by the year 2000, WHO standards for air quality should be mandatory at the EU level. As a concrete action, it stipulated that assessment and measurement systems had to be in place (European Commission 1993, p. 49). This target, as well as the wording of the proposal for the Framework Directive, indicated that the 5th EAP provided the backdrop for the new EU air quality policy. The 5th EAP will be discussed in the following section. The ideas in this programme turned out to be similar to Dutch ideas developed in the NMP. Section 4.2.1 provides a general discussion of the document. In section 4.2.2, the report '1992, the environmental dimension' is considered. This report was a precursor to the 5th EAP, but it also explained why the EU turned to setting environmental quality standards in the latter 1990s. In section 4.2.3, the Dutch influence on European ecological modernisation is discussed.

4.2.1 *The 5th EAP*

The 5th EAP was an ambitious environmental action programme that contained a coherent strategy as well as numerous actions to be taken, and it had a longer time span than the earlier action programmes. The 5th EAP outlined environmental policy for the whole decade, from its inception in 1992 through to the year 2000. It was the first Environmental Action Programme after the establishment of the internal market, and it served as the EU's answer to various important worldwide developments in environmental policy, such as the UNCED's Earth summit in Rio de Janeiro in 1992 and the release of the 1989 Brundtland report 'Our Common Future'. In this report, the concept of sustainable development was articulated systematically, and the 5th EAP was placed squarely in the tradition of sustainable development by its title 'Towards Sustainability'. After Brundtland and the 'Earth Summit' in Rio de Janeiro, an optimistic atmosphere permeated environmental policy in the EU (M. Van Giezen, interview). According to Mirjam van Giezen, a new kind of development seemed possible, where win- win solutions for both the economy and the environment were feasible.

The 5th EAP embodied these new ideas. It was an especially important milestone, because it indicated a shift in the environmental policy of the European Union (Hey 2005; Kronsell 2000; Lenschow 1997). The programme contained many elements of ecological modernisation as defined and considered in chapters 1 and 3.

Firstly, it advocated the positive sum game postulate between environmental protection and economic welfare. It stated: '*The perceived conflict between environmental protection and economic competitiveness stems from a narrow view on the sources of prosperity and a static view of competition. Rather than reduce competitive advantage, stringent environmental requirements can actually enhance it by triggering upgrading and innovation*' (EU Commission 1993, p. 31). In eco-modernistic fashion, the 5th EAP focused on regulation by way of economic and market instruments, such as taxes and other fiscal instruments.

Secondly, it underscored the goal of internalisation of environmental values on the part of polluters, policy makers, and the general public. It aimed at behavioural change. The plan no longer focused on pollution that took place within the various environmental media but on the sectors that caused pollution. This way of organising the plan had a lot in common with the Dutch target group approach. The 5th EAP was divided into five 'target sectors': industry, energy, transport, agriculture, and tourism, and industries within these sectors all had specific tasks in protecting the environment.

The ambition to change behaviour was also evident in the wish to integrate environmental considerations in other policy fields. Where earlier conceptions of environmental policy tended to view the environment as a separate policy domain, the 5th EAP highlighted the interdependencies of community policies that have environmental impacts. The programme stated:

> *'It (the implementation of the strategy of sustainable development TA) requires that environmental protection requirements be integrated into the definition and implementation of other Community policies, not just for the sake of the environment, but also for the sake of continued efficiency of the other policy areas themselves'* (EU Commission 1993, p. 24).

In its quest for sustainability, the Commission saw the public as an ally, and therefore the public had to be mobilised by providing them with information. The document emphasised that each citizen had three roles to play. Firstly, he or she was a citizen concerned with the quality of his or her living environment. Secondly, he or she was a producer of pollution as an employer or an employee, as a commuter, and as a participator in leisure activities. Thirdly, the citizen was a consumer of products and services (EU Commission 1993, p. 27). In order to fulfil these roles in an environmentally benign way, the public should be targeted by awareness-raising campaigns and by environment-related information. The active involvement of NGOs, consumer organisations, and trade unions was considered crucial in this process. The public should also be given opportunities to buy environmentally friendly products, and the price of harmful products could be raised to provide an incentive for making a choice for the environment.

The third eco-modernistic element was the prominent role for consensus policies. The new approach included reaching out to industrial target groups as well, and it aimed at replacing a prescriptive, top-down approach to regulation with a horizontal one. In the words of the Commission, a 'thou shalt not approach' should be replaced with one of 'let's work together', so that industry would become part of the solution instead of part of the problem (EU Commission 1993, p. 28). The Commission should enter into dialogue with industry, and high environmental standards should be combined with incentives to improve efficiency. Key notions involved improving management and strategic planning within the industry, and introducing environmental impact assessment, self-regulation, and covenants (EU Commission 1993, p. 29).

The precautionary elements of ecological modernisation were present in the 5th EAP as well. The bigger role for industry should not be detrimental to high levels of ambition. The 5th EAP contained ambitious environmental legal principles, such as the precautionary approach and the polluter-pays principle. The precautionary principle was named in the same sentence as the notion of shared responsibility: '*The precautionary approach and the concept of shared responsibility*' (EU Commission 1993, p. 24). This is an indication that target groups themselves were considered to internalise the precautionary approach. The rise of this precautionary approach was also evident from the adoption of the precautionary principle in the Maastricht treaty of 1992. It was laid down in the latter EC treaty, in article 130R, and is now found in article 191 EU treaty (Treaty of Lisbon).

Managerial and natural scientific insights came to the fore as well, especially when concrete targets and timetables were discussed. A case in point was the desire expressed to use the WHO guidelines as a legitimation for the air quality targets. The plan allowed for scientific expertise to have considerable clout in the political debate over standard setting.

'Dutch' elements of ecological modernisation were conspicuous. Emphasis was less on prescriptive regulation and more on internalisation, consensus building, and behavioural change, especially on the part of industry and the public. The argument that environmental improvement was a matter of shared civic responsibility was carried through consistently. This theme was also stressed forcefully by the Dutch 'reverend' in national environmental policy.

In summary, the policy was a clear-cut example of ecological modernisation. It contained the positive sum game format, the emphasis on precaution as well as on consensus, the preference for long term-strategic planning, and targets that were legitimated by referring to managerial and natural scientific insight.

4.2.2 *The move towards quality standards in 1992, the environmental dimension*

Many ideas featuring in the Air Quality Directives fit into the new ecological modernistic frame laid down in the 5th EAP. However, there were frictions as well. Proposals for the Air Quality Directives contained a hybrid mix of old and new instruments. Old instruments like quality standards were combined with new measures like information requirements along with new goals such as public participation, prevention, and the greening of other policies. The real complaint regarding the Air Quality Directives, however, had to do with the strict air quality standards. Air quality standards were traditional regulatory instruments, and are in fact one of the oldest instruments around. To see how standard setting featured in the new EU environmental policy, we will have to review another report – '1992, the environmental dimension'.

This document appeared in 1990 as the result of concerns about the environmental effects of a single European Market. On 1 and 2 October 1988, the Environmental Ministers called on the Commission to report on the environmental dimension of the single market. The Commission convened a group of independent experts to identify the key issues and likely environmental impacts of the single market (Taskforce Environment and Internal Market, 1990, section 12.2). The experts pointed out certain significant negative effects of the internal market, due to, among others, the increased transport of goods and increased emissions from industrial sources. They warned that the internal market should not have negative environmental consequences, as this could endanger the long-term sustainability of the economic growth it would offer (Taskforce Environment and Internal Market, 199, section 12.1).

To avoid this scenario, the taskforce presented eco-modernistic ideas such as more policy integration, a favourable climate for improved environmental management, use of fiscal measures, and raising awareness by providing information and education. Ideally, economic growth and environmental degradation had to be 'uncoupled'. This concept means that the supposed link between increased economic growth and increased environmental degradation needs to be broken. It is natural to presuppose this link because increased production and consumption leads to increase strain on the environment, so breaking this link was therefore an important eco modernistic target. However, technical measures alone were considered unlikely to achieve this uncoupling, which was especially considered urgent in the transport sector (Taskforce Environment and Internal Market, 1990, section 12.12).

Contrary to the more market-based direction of the 5th EAP, however, the report pleaded for a system of minimum quality standards for environmental media in order for citizens to enjoy sustained environmental quality. These minimum standards should not be low, because according to the treaty the European citizen had a right to high environmental protection, and the EU's environmental policy should be 'preventative' in nature (Taskforce Environment and Internal Market, 1990, section 12.14).

The Taskforce was quite optimistic about the economic opportunities that quality standards could generate:

> *'Countries adopting comparatively stringent standards will reap benefits from higher environmental quality, which would tend to compensate for higher costs of environmental protection. Moreover, environmental quality can be an important factor in influencing the location of economic activities, including service sectors such as tourism'* (Taskforce Environment and Internal Market, 1990, section. 8.5).

Another key economic argument was the intention to create a 'level playing field'. Divergent quality standards ran counter to the treaty and the free movement of goods and services, because citizens had a right to a decent environment, irrespective of workplace and place of residence (Taskforce Environment and Internal Market, 1990, section 8.2). Moreover, having different standards in different Member States would have the consequence that industries in some Member States needed to comply with stricter regulation than industries in other Member States. This was contrary to the notion of a free market where competition is equal. However, on the basis of the EU treaty, Member States could set stricter environmental quality standards. In order to discourage this, standards needed to be high, which would prevent the more ecology-minded Member States from setting more ambitious standards themselves.

This report is interesting because eco-modernistic arguments, like the economic opportunities in a high-quality environment, were used to legitimate traditional regulation by way of quality standards. The Air Quality Directive contained a mix of eco-modernistic arguments and flanking measures as well as traditional quality standards. This kind of traditional regulation was dictated by the logic of the internal market, and underlying the generic approach of the EU was the important idea of the level playing field. It was considered necessary to impose the same standards on every Member State in order to avoid competitive advantages stemming from differentiation in standard setting. This rigid approach was contrary to Dutch notions that gained prominence in the late 1980s and 1990s, as Dutch policy started to favour flexibility and tailor-made solutions.

The '1992' report '*struck like a bomb shell*' according to Klatte (1997), because up until then EU officials had an eye only for the advantages of the internal market. The message that the internal market could become an '*environmental nightmare*' was highly inconvenient (Klatte 1997, p. 87). This report gave an impetus to environmental awareness on the part of heads of state, and they issued a declaration on the environment in 1990. This declaration and the report by the Taskforce were precursors of the 5th EAP and of the wave of environmental regulation dating from the latter 1990s, including the Air Quality Directives. The directives followed the logic of the report in which quality standards were favoured. Apart from Air Quality Directives, other framework regulation, such as the one on water quality, displayed the same policy philosophy.

4.2.3 Dutch influence on the eco-modernistic shift

The policy discourse laid down in the 5th EAP and to a lesser extent in ''1992', the environmental dimension' bore close resemblance to the Dutch approach laid down in the National Environmental Policy Plan discussed in the previous chapter. This resemblance is not coincidental, as the Netherlands was known at the time as an environmental forerunner. Indeed, the Netherlands, Germany, and Denmark became the EU's 'environmental troika' in the 1980s, and still held that reputation in the 1990s (OECD 1995). The NMP itself argued for active environmental diplomacy and for strengthening international environmental policy (VROM 1989).

A significant number of policy advisors who drafted the 5th EAP were either Dutch or schooled in the Netherlands (Jörgens 2004, p. 162). Arts, Dieperink, and Liefferink also mention that the Dutch Government sent an experienced civil servant to write the 5th EAP (Arts Dieperink & Liefferink 2002). The aforementioned Laurens Jan Brinkhorst in particular played a part in setting up a coalition that shaped EU policy in an ecologically modernistic fashion. Brinkhorst was Director General of the EU's Directorate for the Environment and Nuclear Safety from 1987 to 1994. He argued that stringent environmental standards for products were necessary to ensure that they remained competitive in a post-industrial world economy. If Europe were to be competitive, and intended to gain a position of leadership, it had to enforce strong environmental standards (Weale 1992, pp. 77-78).

Through such arguments, Brinkhorst managed to forge a discourse coalition on environmental and economic issues. The main storyline around which this coalition centred was that environmental policies were the key to future successes in post-industrial economic development and in acquiring a position of leadership in product standards and innovation, again a variation on the themes of win-win scenario's and environment as opportunity. According to Annika Kronsell, Brinkhorst led a group of scientists and policy makers in the EU environmental directorate, who used the Dutch example of the NMP as a blueprint for the 5th EAP (Kronsell 2000, pp. 91-92). There were indeed strong similarities between the NMP and the 5th EAP as noted above. Elements such as the target-group approach, consensus seeking, and the holistic approach to pollution were noted as 'Dutch' characteristics (Weale et al. 2000, p. 61).

The discourse coalition on environmental matters led by Brinkhorst was influential in the early 1990s. In 1992, the Commission chairman Jacques Delors became convinced of the potentially beneficial effects environmental policies could have regarding the environment. The Commission outlined these benefits in a white paper called the White Paper on Growth Competitiveness and Employment. As well as embracing the ecological modernistic tenet that environmental measures could have economically beneficial consequences, it proposed a different model of taxation that would bring about a more environmentally friendly and economically sound path to progress and development. This white paper was a significant success, because environmental projects took centre stage in a White Paper that was essentially concerned with solving economic problems (Weale 2005; De Cendra de Laragan 2009; Lenschow & Zito 1997).

The shift towards an ecologically modernistic discourse created opportunities to draft strict environmental regulation because of the storyline that high environmental standards had competitive potential. The traditional argument made by vested economic interests that environmental regulation harmed competition could be refuted by pointing towards the competitive edge of higher quality products. Instead of causing economic harm, strict regulation could have long-term economic benefits, such as higher potential for innovation, better quality of life, and a healthier, more productive work force.

The Dutch Government followed the proposals made in the NMP to export their eco-modernist ideas abroad and became an ambassador for them in the EU arena. The reasoning behind this is clear; as a small country, the Netherlands profits when its neighbouring states enact strict environmental policies, and the environment in the Netherlands is as vulnerable to foreign pressures as the Dutch economy. Hence, an EU environmental policy that promoted high ambitions without impeding international economic activity was in the Dutch interest. By inducing EU Member States to take ambitious market-based environmental measures, it appeased both the environmental reverend as well as the economic merchant.

4.3 WHO involvement in the EU air quality policy

The 5th EAP and '1992', outlined the philosophy on which the strict proposals for the EU air quality policy floated like a cork on water. Another crucial feature in the emergence of the air quality policy was the involvement of the WHO in air quality policy (Wettestad 2006; Rood 2003; Edwards, 1999). Its participation is of interest because it provides another example of Dutch influence on EU policies; moreover, it demonstrates the disadvantage of using powerful international bodies to reach national legislative aims.

The air quality standards based on the WHO's research was decisive in the negotiation phase of the directive (Rood 2005; Wettestad 2006). The WHO recommendations became a pervasive part of the storyline with which strong measures were justified. According to this storyline, bad air quality was a problematic social condition, and by implication motor traffic was likewise problematic. Reconstructing the reasons for WHO involvement may show us why and how health considerations became an overriding concern.

4.3.1 *Dutch proposal to the WHO*

It was mentioned above that the scientific work of international bodies, and explicitly the work of the WHO, formed the basis of the proposed air quality standards. In the Explanatory Memorandum to Commission Proposal COM(94)109, The Netherlands was praised because it had taken international initiatives that *'have resulted in the past in the drafting of the World Health Organisation Air Quality Guidelines for Europe (1987)'* (European Commission 1994, p. 7). As noted in the proposal for the Daughter Directive, the guidelines were important because the standards would be based on the new updated version of these guidelines. Apparently there was a Dutch connection to another one of the key factors that shaped the Directive – involvement of the WHO.

This involvement began in 1983, more than ten years before the Framework Directive was adopted. Toxicologist Klaas Krijgsheld called it a *'remarkable intervention'* (Krijgsheld 1987, p. 74), because it was rare for the WHO to give such clear recommendations on which to orient policy. For my own reconstruction of the WHO involvement,

I used Krijgsheld's article together with material obtained from an interview with Kees Zoeteman, the civil servant who initiated contact with the organisation and requested that it draft air quality guidelines.

In 1983, Zoeteman had just become Director of the Air Department of the VROM Ministry. Before joining the Ministry, he had been working with the Dutch Institute for Drinking Water. In that capacity, he had been involved in drafting the WHO guidelines for drinking water and was in the subsequent negotiation regarding the European Water Directives. Therefore, when air quality standards were needed, he considered it would be beneficial if the WHO could also draft air quality guidelines (Zoeteman, interview 17-8 2011).

Kees Zoeteman asked the WHO's European Regional Office if it could take the initiative to draft air quality guidelines. The WHO responded positively, and set up a working group. At the time, four pollutants had already been regulated by way of European Directives, and had also been investigated by the WHO in 1972. This new WHO investigation, however, would have a much broader scope, with 28 pollutants on its target list. It was considered that the effectiveness of these guidelines would be strongly increased if the standards were set in numerical values, and the form of a risk assessment or the establishment of a 'no-effect' level were preferred. A 'no effect' was the level of concentration of a pollutant in the air at which no effects to human health can be measured. In the Netherlands itself, this 'no effect level' was the preferred value on which to base standards. It was agreed that the WHO would limit its involvement to establishing the no-effect levels or the risk assessments, and the Dutch Government would bear the project's financial costs.

According to Krijgsheld, the research method adopted was expert consultation. A group of around 130 experts were split into groups, each considering a number of different pollutants. Before the discussion within a working group started, a background document was composed by one of the members. The meeting resulted in a text containing recommendations, and these texts would be further scrutinised by a review panel. The experts hailed from different countries, and various research institutes hosted the meetings. Particulate Matter, for instance, was discussed in Bilthoven, the place of residence of the Dutch RIVM.

Considering the task at hand, the time frame in which the WHO needed to operate was short, as drafting air quality standards for 28 pollutants demanded considerable effort and was a matter of grave complexity. Krijgsheld mentioned that the procedures employed by the WHO in establishing the limit values were *'contrary to WHO tradition of thorough and time consuming research of chemical substances an ambitious planning was adopted to evaluate a number of around 20 pollutants in the space of three years'* (Krijgsheld 1987, p. 74).

The resulting Air Quality Guidelines was a document on which consensus was reached by the experts, which made it attractive as a frame of reference for policy makers. However, the scientific value of this consensus should not be overstated. According to Kees Zoeteman:

> *'There is an international document over which there is consensus, but consensus between whom? The WHO, which means a commission containing fifteen ladies and gentlemen who agree on something. At that time, that was without any consultation. On the basis of their own expertise a document is drafted and these are the WHO guidelines. But there was just nothing better'* (Zoeteman, interview). The project was completed in 1987, and it immediately became an important benchmark in the context of the EU (Wettestad 2006).

4.3.2 *The role of WHO expertise in the adoption of the air quality Daughter Directive*

The WHO air quality guidelines were not the only scientific data on which standards in the EU proposals were based. The EU intended to base its standards on scientific expertise, and it assigned scientific committees to propose standards as well.

In the case of the proposal for the first Daughter Directive, the European Commission created a number of working groups in order to look at the four different pollutants under consideration in the proposal. The most health-relevant pollutant was PM10. The technical working group on particulate matter that had to supply advice on applicable standards for PM10 consisted of delegations from Spain, Denmark, France, The Netherlands, Germany, and the UK, with Germany and the UK co-chairing the group. The task was onerous, especially regarding PM10, because at the time of drafting the proposal and of the subsequent negotiations, the WHO had declined to give a specific numerical recommendation. It considered that a 'no-effect level' could not be established. Instead, it provided the scientific Working Group on Particulate Matter with research findings indicating dose-response relationships.[1] PM10 was a problematic pollutant for both the WHO and the EU experts to assess, because its exact effects and even its composition were subject to uncertainty, as became apparent from chapter 2.

As recounted herein, the Netherlands was one of the first countries to begin measuring PM10, having imported this category from the United States of America. Dutch and American researchers on air pollution were in close contact (Buijsman 2003, p. 116). However, many parts of Europe did not measure PM10 when the proposal was put forward. The information of the Working Group on Particulate Matter contained gaps as well as a large number of other uncertainties. There were limited emission inventories, so it was difficult to know how much PM10 was emitted in the

1. The dose-response relationship indicated by the WHO is provided here in Appendix 1 at the end of the chapter.

European air, but there were additional pressing issues as well. The Working Group felt that uncertainties regarding the effects of PM10 were so large that a fixed numerical standard might not be the best option. Such a standard would present a false sense of certainty, and there was little information about the costs (Technical Working Group on Particulate Matter 1997, p. 33). It was also considered problematic that the final version of the WHO Air Quality Guidelines was not yet fully available (Technical Working Group on Particulate Matter 1997, p. 33). It was questionable as well as to whether the right kind of particulate matter has been targeted. The Working Groups expected that the EU would regulate PM10, but the smaller variety of particulate pollution, PM2.5, was considered more dangerous. Regulating PM10 was a compromise in this perspective. The Working Group on Particulate Matter stated:

> *'At present, however, PM10 as a PM metric seems to be a reasonable compromise between theoretical arguments favouring the measurement of very small particles and the knowledge and practical experience based on existing PM10 and TSP (Total Suspended Particulates TA) (and BS [Black Smoke TA]) measurements'* (Technical Working Group Particulate Matter 1997, p. 1).

Uncertainties notwithstanding, the Working Group recommended the following standards: a 24-hour PM10 limit value of 50 mug/m^3 as a 98-percentile (of daily mean values over a calendar year) in conjunction with an annual mean limit value of 20 mug/m^3. It recommended these standards for health reasons, and considered that the economic evaluation of the attainability of these standards was yet to come. Both US and European studies along with the dose-response relationship provided by the WHO were considered crucial (Technical Working Group on Particulate Matter 1997, p. 2).

In the European Parliament, the Commission proposal was welcomed and in some senses reinforced. The socialist UK Parliamentarian Anita Pollack drafted a report about the Commission proposal to be discussed in the plenary session. She approvingly mentioned the WHO and other scientific involvement, and hailed the Commission proposal as *'an end to the tale of disaster'* (European Parliament 1998, p. 71) facing public health, eco-systems, and cultural heritage. Ritt Bjerregard, Commissioner for the Environment, was present on behalf of the European commission. She also defended the Commission's proposal as having been based on close cooperation with the WHO:

> *'It sets new limit values for sulphur dioxide, oxides of nitrogen, particulate matter and lead. These limit values are based on WHO guidelines adopted in 1996 in the wake of a programme of collaboration with the Commission'* (European Parliament 1998, p. 78).

In Parliament, the proposed Directive met with little resistance. In general, the EP was considered to be more environmental friendly than the Council of Ministers, and the Parliament's Environmental Committee in particular had a reputation as an environmental watchdog in the 1990s (Bomberg & Burns 1999, p. 178).

The final say belonged to the Council of Environmental Ministers. I found no public records of the Council meeting of 16-17 June 1998, but telephone interviews as well as grey and secondary literature shed light on the proceedings. The negotiations were concluded at a fast pace, according to Dutch negotiator Tom Blom (Project Group Air Quality Directives, 1998, on file with the author). Representatives from the northern European countries and especially Austria favoured tough standards, while southern European countries argued for postponement and exceptions to the PM10 standard because of naturally occurring phenomena – such as Sahara dust – that influence the concentrations of PM10 in the air. The proposal was discussed in the last days of the UK presidency of the Council, and the UK negotiators tried to bridge the gap between the conservative and the more progressive nations (M. Williams, telephone interview). In the end, the Council agreed unanimously on a proposal that somewhat weakened the standards for PM10. However, due in part to the Netherlands' insistence, a clause that called for a review of the standards by 2003 was introduced into the document.

Rood et al. (2005) and Jürgen Wettestad (2006) indicate that the WHO air quality guidelines were influential in the Council debates. Most Member States agreed on tough standards for PM10 and NO_2 because they took the WHO recommendations very seriously. Due to the existence of WHO guidelines, representatives of the Member States did not heed the lack of scientific certainty in many respects of air quality (Rood et al. 2005, p. 17). The health threat that was posed by PM10 and the other pollutants had become the overriding argument in the debates, and the seriousness of the threat of air pollution was never called into question.

According to Wettestad, acceptance of the WHO air quality guidelines as targets in the 5th EAP had done much to lessen the Council of Ministers' opposition to tough standards. It had already been accepted that the WHO guidelines would be used as a benchmark, which made it difficult for the Ministers to disregard the air quality guidelines at the time of the negotiations.

Wettestad captured the dynamic between the political action of the EU and the scientific work of the WHO in figure 1 (Wettestad 2006, p. 294).

The WHO guidelines asserted considerable influence on the draft of the air quality Daughter Directive in the expert committees as well as in the negotiation phase. The strategy to involve the WHO to smoothen the eventual negotiations had been a deliberate move by the Commission. It intended to ensure consensus on the principles of the air quality regulation by first securing the adoption of the air quality Framework Directive.

The interdependency between the WHO and the European Commission enforced both parties. The Commission gained the necessary scientific legitimacy for its

Figure 1

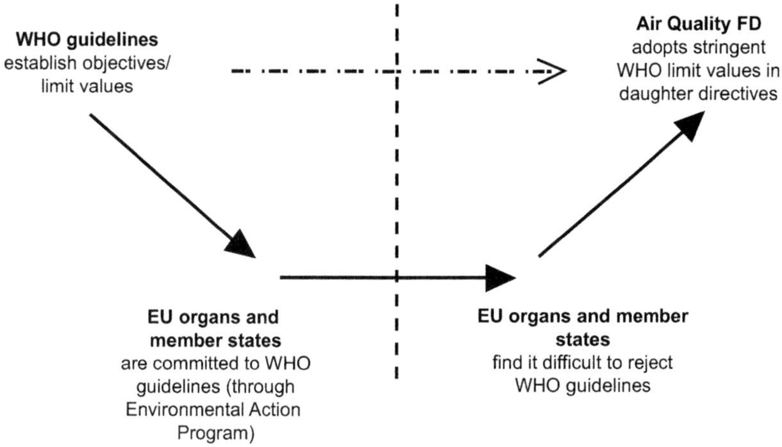

proposal, while the WHO gained real political influence. Involvement of the WHO in the environmental domain reinforced the storyline that health protection and environmental protection were strongly intertwined. Public health arguments reinforced environmental policy and vice versa.

4.3.3 WHO involvement from a Dutch perspective

The WHO involvement is an influential explanation for the eventual acceptance of the Air Quality Directives, and the Dutch Government had proposed enthusiastically that the WHO play an active role. This raises the question as to what the Netherlands had to gain by involving the WHO, and why the Dutch Government paid for the research to be undertaken. The answer can be found in the increasing prominence of environmental matters on the political agenda in the Netherlands back in 1982.[2] The idea of integrating environmental policy into other policy fields was gaining ground, and policy makers were searching for adequate instruments pertaining to environmental protection. In the 1982 IMP Air, environmental quality standards were considered a key element, and – together with standards pertaining to the emission of pollutants – they were the cornerstones of Dutch air pollution policy, which became known as the 'two- track policy'. The EU proposals for its air pollution policy mentioned the same idea, referred to now as 'parallel policy'.

For a number of reasons, it was appealing for the Dutch Government to bring in the WHO. Firstly, a very practical reason was that air quality standards were hard to determine, and setting them required considerable scientific expertise. The WHO took some of the work out of the hands of the Government. Secondly, according to

2. This was mentioned in the previous chapter.

Krijgsheld (1987) and Wettestad (2006), the acceptance of measures increased when it was seen that the international competitive position would not be harmed by unevenly heavy-handed measures in comparison to other countries. The industry feared that if the Netherlands set standards of its own, it would have to incur costs to clean its production while its competitors abroad did not need to face such costs. The WHO lent legitimacy to the standards, and this was useful for the Government in discussions with its own industries. Moreover, it could also persuade other countries to accept similar standards.

Thirdly, the trans-boundary character of air pollution made it difficult for the Netherlands to influence pollution concentrations on its own. Concentrations of air pollution in the Netherlands were also influenced by the emission of pollutants in neighbouring countries. If the WHO guidelines were to become a benchmark, other countries would be more prone to reduce their pollution levels as well.

The air quality guidelines were also appealing to the EU. According to Kees Zoeteman, the European legislator was keen on using knowledge that was already at hand. When there was nothing else, the European Commission was tempted to turn to a document like the WHO guidelines, because apparently there was consensus about it. Discussions with the Member States could be minimised if the EU could demonstrate that it based itself on sound science. When environmental directives were concerned, the Commission started by setting up working groups of experts, and they asked these experts what their proposals were. The Commission proposal was then based on the suggestions of these working groups. These experts also looked at the things that had already been done, and when there were guidelines they took them as a point of reference (Zoeteman, interview).

This kind of action on the part of the Dutch Government was not unheard of. The Netherlands was traditionally active in persuading countries to adopt strict environmental standards, and the mix of idealism and protection of economic interests led the Netherlands to take on an active role in international organisations. The same sentiments were at play here, and the WHO involvement served Dutch interests in a number of ways. The WHO relieved Dutch scientific bodies of conducting time-consuming and costly research, the Dutch Government would have a stronger hand if industry were to oppose the Dutch regulation on air quality, and the WHO document could be used as a benchmark to persuade other countries to incorporate similar legislation.

The Netherlands is generally not averse to applying the tactic of using experts and state-of-the-art scientific knowledge whenever it intends to export its own approach to the European arena; indeed, the Netherlands is known for what Marcus Haverland calls the use of the 'expert strategy' (Haverland 2009). Through the use of scientific expertise, the Netherlands is able to create leverage for its policies while a bigger country can rely on its power within EU institutions. Through its alliance with the WHO, the Dutch Government was able to gain the expertise and clout of the WHO to influence other actors at home or abroad. This internationalisation ensured that competition remained equal and environmental measures did not lead to a worsening of the competitive position of Dutch industry. Again, we witness the dual focus on idealistic initiative and competitive position: the well understood self-interest of the merchant / reverend. Although one could never predict such a document would eventually become EU policy (Zoeteman, interview), the advantages for the Dutch Government of having the WHO on its side were evident.

In the Netherlands itself, enthusiasm regarding the air quality guidelines died down. Dutch involvement with the WHO Air Quality Guidelines was mentioned in the IMP air 1984-1988 (VROM 1983, p. 114), but was no longer mentioned in the later IMPs or the NMP. Air quality standards lost their political saliency, and the Netherlands did not regulate air quality further on its own. The European Commission took over the initiative at the beginning of the 1990s.

When the Daughter Directive was under discussion in EU circles, the situation in the Netherlands was very different from that in the 1980s. Memorandums from a project group formed during the negotiations indicated there was increasing alarm about the consequences of air quality standards. In order to monitor negotiations on the Daughter Directive, this project group was set up by the VROM Ministry, and included scientists within the RIVM, representatives of the VROM Ministry, and other stakeholders. This group became increasingly worried about the validity of the scientific data used, and privately considered the PM standards very difficult to meet. In an internal Memorandum, toxicologist Peter Rombout complained that if it had been known beforehand how the WHO air quality guidelines would be used in legislation, their conclusions might well have turned out to be very different (Internal Memorandum Project Group Air Quality, Sept. 1998, on file with the author).

However, WHO involvement had already infused debates on the EU air quality policy with a powerful storyline: namely, that air pollution was an important health threat that needed to be tackled forcefully. Subsequent European debates displayed that the involvement of the WHO provided a powerful sense of urgency to the negotiations (Rood et al. 2005).

4.4 Cost-benefit analysis of proposal COM(97)500

The third factor of importance was the cost benefit analysis supporting the proposal for the Daughter Directive. In the Netherlands, the overwhelming costs of implementing the directive led to debates between the Minister of VROM and Parliamentarians.

A figure of 30 billion Euros was mentioned in those debates.[3] Yet how could such high costs be caused by a policy that was supported by an economic evaluation and should have been highly cost-effective?

The evaluation was conducted jointly by researchers from the Free University of Amsterdam, the Austrian Institute for Applied Systems Analysis, and the Norwegian Institute for Air Research (NILU). The research demonstrated that the air quality policy in regard to PM10, NO_2, and SO_2 would provide benefits that far outweighed the costs. The research was important to persuade Member States to accept the proposal, because it showed that the air quality standards could be implemented without high costs. In the following sections the assessment and its conclusion that the air quality regulation would not lead to high costs will be examined.

4.4.1 *Economic forecasts for the air quality standards*

The purpose of the impact assessment was to determine what additional action would be needed in order to meet proposed standards. The researchers calculated the costs of such additional action by using the most effective solutions, and assessed the benefits that could be expected from meeting the standards (European Commission 1997, p. 3). Where possible, the benefits were quantified in monetary terms. Since health benefits were the main advantages of the regulation, a monetary value was given to a statistical human life. Based on various studies, the value of a human life was estimated to lay between 2.6 to 4.2 million ECU (European Commission 1997, p. 4).

The figures appeared highly favourable for air quality regulation, as the economic evaluation showed that the benefits were very high and lay between 5 and 50 billion ECU.[4] The large gap between the figures was explained by the range within which a human life had been quantified, as well as by uncertainties in relation to mortality rates attributable in particular to PM10. Basing their conclusions regarding the mortality rates on the same US research that the WHO had used, the researchers pooled the estimates of the various studies provided by the WHO, in essence combining the figures of separate studies. The WHO itself cautioned against pooling estimates. The institutes resorted to pulling the estimates because '*at present it is not possible to take into account the local factors contributing to heterogeneity and to develop locale detailed estimates of impacts*' (Olsthoorn et al. 1997, p. 66). Admittedly, this procedure increased uncertainty, and was the reason that the uncertainty interval on the benefit side was very high.

It was projected that compliance with the directive would cost the EU countries taken together between 50 to 300 million Euros per annum. These costs were low for a

3. Kamerstukken II 2000-2001, 27 400 XI, nr. 56, p. 1/2.
4. European Currency Unit: ECU. The value of the ECU is equivalent to that of the Euro.

directive that purportedly set stringent air quality standards. The researchers considered that the standards could be met against relatively little cost because other EU legislation would also lead to lower emissions, especially due to cleaner cars. The cost-benefit assessment was drafted under the assumption that another EU programme – the auto-oil programme – would yield improvements for air quality:

> *'The assumptions with respect to the options for control of PM10 emissions from mobile sources are based on the results of the auto-oil study. This study assumes an emissions reduction of 50% relative to the current 1996 level of emission standards'* (Olsthoorn et al. 1997, p. 46).

It was envisioned that emissions from transport cars and motorcycles would decrease by 80% in the 1990s due to planned EU regulation. It was estimated that emission from heavy-duty vehicles would be reduced by about 50% (Olsthoorn, 1997, p. 24).

The researchers were also optimistic about non-technical market-based measures such as introducing road pricing and the introduction of LPG busses. They foresaw a reduction of emissions of 20% by such measures alone. However, the researchers noted that if these measures did not have the desired effect, the cost curve would rise sharply. The results of the auto-oil programme were especially crucial. The auto-oil programme was the name of a covenant between the European Commission and the main bodies representing the auto industry and the oil industry at the EU level. The environmental Commissioner Ritt Bjerregard had high expectations of this programme as well, and endorsed it by stating: *'we should see cleaner air in our cities and we shall have fewer ozone episodes in the summer'* (Wettestad and Farmer 2003, p. 19). In summary, the researchers argued that only a modest additional effort needed to be made, because the lion's share of the work had already been done, owing to the reduction of emissions resulting from other EU regulation.

4.4.2 *Problems with the economic evaluation*

In hindsight, figures for the costs presented in the economic evaluation were much too low. The Dutch Court of Audit, the Algemene Rekenkamer, drafted a report about the rising costs of compliance with the air quality standards. In 1997, at a time when the Dutch Government still relied on the Commission's figures, it estimated costs from 1.3 million to 3.2 million Euro per year, an amount that could be incorporated without much difficulty into the roughly 1 billion Euro budget for environmental management.[5] However, it soon became clear to the Dutch Government that expenses would rise significantly, and a year later the costs of meeting the standards were considered to be 340 million Euro per year. But at the time the policy was implemented in 2001,

5. Based on the VROM budget for the year 2000, Law of 14 September 2000, regarding changes in the budget of the Ministry of Housing Spatial Planning and Environmental Management for the year 2000; Stb. 2000, 393.

various assessments of the costs had been made that ranged from only 21 million Euros to an estimate of 16 billion Euros (Algemene Rekenkamer 2004, p. 48). In 2002, Buringh and Opperhuizen concluded that reducing PM10 concentrations by a modest 1.1. mug/m^3 countrywide would cost 6 billion Euros. They concluded that even with all possible reduction policies, compliance with the daily standard for PM10 as prescribed in the EU directive was not feasible (Buringh & Opperhuizen 2002b, section 6.14).

The reason the costs rose exponentially was that calculations in the economic evaluation relied to a great extent on projected advantages emanating from other policies; for instance, the calculations included the supposition that the auto-oil programme would generate significant benefits. However, these did not materialise, as the auto-oil negotiations were mired in conflicts between the European Parliament, the oil industry, and the automotive industry. The programme was based on a novel approach that fit with the philosophy of the 5th EAP, but it was criticised for its lack of transparency. Moreover, the German Government demanded the inclusion of many more stakeholders in the second stage of the programme. This wrangling caused significant delays and brought the policy process to *'all but a halt'* (Würzel 2002, p. 175). Moreover, in a review of the Auto-Oil Study it was mentioned that only modest reductions of PM10 could be expected by 2010 (European Commission 2000a, p. 11).

High yields were also expected from non-technical measures such as road pricing, but road pricing did not emerge in the European Union. In the Netherlands, it was considered highly controversial, as we saw in the previous chapter.

This reliance on other policies was the central problem distorting the economic evaluation. These environmental policies were designed to cooperate with other policies, and to spill over into other sectors like transport or agriculture. However, when policies are not as compatible as hoped, and if the projected benefits from one policy do not materialise, the whole system becomes unstable. At the time, his instability was exacerbated by uncertainties that were rife in the air pollution case, and there was little in the way of reliable measurements and models, little information about the exact health impacts of PM10 in particular, and little knowledge as to how concentration levels could be reduced. When policies are designed in response to other policies, the problem of uncertainty becomes doubled because these policies are subject to uncertainties, thus exponentially increasing the amount of uncertainty.

This kind of cost-benefit analysis exhibited another problem. The benefits were presented in monetary terms, but they did not constitute real financial revenues – they constituted life years gained by the population. These statistical lives saved could not be added to the balance sheet of any company or Member State. The costs were felt immediately, and by some sectors and states more than by others, while the benefits remained diffuse and spread over the whole population. To calculate the benefits, the working group valued a statistical life at 3.2 million Euros, making pollution reduction cost-effective, but who earned that amount of money? In the end, health was gained in exchange for money. This increase in health could well have been bought for a good price, but presenting it as an investment that would produce revenues in monetary terms distorted the economic picture.

The cost-benefit analysis had been so influential that the Dutch delegation, for instance, did not conduct an alternative analysis; it simply took at face value the numbers presented by the Commission (Rood et al. 2005, p. 19).

4.5 The UK as a 'champion' of the Air Quality Daughter Directive

Drafting new directives is a political process in which Member States play an important part. They have a say in the Council of Ministers, but they can also choose to influence the Commission when drafting the proposal by sending their own personnel to aid the Commission, or by presenting their own policies as an example for the Commission. The role of Member States in the policy process is crucial. According to Adrienne Héritier, Christoph Knill, and Susanne Mingers (1996), Member States can act as 'champions' of a certain European regulation and play a front-runner role in the process of its adoption. I argue that in the air quality case, the UK acted as a champion for the Daughter Directive.

The position of the United Kingdom was of importance, not only for adoption of the air quality regulation but because a more in-depth review of the UK's involvement demonstrated that similar problems had erupted in the UK about ten years earlier than in the Netherlands. This points towards an interesting dynamic regarding the construction of environmental problems within the EU. Countries export these problems to the EU level for regulation, and this regulation in turn creates similar problems for other Member States. Moreover, the UK example reveals what was at stake in the issue of ambient air quality. As the subsequent analysis will show, the interests of transport and auto-mobility were pitted against public health interests.

The importance of the UK emerged during the investigation, and is based on interviews with people involved, on the literature, and on grey literature. Especially important are references from the following: Rob Maas and Hans Eerens from the RIVM, Marjan Van Giezen from the Ministry of VROM, Richard Mills, a civil servant from the UK, the Dutch Project Group for Air Quality Directives, and a report from the Clingendael institute (Rood et al. 2005). UK Parliamentary documents also point to leadership on this issue.

There are various reasons for considering the UK a champion in the political process that led to the Air Quality Directives, especially the Daughter Directive. The UK supplied scientific and political expertise in the negotiation phase as well as in the expert phase, and this influence caused a convergence between the EU and UK strategies (Elsom 1999). It sent a national civil servant, Lynne Edwards, to the EU Commission to oversee negotiations, and she was extremely influential (Maas & Eerens, interview; R. Mills, interview). Moreover, negotiations between the Member States were finalised in June 1998 under UK Presidency. The rotating EU Presidency of the Council of Ministers is an important office because the President can shape the policy making agenda (see also box 4). The President acts an important broker between the various Member States and their respective positions on the negotiation table. The negotiations were finalised hastily in the last days of the UK presidency, indicating that the UK attached great significance to achieving a deal under its leadership. UK officials and scientific expertise from UK institutes played key roles, with the UK acting together with Germany as chair of the Working Group on Particulate Matter. The scientific data used to underpin the standards were gathered predominantly from UK research institutes, and the rapporteur to the EP in the case of both the Framework and the Daughter Directive was from London.

The UK was a prima facie unlikely candidate to have such a strong involvement in the air quality Daughter Directive, as it was not known as an environmental front-runner. There were, however, two main reasons the UK attached importance to the proposal for an ambient Air Quality Directive. The first was the fact that the UK already had an air quality strategy in place when the EU negotiations over the Daughter Directive began in 1997. Advanced regulation by a Member State acts like a benchmark for the EU, and is frequently incorporated by the EU itself (Knill & Liefferink 2007; Holzinger et al. 2008). It was in the interest of the Member State to have the EU regulation match the national regulation as closely as possible so that no costs would be incurred by having to change the national regulation to any extent. Moreover, by succeeding in having its national policy 'raised' to the level of European policy, it ensured that its own industries would not be alone in being burdened by regulation (Héritier, Knill & Mingers 1996).

The second reason was that the UK Government of the young Labour leader Tony Blair was intent on showing a green and pro-European face during its 1998 presidency. During his presidency, Socialist leader Blair made the environment one of his priorities, and the Air Quality Directive fit this overall aim.

Both of these reasons for UK involvement had to do with a domestic conflict over air quality in the early 1990s. An examination of the policy documents relating to air pollution showed that during the 1990s air pollution had become a salient political topic because of concerns over childhood asthma. This finding was important, because ten years later Dutch environmental movements raised a similar concern over children who lived close to highways, therefore I will take a closer look at the emergence of this strategy here.

4.5.1 *Emergence of the UK National Ambient Air Quality Strategy*

In the UK, air quality was of concern in the first programmatic approach to environmental management launched in the country; it was called This Common Inheritance. The approach marked the start of a greener phase in the UK's environmental policy, which up until then had been paltry. With regard to air pollution policy, the document stated that the Government was interested in developing standards for air quality and that the WHO guidelines could act as a guide. The document did not include any policy initiatives on this front, however, but advised an increase in monitoring stations and set up an advisory council that would inform the Government on air quality standards.

The UK was known to base its environmental policy on sound scientific insights (Hajer 1995) and the air quality standards were no exception. A number of scientific bodies were dealing with air pollution, and during my investigation I identified five different ones in total. The first was the Expert Panel on Air Quality Standards (EPAQS), instituted shortly after This Common Inheritance was issued. EPAQS in turn based its recommendations on the work of other expert bodies, most notably COMEAP (Committee on the Medical Effects of Air Pollutants; independent but connected to the Ministry of Health).

Not mentioned in This Common Inheritance but nevertheless important was the advisory body called the Quality of Urban Air Review Group (QUARG). The highly regarded Royal Commission for Environmental Pollution investigated air pollution as well. This was and remains a reputable institution that had been sworn in already back in 1971. Finally, the research body, 'Medical Aspects of Air Pollution Episodes' (MAAPE), was set up in 1990 as a subdivision of COMEAP. Its explicit terms of reference were to determine whether public advice needed to be given during air pollution episodes, and, if so, what kind of advice (Website COMEAP, last accessed 24-06 2015).

All these research institutes released worrying scientific reports in the 1990s. In July 1991, a report by MAAPE appeared, regarding the pollutant ozone. Though not alarming on the whole, the MAAPE report concludes that the effects of ozone could be more troublesome for people with asthma. Children were mentioned separately. General measures did not need be taken, but extra information should be supplied to groups that might be more sensitive. The UK Government led by the Conservative party responded by providing information bulletins on television and by instituting extra help-lines that could be called free of charge.

During the Christmas period of 1991, the UK experienced a week-long episode of intense air pollution. In January of the following year, QUARG was assigned to examine urban air quality, and published its report in 1993 (QUARG 1993a). The researchers pointed their fingers at transport as being responsible for bad air quality in British cities, especially in London. The issue led to a number of Parliamentary debates and to written questions about air quality (for example: Hansard 28 Jan., 10 Feb., 17 Feb., 2 March, 22 March, 23 March and 25 June 1993). The press meanwhile picked up on the story of childhood asthma, in 1993 as exemplified by publication of the heart-wrenching story of the wheezing baby, Lydia Wolfe. The dramatic headline proclaimed: '*Exhaust fumes are suffocating our children*' (The Independent 10-10 1993).

By December 1993, QUARG had issued its second report. This time diesel vehicles were considered the guilty party in environmental pollution (QUARG 1993b). Air quality came under closer scrutiny when data about the potential damage of PM10 emerged. During the spring of 1994, Parliamentarians responded to a short article in the New Scientist in March. According to the article, PM10 posed a much more significant threat than had previously been thought (Brown 1994, pp. 12/13). On 10 March 1994, the Independent quoted from the article and concluded that according to epidemiologist Joel Schwartz, 10,000 people in the UK could be dying annually because of exposure to PM10 (The Independent 10-03 1994).

On 17 March 1994, Under Secretary of State for Public Health, Tom Sackville suggested that the Committee for the Medical Effects of Air Pollution was of the opinion that there was evidence to suspect PM10 of being responsible for having an effect on health (Hansard, 17 March 1994). This represented a change of position. Back in 1992, MAAPE had still suggested that PM10 caused little harm (MAAPE 1992).

By then, the link between air pollution and transport was by now firmly established in the public eye, and these concerns had started to pick up in Parliament as well. Labour MPs from London, notably Jeremy Corbyn and Austin Walker, began to campaign on the issue in December 1993. In Parliament, Secretary of State John Gummer was confronted by questions as to why there were so few monitoring stations, especially considering the link with asthma (Hansard 1 Dec. 1993). Minister of State Tim Yeo had to reply to questions in writing (Hansard 17 Dec. 1993).

In the summer of 1994, a new smog episode appeared to be on its way, and air quality was discussed in the press in increasingly jittery terms (The Guardian, 5-07 1994; The Independent, 5-07 1994). The issue reached its zenith after 17 July. On that day, the Independent reported an asthma incident following the evening of 24 June, three weeks earlier. On its front page, the newspaper wrote that after a period of severe thunderstorms, hospitals had become overcrowded at night with people suffering from asthma. The doctors questioned by the journalists compared the situation to '*a plane crash happening near every hospital*'. The newspaper itself featured the dramatic headline: '*The day Britain choked*' (The

Independent, 17-07 1994).[6] The next day, The Guardian called for *'a politician that can clear the air*, because *'we were told that legislation had banished smog, but now it is stiflingly, chokingly, sometimes murderously, back'* (The Guardian 18-07 1994).

The conservative Government came under mounting pressure when concerns over asthma were intertwined with ongoing protests over road expansions. In 1990, the conservative Government announced one of the most ambitious road-building schemes in UK history, but protest against it was fierce. In 1994, the issue of road expansion and asthma became connected. In July 1994, the month of the smog panic around asthma, Labour MP Joanne Walley, who had been in the frontline from the start, attacked the Government's transport policies together with Labour MP Denham. Denham accused the Tory Transport Minister of 'complete complacency' regarding the fate of *'one million sufferers of asthma'* (Hansard 18 July 1994). She demanded a halt to the road-building scheme.

Combined with the anti-roads protests, the asthma issue became a campaign issue for Labour. On 20 July 1994, a Labour document denouncing the roads programme featured in a long debate, during which different Labour MPs lambasted the initiatives taken by the Environmental Minister of State, Robert Atkins. Among the critical MPs were Austin Walker, Keith Hill, and Paul Flynn. All the emblematic cases were cited, such as the WHO guidelines, the 10,000 deaths due to PM10, the pollution incident in 1991, the deadly smog episodes in London's past, and the large increase in asthma reported by the Independent in its article on how Britain had choked (Hansard 20 July 1994). In October, Michael Meacher, Labour Shadow Minister for the Environment, held a long speech in which he condemned the roads programme on behalf of the Labour Party (Hansard 31 October 1994).

The Government finally budged early in 1995 by releasing a blueprint for an air quality strategy (DOE 1995), and it was included in the new environmental bill. The actual standards were not known at the time, but they would be supplied by COMEAP. Conservative Secretary of State John Gummer's strategy was promulgated on 12 March 1997. The values were indeed strict, even more so than the European standards that would be promulgated in 1999. However, the plan also included provisions that were more lenient than the EU proposals. The standards were mandatory, but they were only considered relevant for areas having public exposure. Moreover, a weighing of interests between the air quality standards and economic interests was possible (R. Mills, Telephone interview). The strategy was to be Gummer's last act as Secretary of State, because a Labour Government took over two months later.

6. Incidentally, the Independent's report was taken over uncritically in the Nederlands Tijdschrift voor Geneeskunde, 1994. 6 Augustus, 138 (32), under the heading of foreign news.

4.5.2 *UK 1998 Presidency*

After the 1997 elections ended in a landslide victory for Labour, the party maintained its positive attitude towards air quality for at least the time being. Deputy Prime Minister John Prescott declared that Labour Administration would be *'the Greenest Government yet'* (Carter 2001). Labour promised to deliver environmental improvements, and got off to an environmentalist start. Prescott outlined new proposals for transport, and cut the already decimated road programme even further (DETR 1997). He also outlined his progressive and somewhat eco-modernistic transport programme in the white paper 'A new deal for transport' (Department of the Environment, Transport and the Regions 1998). In order to integrate transport and environment, the Labour Government combined the Ministries of Transport and the Environment. The Department of Transport (DOT) now merged with the Department of the Environment (DOE), and together they formed the Department of the Environment, Transport and the Regions (DETR). This merger had the explicit aim of integrating environmental and transport policies (House of Commons 1998). Michael Meacher became the Minister of State for the Environment, but John Prescott held the actual power.

In this light, it was natural for Tony Blair to prioritise environmental subjects for the 1998 Presidency of the European Council. Labour announced that the UK would use the Presidency to embark on a significant greening of the EU policy. It made the environment one of the top three priorities of its presidency alongside crime and jobs, and would look in particular for integration of the environmental policy with other policy fields. The Labour presidency was successful in, among others, establishing a common position on the air quality Daughter Directive, and in reaching an agreement in the auto-oil programme. It also initiated the Cardiff Process, designed to integrate environmental considerations in other fields.

According to Rudiger Würzel (2003), the UK was bent on maintaining a high profile for environmental regulations in order to involve Euro-sceptic Britons more with the EU. The British Prime Minister, Tony Blair, declared that he wanted Britain to be at the heart of Europe. Holding the Presidency in early 1998 constituted an important opportunity for 'new' Labour to show its European credentials and to convince a largely Euro-sceptic British public of the merits of EU membership. Though the environmental flamboyance of the new Labour Government withered away swiftly (Carter 2001, p. 133), it had a considerable impact on the EU's air quality regulation. The UK had an interest in having the EU clean air policy modelled after its own, and for the European Commission as well the strategy was useful as a blueprint. UK civil servant Lynne Edwards was sent to advise the Commission. Edwards was part of the team that had drafted the UK strategy, and according to RIVM researchers Rob Maas and Hans Eerens, she *'constantly explained which route to take to arrive at the right direction* (interview, Hans Eerens and Rob Maas). The final EU Directive ended up being stricter than its UK counterpart, and did not allow for a weighing of interests in the same manner as that of the UK strategy. According to Richard Mills, the EU possibly intended to 'one up' the UK strategy, showing that it was prepared to go further (R. Mills, telephone interview).

The UK debate preceded the Dutch debate by roughly ten years, but the clash that took place in the UK over childhood asthma displayed a number of discursive elements that played a part in the Dutch case as well. Particulate Matter was connected discursively to traffic and transport, roads became considered dangerous places because of bad air quality, and the 'car economy' was portrayed as unsustainable and in need of an overhaul. The UK case showed the potency of health-based claims against the expansion of transport and mobility interests.

4.6 Promulgation of European air quality standards

Now that four crucial actors and factors for the promulgation of the directives have been analysed, we will turn our attention to the eventual adoption of the proposals. The proposal for the Framework Directive COM(94)104 eventually became 'Council Directive EC/96/62 on Ambient Air Quality Assessment and Management', the Framework Directive for short, promulgated in 1996. The Directive faced little political opposition, because it contained only programmatic aspects that did not place large burdens on the Member States. The Daughter Directive, proposed in 1997, was politically more salient, and was officially adopted on 22 April 1999 as Council Directive 99/30/EC, 'relating to limit values for sulphur dioxide, nitrogen dioxide and oxides of nitrogen, particulate matter and lead in ambient air'.

Negotiations over this directive in the Council of Ministers were much more difficult. The reason was that this proposal contained the actual air quality standards, and these had a significant impact on Member States. A rift opened between southern Member States, which aimed at having exemptions from and postponements to the standards, and northern Member States, which felt the proposed standards did not go far enough. The southern Member States invoked the argument that the regulations should be responsive to national characteristics.[7] However, this argument in favour of flexibility was frowned upon by the more ecologically inclined Member States, including the Netherlands (Rood et al. 2005, p. 18).

The European Parliament was in favour of setting strict standards, and the UK influence was present in these Parliamentary debates as well. The rapporteur, Ms. Anita Pollack, was a MEP from the Parliamentary Commission of Environment, Public Health, and Consumer Protection, and was from London. Her job was to draft a rapport on the proposal to inform her fellow MEPs. Ms. Pollack was a Labour representative, and when the proposal was debated, she referred explicitly to the plight of asthmatic children and the threat posed by road traffic. She stated: '*City dwellers suffer*

7. Kamerstukken II 1997-1998, 21 501-08 nr. 72, p. 3.

most, of course, since pollution from vehicle traffic is becoming a bigger problem every year' (European Parliament 1998, p. 71).

The eventual standards promulgated in the Daughter Directive were somewhat more lenient than those proposed originally by the Commission. This concerned especially the limit values for PM10. The original proposal contained two separate standards for PM10: a yearly one and a 24-hour one. The yearly standard stipulated that the air should not contain more particles than 30 micrograms (Mug) per cubic metre (m^3) as a yearly average. It also proposed a maximum daily concentration of 50 mug/ m^3 of PM10. This maximum concentration could be exceeded 25 times per year. The Directive as promulgated contained a yearly and a 24-hour standard as well. The yearly standard was raised to 40 mug/m^3 of PM10 per year, while the 24-hour standard remained the same, but instead of 25 exceedances, 35 were allowed. The standards for PM10 had to be met by 2005. It was foreseen that by 2010 the standards would be lowered to a yearly average of 20 mug/m^3 PM10 and to a daily average of 50 mug/m^3, but only allowing for 7 exceedances.

The values for NO_2 remained unchanged from the Commission proposal, and they had to be met by 2010. The Framework and Daughter Directives included other notable features, such as extensive information rights for the public (Daughter Directive, article 8) and stipulations on the harmonisation of measuring methods and locations; Directive 99/30/EC also contained a review clause, and the standards were to be evaluated by 2003. At this time, the Commission would also assess whether meeting the indicative standards for 2010 would be feasible. The southern Member States managed to obtain some exemptions for naturally occurring PM10 (Daughter Directive, article 5, sub. 4), but the directive did not allow for a weighing of interests. The standards adopted were termed limit values, and that indicated they were result-oriented obligations. The air quality standards concerned the outdoor air in the territory of the whole of the EU, except where the workplace was concerned.

Even though the standards were weakened somewhat, the directive was still considered an ambitious piece of environmental legislation. The European Environmental Bureau (EEB), which acts as a pressure group for the environment at the EU level, was pleased with the policy making process. It concluded: *'The level of environmental policy demands is high'* and there was *'even an excess of environmental representation'*, especially during the expert phase when representatives from the industry were outnumbered by representatives from the environmental movement and from environmental and health departments (Hey 2000, p. 12).

The position of the Netherlands during these final negotiations is of interest, though the country itself did not play a crucial role during the negotiations. Despite scientific knowledge that meeting the standards would be very difficult, it chose to side with the northern camp. Rood et al. (2005) state that from a pragmatic standpoint this choice was difficult to defend and that an alliance with the southern block would have been a more obvious choice (Rood et al. 2005, p. 18). In retrospect, this is certainly the case, but in the light of the time, the Dutch position is understandable as well. The Dutch Government had a hand in many of the developments that eventually led to the Framework and Daughter Directives, and it consciously lobbied for ambitious EU policies based on eco-modernistic ideas. Getting the WHO involved in air quality standards had been a Dutch idea as well. At the time, the Netherlands was still clinging to the role of front-runner in European environmental politics, and intended to go further than the Council of Ministers eventually had done (Jan Pronk, interview).

On the whole, the Dutch Government was not simply a marginal player in the air quality policy field. Although Dutch politicians painted this picture during the air quality clash of 2004-2008, based on the current analysis, that depiction is incorrect. On a number of fronts, the Dutch Government played a decisive role in the emergence of the directive. In essence, the rhetorical talent of the Dutch 'reverend' was influential in European environmental policy making. Nor did the Netherlands forget its merchant side, but in this case the merchant was overruled by a highly influential WHO doctor that delivered its troubling diagnosis to an audience of worried European policy makers.

4.7 Concluding remarks

In the conclusion to this chapter I will make a number of comments on four topics that came to the fore. The first topic of interest is the policy making process in the European arena. This process should be characterised as a process of bricolage. Many interests and stakeholders are present, and they all intend to influence the policy making process. The directives reflect this diverse set of interests and actors, and consequently they contain a mix of elements, sometimes haphazardly put together. Secondly, the emergence of the Air Quality Directives shows the importance of involving doctors in discourse coalitions over environmental problems. When an environmental problem is recognised as a health threat, even vested economic interests have to back down. Thirdly, environmental problems facing one country become problems concerning all the European Member States. When the public of a Member State experiences a situation as problematic, it makes sense to export these problem definitions to the European arena. In this way, a Member State makes sure it becomes a European problem and subject to European regulation, thus avoiding having to take domestic measures. Fourthly, I will address the question of whether the Air Quality Directives are examples of precautionary legality. It will be argued that although the air quality policy in the EU has the hallmarks of precaution regulation, the policy process also contained elements of the legality of risk and compensation, notably the presence of an economic evaluation of the proposed Directives.

4.7.1 *Main influences on the EU Air Quality Directives from the 1990s*

In theory, the European Council is the most important policy maker in terms of Joel Best's scheme. Formally the Council of Ministers and the European Parliament enact legislation, however they do so on the basis of Commission proposals, giving the European Commission a key position in the policy making process. The European Commission is a policy maker that is influenced by many different external pressures, as this research indicated. These different pressures cause legislative proposals to become cobbled together as it were by the interests of various actors and various policy philosophies in vogue at the time at the time of their preparation.

Four aspects played key roles in the shaping of the Air Quality Directives. The WHO and the UK Government were key actors that decisively influenced the policy process. The rise of eco-modernist ideas and the use of a cost-benefit analysis were pivotal discursive factors that shaped the eventual form and adoption of the directive.

My research indicates that within the European regulatory institutions such as the European Commission, the Parliament, and the Council of Ministers, policy making was influenced by the ideology that held sway at the time. Ecological modernisation discourse was on the rise in the EU during the 1990s, and – after the single market, the Earth Summit in Rio, and the Brundtland report that popularised the term sustainable development – the EU was in an upbeat mood about environmental regulation. It had high hopes of the possibilities it provided to steer both the environment as well as the economy to prosperity. Environmental leadership could pay off, and therefore it was possible to table ambitious legislation.

Policy making was influenced further by institutions that had no formal political stake, such as the WHO. The scientific legitimacy that it provided to the proposal for the Directive made sure it encountered less opposition from Member States than if it were simply a proposal from the Commission based on political justifications and 'in-house' scientific expertise. The Commission used the WHO guidelines explicitly for that reason, and the WHO was so influential that dissenting voices stood no chance. These standards, however, were drafted by doctors and epidemiologists. Members of these professions have considerations that are quite different than those of politicians. Their duty is to propose standards they consider best from the perspective of public health, and they are not concerned with political considerations like feasibility of costs. The air quality guidelines drafted by the WHO did become regulatory benchmarks however, quickly after their inception in 1987.

The influential economic evaluation served as an argument for the Commission as well. The proposal under consideration was related to other policies, and natural scientific findings were used to draw conclusions about the economic feasibility of the proposal. The package was presented as a matter of calculation, but in the end health was traded off for money. Human lives saved were monetised and presented as economic gains. The payoff would never be financial, however, but would be of a diffuse nature, such as a healthy population. Within the confines of cost-benefit analysis, such considerations were obscured. Moreover, the fact that costs could vary between Member States was left out as well. Not all countries had the same problem with PM10 in their cities, and those countries faced fewer costs than a country such as the Netherlands.

UK involvement was a critical factor as well, since the EU strategy was influenced by that of the UK, and the UK supplied information and know-how during various formative stages of the process. Tony Blair needed to score green points as well to secure his image as a young progressive leader and he needed to 'sell' the European Union to a sceptical UK audience as well. Air quality is a subject close to heart of the British public, making it an ideal topic for the new UK Government to profile itself on.

4.7.2 *The Air Quality Directives as a product of 'bricolage'*

Member States have always been considered important players in the formation of European regulation, but their influence goes further than their role in the Council of Ministers, or in supplying experts in the preliminary phases. Policies from Member States that the Commission considers advanced act as blueprints for Commission proposals. Member States send their personnel to aid the Commission, but these Commission aides also look after the interests of the Member State in question.

The European policy arena is open to Member States that have the ability to act as 'insider claims makers' in Bests terminology and to push their own policies on the Commission agenda. It is suggested in the literature that for the Member State it is attractive to do so, and to bring its own regulation to the attention of EU legislative bodies. Nevertheless, the export of policy to the European arena is bound to have unintended consequences.

My conclusion is that the content of European policy depends on a process of what can be considered bricolage. Bricolage refers to a construction process in which elements are put together in a willy-nilly fashion from materials that happen to be available. Member States and other institutions provide ideas and policies, but elements of these are used and combined with other elements, creating something that is totally different from what was envisaged.

The European arena is a policy making arena in which it takes a long time for a policy initiative to mature and eventually be enacted in the form of a regulation or Directive. First Member States are consulted about a certain intension of the Commission, than experts are sought and they work out the scientific and economic details of the proposal. Subsequently, rounds of negotiation between the Commission and the Council of Ministers and the Council of Ministers and the European Parliament take place. During all these phases European institutions, lobby groups, Member States and other pressure groups try to make their mark on the eventual regulatory end product.

This process becomes clear when we view the long trajectory during which the European air quality policy from the 1990s came to fruition. The Netherlands introduced the idea of asking the WHO for air quality guidelines, subsequently they became a benchmark for EU air quality policy. The Netherlands also helped to introduce eco-modernist ideas, which found their place in the 5th EAP and through the 5th EAP influenced the Air Quality Directives. Meanwhile the idea that policies should be supported by cost benefit assessments and impact assessments gained ground and such a document was duly produced. UK civil servants had their own designs for the Air Quality Directive and it was moulded after UK, but also Dutch air quality policies on the national level. Eventually Directives emerged that were concerned with sustainable development and public health, but also with research into the effects of air pollution, the role of public health institutes, the harmonisation of measuring equipment, reporting to the Commission, and making pollution data available to the public. These various aims of the directives reflect the interests of the actors involved and their underlying ideology.

In the air quality case, WHO input, eco-modernistic ideas, UK influence, and economic data were all combined and mixed, and the result was a directive that contained elements of all these disparate aspects. This process of bricolage is of interest for Member States that intend to export their policies or their policy ideas to the European arena. Such a process is highly unpredictable, and is influenced by factors that are difficult to take into account and foresee in advance. Over time, the original ideas will be mangled, recycled, translated, and embellished in unforeseen ways, and by a number of different actors. It is questionable whether the original Member State will still be happy with the regulatory end-product.

The case of the Netherlands in the field of air quality is a case in point. When the Netherlands asked the WHO to set air quality guidelines, the issue was taken out of Dutch Government's hands. Years later, the WHO guidelines became an instrumental benchmark, and would eventually cause the Dutch Government a considerable number of political and economic problems.

The same scenario is applicable to the Dutch export of eco-modernist policy ideas. Public mobilisation was an eco-modernistic strategy, and was applied in the Air Quality Directives. In the Netherlands, the public was indeed mobilised, and went to court over noncompliance with the directives. All was in conformity with eco-modernist policy philosophy, but it proved to be a major economic problem. The explanation for this mishap may be found by comparing the Dutch policy arena with the EU arena. In the Netherlands, ecological modernisation was based on a tacit consensus between market parties and the administration. Market parties would not be overburdened, and in exchange they would adopt environmental measures considered reasonable by both parties. It worked in an atmosphere in which the parties knew each other and relations were not polarised. The Dutch strategy combined the target group policy with mobilisation of the public. In the European Union, however, things worked out differently. The Commission focused on mobilisation of the public to act as its 'eyes and ears' in a rather antagonistic fashion. Information rights for the public in the Air Quality Directives were included to make the public aware of air pollution problems, and to create pressure on the Member States to comply.

This information- and mobilisation-based strategy was combined with top-down standard setting in the Commission proposal, which was not in conformity with Dutch ecological modernisation, but followed from the ideas outlined in the report '1992, the environmental dimension'. Quality standards were the preferred British way to regulate air quality. Both public mobilisation as well as the top-down character of the air quality regulation proved to be thorns in the side of the Dutch Government when attempting to break the deadlock over air quality in 2005. The combination of public mobilisation and top-down standard setting together in one directive caused unforeseen problems, because in an environmentally aware public began to use the rigid standards to halt infrastructural development.

The importance of eco-modernistic arguments may also be illustrated by the importance attached to the cost-benefit assessment underpinning the Daughter Directive. On the basis of overly optimistic assumptions, economists concluded that the directive was cost-effective. The emphasis on the economic aspect of environmental policy was typical of ecological modernisation, but also typical of the multitude of considerations that come into play when European proposals are under consideration. The cost-benefit assessment took into account the effects of many policies that still had to deliver air quality benefits, heightening the uncertainties. Nonetheless it was used to argue in favour of strict standards.

These standards were the unintended result of initiatives and policy ideas that the Netherlands helped to shape. In the end, they turned out to conform neither to Dutch economic interests, nor to its policy philosophy.

4.7.3 *Speaking health to power: the discursive force of 'doctor knows best'*

The WHO involvement in air quality regulation and the asthma case in the UK in the early 1990s contributed to the re-emergence of a storyline that bad air quality was a serious threat to human health. It was a re-emerging storyline, because the consideration that bad environmental hygiene, including air pollution, damaged public health was a pervasive rationale behind environmental regulation in the 1970s. However, in the 1980s this connection receded to the background, and environmental problems became linked to eco-system threats.

From a discourse analytic perspective, WHO involvement in air quality regulation was of interest. Many respondents and the reviewed literature indicated that the involvement was crucial and that it managed to silence otherwise critical Member States. As Krijgsheld noted, the WHO was invited by Dutch civil servants to produce air quality guidelines. A critical factor was the insistence that the guidelines contain numerical values of risk estimates. Numerical values are easily usable for policy makers, and that was the idea behind the WHO involvement. The WHO produced a document that could be used easily for political purposes, and the Commission could refer to the WHO document to legitimate its regulatory aims. We find a discourse coalition between the WHO scientists and the Commission, around the storyline that bad air quality was a highliy significant health threat that proved difficult to counter.

This discourse coalition succeeded in silencing dissenting voices. No one asked critical questions about the necessity of setting health-based standards. The Dutch had cautiously raised the issue of scientific uncertainty, but were quickly argumentatively outgunned according to Rood et al. (2005). The policy process of the Air Quality Directives displayed the force of health-based arguments, and the WHO medical professionals managed to shift the discussion in the direction of health protection, which succeeded in overriding traditional economic concerns. With a play on words regarding the classical phrase – that one may speak truth to power – we witness here a situation of speaking health to power.

The decision to bring in the medical experts in the early 1980s turned out to be a powerful discursive move by the Dutch Government, and its influence in the 1990s shows that the alliance between the fields of environmental policy and public health was still a potentially strong one. This is an intriguing finding, because in the 1980s a more ecocentric approach held sway, in which the importance of eco-systems was underlined. The environment was considered important for its own sake, and the emphasis on human health was considered a narrow-minded approach. Apparently an emphasis on public health could still strengthen considerably the environmental cause. In fact, the WHO's influence was stronger in the 1990s than in the 1980s, a sign that public health was an increasing public concern that even managed to override traditional economic considerations.

4.7.4 'Pet fears' and the Europeanisation of environmental concerns

The storyline that bad air quality was a health threat was developed further by the asthma episode in the UK. UK representatives played an important role at the inception of the Daughter Directive and they iterated that air pollution resulting from exhaust fumes was particularly dangerous. The connection between asthma and air pollution caused by road traffic was made explicitly in the discussion in the UK, but it became a wider issue through the promulgation of the directives. The UK discussion preceded the one in the Netherlands. Worrying scientific results were published here as well, but they raised little alarm until 1999. In the UK, the findings of epidemiologists sparked public protests much earlier, causing the UK Government to side with the Commission and support the proposal. Discursively, the involvement of the UK directed attention to the harmful effects of road traffic.

UK involvement was no surprise when one considers that the UK was historically prone to be fearful of air pollution. Fog in London is a peculiar British phenomenon, but it is more than an idiosyncrasy. In 1952, the Great London Fog killed thousands due to a combination of fog and smoke that indeed choked the city. Air quality standards were set in order to combat this type of smog in the future. The Great London Fog is still a part of the collective British memory; it was recalled dramatically in the asthma crisis of the 1990s, and is part and parcel of UK presentations delivered on air pollution (Marjan Van Giezen, interview).

I am borrowing the description 'pet fear' from UK researcher Ragnar Lovstett to describe a peculiar social problem that is culturally defined, and to which a particular country or group is subject.[8] Urban air quality is a British pet fear, and it is therefore understandable that childhood asthma caused by exhaust fumes could become a social problem in the early 1990s. This problem coincided with recent scientific results on the danger of PM10 and the EU regulatory preparations for a new air quality policy. This presented the opportunity to the UK Government to export its own pet fear to the European arena. Not every country of course has the same pet fears. According to Lovstett, people in Denmark are fearful of the harmful effects of radiation from power stations, while to the German psyche, the dying of forests is deeply troubling (Bramwell 1989).

I hypothesise that we are witnessing a dynamic that leads to the proliferation of environmental fears in the European Union. The European policy arena provides opportunities for Member States to export their own pet fears to the European level, and it is tempting to do so. Member States are under domestic pressure to propose strict regulatory measures to curb the pet fear, just as was the UK in the case of air quality. Because strict domestic regulation leads to a comparative disadvantage for domestic industries, it makes sense to export one's policy solutions to the European level. This may not always lead to satisfactory outcomes, as discussed above, but it is tempting nonetheless.

This export of policy is done through the discursive device of the storyline. To convince the Commission and other Member States, a storyline must be presented about why this problem needs urgent regulatory attention. Through the use of storylines, problems are exported together with the policy solutions for them. UK representatives will inform us about the potential catastrophic impacts of urban air pollution; Danish representatives will tell us about the harm of radiation; Swedish representatives push concern for chemicals in consumer products; and the Germans will warn us about forest decay.

8. I heard the phrase in his keynote lecture, 'Risk Communication in the 21st Century', at the conference 'Grenzen aan de risicobenadering', 13 May 2009, Scheveningen, the Netherlands.

The threat of bad air quality was similarly exported to the Netherlands. Before the promulgation of European directives and their outcomes in court cases, air quality was not a topic of great concern in the Netherlands. Through European policy making, however, it became a concern for Dutch residents as well, and the storyline took hold. In the air quality clash, the storyline that air pollution caused by motorised traffic leads to premature mortality is one of the most important ones. It started with the claims made by epidemiologists and was reinforced by UK representatives for domestic reasons and the WHO and finally embraced by EU institutions which fed it back to the public by way of regulation with an explicit awareness raising aim.

4.7.5 *Air quality standards as examples of a legality of precaution*

The standards for air pollution promulgated in 1999 were European law, and they had to be implemented in Dutch law. The Netherlands did so in 2001. Air quality standards were in fact not new; earlier EU Directives contained air quality standards as well, but their practical importance was limited because the Netherlands generally complied with these standards. The situation was different for the 1999 standards, and in many parts of the country the standards would be exceeded. The question is whether the promulgation of these standards should be considered as fitting within a change towards a legality of precaution.

A number of considerations led to the conclusion that the air quality standards were indeed precautionary. The main argument for considering the Air Quality Directives as examples of precautionary legislation is that the standards for PM10 were set without much scientific certainty. The EU Technical Working Group on Particulate Matter highlighted many uncertainties and even considered whether a limit value should be set or whether a range should be given. It was also unsure as to whether PM10 was the right fraction of PM to regulate, or whether PM2.5 should be targeted instead. The Working Group opted for PM10 because the knowledge regarding PM2.5 was even more incomplete. PM10 was a pollutant measured in only a few countries at the time, and the Dutch Government considered that the information about emission inventories was insufficient. Moreover, there were serious doubts regarding the feasibility of the air quality standards. One may argue that the European Union acted early on the basis of new scientific data such as the epidemiological studies discussed in chapter 2. However, this discussion also revealed that much was still unclear. Policy making to protect health or the environment without complete scientific certainty is a hallmark of a precautionary approach because it conforms to the precautionary principle in environmental law.

The precautionary principle is of course a cornerstone of a precautionary legality. In the air quality case, the precautionary principle was hardly ever mentioned as such. However, precautionary considerations lay at the heart of the air quality standards.

They certainly lay at the heart of the UK strategy, which influenced the eventual EU strategy. When confronted with the possibility that PM10 caused large-scale health problems, the UK's air pollution Committee, COMEAP, stated in 1995: '*It would be imprudent not to regard the associations as causal*'. Such considerations are a clear indication of precautionary logic, because even before it is known whether an association found is indeed a causative relation, it is treated as such. This implies that measures need to be taken before the scientific evidence is settled on the issue. Moreover, respondents from the VROM Ministry considered that these EU policies were a form of 'no regret policy', as such policies prescribe taking preventative measures early in order to avoid the possibility of acting too late.

The information requirements included in the Air Quality Directives were in line with the legality of precaution as well. The public had to be made aware of the remaining risks of air pollution so that it could take preventative measures, but also to enable the public to make its voice heard.

The primary aim of the directives was to reduce the effects of long-term exposure to air pollution. Such attention to damage becoming manifest in the long term, and with a considerable margin of uncertainty, is the type of damage that takes centre stage in a legality of precaution.

These reasons demonstrate that the Air Quality Directives themselves fit neatly within the legality of precaution. However, the policy process that led to them did not fully subscribe to this legality. Elements from the legality of risk and compensation are also conspicuous in this process, and most conspicuous is the presence of an economic evaluation. The existence of such a cost-benefit analysis points towards the calculation of risks, and this is indicative of a legality of risk and compensation in which the concept of risk rather than precaution takes centre stage. Reliance on expert data from the WHO fits within the ideal typical picture of the legality of risk and compensation as well, because in this type of legality, expert knowledge is considered the best justification for policy, while the legality of precaution also considers the layperson's subjective judgment to be crucial.

An analysis of these different elements indicates that the results of the policy making process were precautionary, but that the policy process itself also contained elements belonging to the legality of risk and compensation. Justification for the strict standards was not the outrage of civilians over the occurrence of damage, as one would expect in a legality of precaution. Instead, it was an expert-driven process in which health considerations dominated. In this case, even economic rationality, such as the rationality to choose air quality standards in service of the common market and the presence of a cost benefit analysis, led to precautionary legislation.

Timeline EU air quality policy in the 1990s

1983	Dutch civil servant approached the WHO in order to ask for air quality guidelines
1987	WHO published first air quality guidelines
1989	NMP released
1990	Publication of "1992" The Environmental Dimension'
1993	First contacts with Member States over new air quality policy
1994	Major Concerns over asthma in the UK
1994	Proposal COM(94)104 for the Air Quality Framework Directive
1996	Adoption Framework Directive 96/62/EC
1997	UK air quality strategy unveiled
1997	Labour wins UK elections
1997	Proposal for First Air Quality Daughter Directive COM(97)500
1998	Common position on proposal COM(97)500
1999	First Daughter Directive on ambient Air Quality Directive 99/30/EC promulgated

Table 6: PM10 risk tables in 1996 provided by the WHO

Table Summary of Relative Risk estimate for bronchodilator use, cough and LRS reporting, PEF changes and respiratory hospital admissions and daily mortality, associated with a 10 µg/m increase in the concentration of PM10 or PM2.5 (results of meta-analysis of available studies)

Endpoint	**Relative Risk for PM2.5 (95% C.I.)**	**Relative risk for PM10 (95% C.I.)**
Bronchodilator use		1.0305 (1.0201-1.0410)
Cough		1.0356 (1.0197-1.0518
LRS		1.0324 (1.0185-1.0464
PEF change (relative to mean)		-.13% (-.17% - -.09%)
Respiratory hospital admissions		1.0080 (1.0048-1.0112)
Mortality	1.015 (1.011-1.019)	1.0074 (1.0062-1.0086)

Table Summary of Relative Risk estimates for effects of long-term exposure to PM on morbidity and mortality associated with a 10 µg/m increase in the concentration of PM10 or PM2.5

Endpoint	**Relative Risk for PM2.5 (95% C.I.)**	**Relative Risk for PM10 (95% C.I.)**
Mortality (1)	1.14 (1.04, 1.24)	1.10 (1.03, 1.18)
Mortality (2)	1.07 (1.04, 1.11)	n.a.
Bronchitis (3)	1.34 (0.94, 1.99)	1.29 (0.96, 1.83)
% change in FEV_1, children (4)	-1.9% (-3.1%, -0.6%)	-1.2% (-2.7%, -0,1%)
% change in FEV_1, adults (5)		-1.0% (n.a.)

(1) **Dockery, D.W. et al.** An association between air pollution and mortality in six U.S. cities. *New England Journal of Medicine* 329: 1753-1759 (1993).

(2) **Pope, C.A. III. et al.** Particulate air pollution as a predictor of mortality in a prospective study of U.S. adults. *American Journal of Respiratory and Critical Care Medicine* 151:669-674 (1995)

(3) **Dockery, D.W. et al.** Health effects of acid aerosols on North American children: respiratory symptoms. *Environmental Health Perspectives*, 104: 500-505 (1996)

(4) **Raizenne, M. et al.** Health effects of acid aerosols on North American children: pulmonary function. *Environmental Health Perspectives,* 104: 506-514 (1996)

5 DUTCH AIR QUALITY LEGISLATION AND THE IMPLEMENTATION OF THE AIR QUALITY DIRECTIVES 1982 – 2001

INTRODUCTION

In Europe the new scientific findings related to air quality and the turn to ecological modernisation led to an ambitious new air quality policy in 1999 that contained tough standards as well as a program to raise awareness among the population. Meanwhile, also in 1999, Dutch public health agencies were claiming that air quality caused significant health damage in residential areas near highways such as Overschie. These dual developments impacted Dutch environmental policy. The Dutch Government was under a European obligation to implement the European Directives in Dutch law and the new scientific findings in combination with the public stir in Overschie also invited a policy response. In this chapter we will look at the development of Dutch air quality policy up to and including the transposition of the Air Quality Directives.

From my interviews and the policy documents it turned out the Dutch policy makers were aware that meeting the air quality standards in the Netherlands in time would be very difficult and very costly. The main question asked in this chapter in why the Dutch Government chose a strict implementation of the Air Quality Directives in Dutch law, in the face of this knowledge. In this chapter two conclusions are provided. One explanation focusses on political and strategic choices made by Minister of VROM Jan Pronk (PvdA) at the time of the implementation. A second more structural explanation relates the strict transposition to the failure to find a political answer to the new scientific findings. The ecological modernistic consensus in the Netherlands entailed that standards that were perceived as being scientifically not well supported and that placed unreasonable demands, could not be considered legitimate standards and the Dutch Government could not be held to account for them. In light of the events in 2004, that proved to be a costly miscalculation on the part of the administration.

First we will examine early Dutch air quality policy from the 1980s and 1990s. The policy process regarding air quality and its lack of urgency in these decades helps to understand why the Government underestimated the potential regulatory problem the air quality standards could pose. Subsequently the implementation of the EU

Directives in Dutch law is considered. In the final section, a number of explanations for the strict implementation are provided and the role of ecological modernisation and the legality of precaution in this process are considered.

5.1 Policy development in the 1980s and 1990s

Before discussing the eventual implementation of the European air quality standards in Dutch law, it is necessary to dwell on air quality policy in the 1980s and 1990s. The policies from those decades form the political context within which the eventual implementation of the Air Quality Directives may be understood. An investigation into air quality policy in the 1980s is undertaken in section 5.1.1 and this reveals that while the Dutch Government voiced ambitions in the field of air quality, in practice it trailed European developments. In the 1990s PM emerged on the scene as a potentially serious health problem (5.1.2.) and Dutch policy in response to this problem is investigated in 5.1.3. Finally, in 5.1.4., the emergence of policy – or rather the non-emergence of policy – in the latter half of the 1990s is under scrutiny.

5.1.1 *Early Dutch air-quality regulation*

In the 1980s, a policy regarding air quality received its first serious consideration as a possible means of combatting air pollution. Quality standards became popular as a way to integrate environmental concerns into other areas of policy, most notably spatial planning.[1] At the beginning of the 1980s, just before the takeover by Pieter Winsemius and the change of Ministry from VOMIL to VROM, air-quality standards gained political prominence.

When VROM Minister Winsemius took over in 1982, he considered air-quality standards as well, and in the plans drafted under his auspices, a list of priority substances was provided for. These substances were to be regulated within an encompassing system of air-quality standards. It was clear, however, that such standards could not be formulated in the short term. In principle, they had to be set at a level that could safeguard public health, the ongoing existence of certain plants and animals, and even economic goods. In many cases, scientific knowledge was not yet sufficient to formulate such standards, making the establishing of dose/effect relationships an arduous task. Hence, the development of WHO guidelines was considered to be helpful (VROM 1983, p. 28).

However, practical policy moved in a different direction. Targeting the effects of pollution generally became subordinate to changing the behaviour of polluters, and Winsemius and his successor Ed Nijpels focused increasingly on target-group policy.

1. See chapter 3.

Nijpels noted that it was difficult to formulate quality standards for all the different substances, and that sources of pollution could also be tackled without those demands (VROM 1986, p. 9). In later IMPs up to the NMP, the WHO guidelines were no longer mentioned.

The IMPs contained air-quality standards for a host of polluting substances, but because there was no basis in Dutch law for setting such standards (Spaans & Michiels 2000, p. 102), they remained simply as extra-legal guidelines for policy makers. In practice, the Netherlands imported European air-quality policies; standards with a legal basis were set only for a certain number of pollutants because they became mandatory as a result of EU law. In 1980, the EU proclaimed standards for particle pollution and SO_2 (Directive 80/779/EC). In 1982, standards for lead were promulgated, and NO_2 was regulated in 1985. These early European air-quality Directives were not very strict, however, and required only minimal policy interventions from Member States (Hey 2005, p. 19). After setting standards for air quality became possible under Dutch law, European air-quality Directives were implemented for SO_2 and Particulates (not as PM10, but as black smoke), carbon monoxide, NO_2, lead, and the last one, benzene, in 1993. Only in the case of benzene was the Dutch Government not under an obligation imposed by previous EU directives.

Within the scope of the inquiry into the PM clash, the Directive that regulated SO_2 and particles together is of most interest. The Air Quality Order[2] on SO_2 and Particulates (Besluit Luchtkwaliteit Zwaveloxide en Zwevende Deeltjes (Stb. 1986, 78), henceforth AQO SO_2 and Particulates) regulated particles and SO_2 together. The AQO on SO_2 and Particulates included the terms 'limit value' and 'target value'. These terms would remain in use later on and throughout further air-quality regulation. Target values needed to be taken into consideration when policy was formulated, which meant that although policy had to be aimed at reaching the values, failing to reach them did not have immediate legal consequences. This target value represented a long-term aim. The limit value was stricter, and needed to be observed ('in acht nemen', in Dutch). Observing a limit value meant that policy should be adjusted to reach and maintain those values and the administration was under a legal obligation to reach the limit values in time, and to guarantee that they were maintained.

While use of the term limit value suggested a strict implementation, the wording of the Explanatory Memorandum demonstrated that the chance of this law having social economic repercussions was considered minimal (Stb. 1986, 78). The Netherlands was already in compliance with the set of limit values. For particles, it was not expected

2. The General Administrative Order is an instrument in Dutch administrative law by which the administration can execute certain competencies granted by law, without needing the approval of the First and Second Chamber in Parliament. The General Administrative Order has the status of law see also section 3.1.1.

that any additional measures would need to be taken, and it was anticipated that the Netherlands would comply relatively easily with the Directive and the Dutch implementation.

In later documents such as VROMraad (2008, p. 9) and Rood (2005, p. 8), it was clear that the Netherlands was labouring under considerable pressure relating to the environment, but this notion was absent in the E.M. to the Air Quality Order SO_2 and Particulates. On the contrary, environmental pressure was considered to be low in the Netherlands. On page 12 of the memorandum, it was stated that the limit values were not exceeded in the Netherlands, because it was a flat country, because certain climatologic conditions prevailed, and because gas was used on a large scale (Stb. 1986, 78). The memorandum did not indicate what to do when limit values were exceeded, and this lack of precision reinforced the impression that it was not considered likely that the standards in the Netherlands would become problematic.

The Dutch AQO on SO_2 and Particulates was evaluated in 1992, and it was made clear that even those modest obligations had not been met. Exceedances did occur occasionally, but were rarely reported, and administrative commitment to comply with the directive was low, as lower administrative bodies did not attach great weight to the air-quality obligations stemming from the 1980s (VROM 1992). The easy acceptance with which EU regulations were imported, and the few consequences attributed to them, indicate that air-quality standards were largely ignored in practice, due to administrative indifference and lack of commitment. Such standards were generally not considered by Dutch policy makers to be 'painful' (Maas & Eerens, interview).

5.1.2 *The turn to PM in Dutch policy from the 1990s*

Scientific awareness with respect to air quality flourished in the Netherlands, and state-of-the-art information about pollutants was laid down in 'criteria documents'. In these documents, pollutants were described and their effects considered, and the compilations were used to aid in policy making. In early 1994, as a result of the European Union's regulatory activity, a new criteria document on Particulate Matter appeared (Annema et al. 1994), which was geared to the issuing of more rigorous air-quality regulation, including PM10 standards.[3] The Dutch Government was forced to form an opinion regarding PM10 policy. The document indicated that considerable uncertainty regarding PM still existed, and urged that the findings be treated as indicative. In most cases, it was not fully known what type of particle was being emitted, which made it difficult to say much about PM10's general toxicity. The RIVM estimated at the time that industry was primarily responsible for the emission of PM with 50-60% and traffic with 10-25% (Annema et al. 1994).

3. The first criteriadocument on PM10 dated from 1987 (Van der Meulen et al. 1987).

This criteria document was sent to the Dutch Health Council, which reviewed the findings and compiled a final advisory policy concerning a quality standard. The Health Council considered that there was evidence that particulate pollution could cause chronic health problems, but it was still not known how it led to ill health, or what components were responsible for its effects. This made it difficult for policy makers to recommend a limit value. PM did not appear to exhibit a no-effect level – one at which no adverse effects occur. Due to the lack of data and the absence of a no-effect level, the Council was hesitant to advise a specific value, but considered that any level of PM10 was probably harmful.

This conclusion had implications for policy. The Health Council advised that it was not particularly useful to reduce the number of days in which PM10 levels were high, but structural and technical measures reducing industrial and traffic emissions should be taken. These measures should, however be considered in an international context. In the Netherlands, the background concentrations were structurally high, and had to do with the amount of PM10 in the air regardless of any source of PM10 nearby. The high level of background concentration caused more harm than the limited exposure in peak periods, as it led to chronic exposure.

Moreover, if there was no 'no-effect level', any standard was in principle arbitrary. There was no standard below which no or far less mortality occurred; it increased incrementally. Whether actual concentration levels were 39 mug/m^3 or 41 mug/m^3 did not greatly matter in terms of health. This implied that taking measures to reduce concentration levels from 41 mug/m^3 to 40 mug/m^3 simply to meet the standard did not result in large gains for public health.

Owing to such considerations, the Health Council recommended not informing the general public. People who were particularly vulnerable were to be advised by their general practitioner, but public information campaigns were not considered prudent, because they could cause people to modify their behaviour unnecessarily (Clarenburg 1995, p. 68).

5.1.3 *PM10 as a serious problem for Dutch policy*

PM10 presented a tough challenge for Dutch policy. Nationally and internationally, epidemiological scientists pointed out the possibility of serious health effects, but the level of uncertainty remained high.

The political quandary was worsened by the fact that abatement policies were difficult and costly. Firstly, roughly two thirds of the pollution found in Dutch air was of foreign origin. The export of air pollution from the Netherlands was on a similar scale. That meant that a reduction of emissions in the Netherlands would lead to drops in the total number of pollutants emitted in the air, but since most air pollution was imported anyway, or exported out of the Netherlands, it would have a much smaller effect on concentrations in the Netherlands itself.

Secondly, due to the high background concentrations in the Netherlands, it turned out to be very difficult to achieve any drop in concentrations at all, and almost impossible to force air-quality concentrations below the European standard. In 2005, the background concentration was already approaching the yearly standard of 40 mug/m^3 proclaimed in the EU Directive, and exceeding the average daily standard. Dutch policy could only be marginally effective in the case of PM10, because the contribution of Dutch sources of pollution to the background concentrations in the country was limited. European sources contributed a greater share. Moreover, the origins of a sizable fraction of PM concentrations were unclear. According to the MNP in 2005, the greatest share of background PM was due to 'other sources', meaning that they could not be specified (MNP 2005a, p. 14, table). Since we had little knowledge regarding the origin of this amount of PM, it was unclear as to how those concentrations could be effected. Even if peaks were to be greatly reduced – for instance, by closing the roads – the standards could not be reached.

The limited influence that the Netherlands had on its own concentrations of PM, made it very costly to achieve extra reductions. Buringh and Opperhuizen concluded in a survey for the RIVM in 2002 that the maximum feasible reduction that could be achieved by reducing PM10 concentrations was 1.1 mug/m^3. This package of measures would cost the Dutch state 6 billion Euro per annum (Buringh & Opperhuizen 2002a, p. 14). Since annual regional background concentrations averaged between 32 and 42 mug/m^3 in 1995 (Van Velzen et al. 2000, p. 11), one may conclude that such a reduction constituted only a small fraction of the total concentrations of PM in the Netherlands. The researchers did note that higher reductions were predicted locally, also with a less expensive package.

The question as to what to do in this situation presented a challenge for Dutch environmental policy throughout the 1990s, but not one that policy makers were eager to confront head on. The Netherlands had a 'smog regulation', which controlled the provision of information. In 1995, the Government issued a 'preliminary policy opinion on Particulate Matter and winter smog' (interim beleidsstandpunt fijnstof en wintersmog). It was proclaimed that reducing PM was considered to be of the highest priority, and that it was possible that health effects could still occur, even with the low levels of PM10 experienced at that time. The document mentioned the paucity of information with respect to the PM's mixture of toxic components, and that made it unclear as to what substance should be targeted. Reducing PM10 was considered to be well-nigh impossible for the Netherlands to do on its own. The letter accompanying the preliminary policy opinion noted: *'The key to abate harmful effects on public health is reducing emissions of PM10 on a European scale'*.[4] The political plan was to wait for Europe to act.

4. Kamerstukken II 1996-1997, 25 005 nr. 3, p. 1.

However, this option of waiting for Europe carried political risks, as policies from Europe might not fit within the Dutch national context. Warnings came from the Provinces that the new rules should fit the Dutch decentralised system, in which lower administrative bodies were responsible for conducting local environmental policy (Van der Werf 1997, p. 66). The VROM Ministry promised to tackle air quality in tandem with the lower administrative bodies in a plan in which they would be granted enough flexibility to deal with the coming standards. It promised an urban air-quality action plan that was especially tuned to the needs of lower administrative bodies. The Environmental Programme 1998-2001 stated:

> *'One subject in the action plan is the way in which the aim of decentralisation on the national level and centralisation on EU level will take shape. [..] The intention is to create as much discretion for local administrations as possible and come to terms over conditions (over procedures, among other things)'.*[5]

The quote indicated that policy makers were aware early on that these standards could conflict with the Dutch approach to environmental regulation.

In Europe, the Dutch Government supported ambitious policy development, and Dutch negotiators associated themselves with the green camp (Rood 2005). Environmental Minister Margreeth de Boer was positive as well. Her successor Jan Pronk stated that it was Government policy at the time to be positive towards the European regulation; the same conclusion was drawn by the Social Economic Council, which reviewed Dutch policy on European environmental regulation (SER 2006, p. 54).

5.1.4 *Air Quality policies in the latter 1990s*

In the Netherlands itself, things remained mostly quiet on the policy front. Environmental policy for the long term was laid down in National Environmental Policy Plans, the NMPs, and for the short term in Environmental Programmes. In 1998, the Ministry of VROM promised to come up with a new policy document on PM to replace the interim policy opinion. It was expected that more would be known about the continuing European negotiations. In the meantime, more research had to fill the existing gaps in information. The third NMP, also issued in 1998, outlined Dutch environmental policy from 1998-2002. It referred to this coming policy opinion (VROM 1998, p. 303), but other than that, little was said about PM10. In these NMPs, PM10 was named as a priority, and road transport was indicated as a major contributor to PM10 emissions (VROM 1998, p. 150), but extra reduction measures were not proposed.

In 1998, the policy statement on PM10 did not appear, nor did the plan of action on urban air quality. According to the Environmental Programme stating the actions for 2000 to 2003, these policy documents were to be combined into one big Air Quality

5. Kamerstukken II 1997-1998, 25 605, nr. 2, p. 102.

Memorandum.[6] However, this Air Quality Memorandum also met with delays. According to an internal memorandum (Project group EU Daughter Directives, Minutes meeting 23-09 1998, on file with the author), these delays were due to the fact that new scientific reports on PM10 from the RIVM were not available.

After the air-quality Daughter Directive was adopted in 1999, the uncertainties remained considerable – so much so in fact that because of a lack of insight into the causality between the components in Particulate Matter and health effects, it was considered impossible to formulate any cost-effective policy on PM10. The Environmental Programme 2002-2005 stated: In contrast *to other substances, according to the RIVM it is not possible to formulate any cost-effective reduction policy, due to lack of insight into the causal relation between the composition of PM10 and health effects'*.[7] The smog regulation was updated in 2001.

The promised memorandum on air quality was also not released. In 2001, political developments caught up with the Government, because the air-quality standards laid down in the Daughter Directive had to be implemented. That meant that in the explanatory memorandum issued with the law, and in this law itself, the new air quality policy had to be laid down. It was already known that achieving the standards would be very difficult, and the strategy of the Government was to wait for an evaluation of the Directive as had been promised in article 10 of the Daughter Directive, and if possible to renegotiate with the European Commission. The political line regarding the pollutant NO_2 was summarised as follows in the Environmental Programme 2002-2005:

> *'Until the evaluation, the Government takes the following line: even though the standards for* NO_2 *in the directive are binding, this does not imply that in all existing situations measures need to be taken immediately. The first plans need to be made by 2003, and they need to be executed in the years immediately following. Definitive decision making about taking expensive clean-up measures needs to be delayed as long as possible until the outcome of the evaluation of the Directive becomes available'*.[8]

The mandated reduction of PM10 concentrations was considered to be in all probability not even feasible: '*Concerning PM10, the situation is that concentrations in the whole of the Netherlands are so much above the limit values that it cannot be expected that they can be brought down enough to comply with them within a reasonable time frame'*.[9] This expectation was in line with RIVM's expectation. It noted: '*For PM10, the EU mandated daily limit values are also not yet feasible for 2005* (RIVM 2000, p. 124).

Before the implementation, the following rule of thumb was formulated: there would be no large-scale clean-up operations; for instance, housing in areas that were in

6. Kamerstukken II 1999-2000, 26 804, nr. 2, p. 133.
7. Kamerstukken II 2001-2002, 28 003 nr. 2, p. 13.
8. Kamerstukken II 2001-2002, 28 003 nr. 2, p. 58.
9. Kamerstukken II 2001-2002, 28 003 nr. 2, p. 56.

breach of the air quality standards would not be demolished. However, new projects would only be realised in locations that were considered to be non-sensitive – namely, areas in which people did not spend a significant amount of time. Pronk reasoned that this strategy would control the damage done to public health without making construction impossible.

Policy started moving in earnest only after 'Overschie' attracted attention in 1999. In Parliament Minister Pronk tried to raise awareness and proposed measures, among which a moratorium on building close to highways. He also opposed his colleague Tineke Netelenbos, Minister of V&W, on the issue of road construction and upon his instigation the speed limit near Overschie was lowered. Initially though his proposals were not heeded.

I conclude that with regard to air quality, the Government was caught between a rock and a hard place. National measures had little impact on actual air-quality concentrations, because of the strong influence of foreign sources and high background concentrations. Reduction policies were costly, and it was questionable as to whether they would be effective. Scientific data was lacking as well. Internally, PM10 and NO_2 were subjects of discussion, for instance, by the project group and by the Directorate General of the Environment at the VROM Ministry. These internal discussions were confirmed by internal memos, such as the one issued about the Action Plan Urban Air Quality and memos made by the project group EU Daughter Directives. Despite the discussions, however, an effective policy was not formulated in the 1990s. The Ministry bet on the evaluation of the Directive, and opted for a strategy that could be summarised as 'waiting for Europe'. In the meantime, it tried to contain the problems by formulating a pragmatic strategy.

However, the Government was also not well prepared, as those strategies were not laid down anywhere, and concrete proposals to tackle air quality did not emerge either. In fact, after leaving the hands of scientists and entering the political realm, the topic of air quality faded into the background. From a political perspective, the subject of air quality remained a strange issue, because politicians did not foresee it as having strong social ramifications. It was considered a European problem, and should be dealt with on that level.

5.2 Implementation of air-quality directives in 2001

The implementation of EU air-quality standards in Dutch law ended the political process that began in the 1990s, as a result of the emergence of new scientific information. In the absence of other policy documents, the new to be drafted Air Quality Order 2001 (Stb. 2001, 269). Hereafter AQO 2001, and its explanatory memorandum would be the Dutch answer to these findings and to the regulation imposed by the

EU. The Dutch Government was in a difficult situation at the time, because although scientific findings signalled a potential threat, there were few credible interventions to counter it.

Before looking at the transposition itself, this section is devoted to the place of environmental quality standards in Dutch environmental law. The implications of the implementation can only be fully understood when it is clear what the legal consequences of air-quality standards were. Moreover, in order to understand the function of those standards within the context of the Dutch system of environmental law, we will examine the question of when it was incumbent to set environmental quality standards in accordance with the Dutch policy philosophy.

5.2.1 *The link between environmental standards and spatial planning*

From early on, environmental quality standards played a role in Dutch environmental policy, even though specific air-quality standards had less impact. The use of environmental standards resulted in a far-reaching integration between environmental policy and spatial planning. As recounted in chapter 3, the law on noise nuisance resulted in the blueprint for the use of environmental quality standards in Dutch law. These standards had spatial implications, because in areas where standards were exceeded, no new activities could take place that would cause further pollution.

Air quality standards had the same legal implications as any other standards. In the Dutch Law on Environmental Management dating from 1993, the possibility of setting quality standards had been retained, and the distinction discussed earlier between target value and limit value had been made in this law as well. Target values only had to be taken into consideration, but limit values had to be observed (Art. 5.2.1 sub a, Law on Environmental Management [Stb. 1992, 551]). An environmental quality standard with the status of a limit value could not be exceeded, except in the case of 'force majeure'.[10] Those quality standards were mostly directed at lower administrative bodies, because they became linked to the granting of permits and to policy decisions regarding land use plans.

Environmental quality standards essentially 'sealed off' areas for further development, because no new polluting activities could take place. In Parliament, these consequences were realised when the status of quality standards was under discussion. It was considered: '*Such quality standards are the result of a weighing of what is desirable from an environmental perspective and what is technically, economically, financially or otherwise (for instance planologically) possible*'.[11]

10. Kamerstukken II 1988-1989, 21163 nr. 3, p. 47.
11. Kamerstukken II 1988-1989, 21163 nr. 3, p. 43.

It was felt that quality standards should not be based on scientific data that were not considered to be strong enough, and in the case of insufficient data, legal status should be withheld.[12] Article 5.1.1 of the Law on Environmental Management granted the administration the right to set environmental quality standards by way of a general administrative order, but only if the administration had properly weighed interests, including economic ones (Art. 5.1.2 sub E).

Legal scholars tended to take a hard line regarding the phrase 'observing the limit values'. Drupsteen and Koeman considered that a permit needed to be refused by the relevant administrative body if the environmental quality standard was likely to be exceeded. If the administrative body refused the legal obligation to observe the limit value, a decision could then be annulled by the administrative court because it had been taken unlawfully or because principles of good governance were at stake (Drupsteen & Koeman 1996, p. 77). Environmental quality standards bound administrative bodies as well when they acted on account of competences not based on environmental law, but on other laws. Michiels concluded that this included establishing legally binding plans involving land use (Michiels 1998, p. 135). This opinion is shared by Van der Geest and Lam (2006, p. 121), and these consequences made quality standards a highly invasive kind of regulation. In fact quality standards are an old instrument of environmental regulation. They were discussed predominantly in the 1970s, the era before ecological modernisation. The limit values should be viewed from the perspective of the limits to growth discourse, In the Dutch system they set limits on the amount of pollution an area may legally endure.

The link between quality standards and lower administrative decisions was an important and unique feature of Dutch environmental/spatial planning, and it featured strongly in debates on the air-quality clash. From this point, I will speak of 'the link' when I mean the link between spatial planning and environmental rules. Historically, the link is easy to account for, as it provided the much-desired integration of environmental considerations into other fields of policy. Lower administrative bodies as well as the Central Government had to apply environmental rules when taking spatial decisions, and quality standards strongly influenced spatial planning, the development of infrastructure, industrial areas, transport, and so on. The link was therefore an important instrument by which VROM was able to influence the development of the Netherlands. The corollary of this is that the EU quality standards had potentially more significant implications in the Netherlands than in other European countries.

5.2.2 *Implementation: the legislative procedure*

The European air-quality regulation was to be implemented in 2001, a pre-election year in the Netherlands. It was the last year in which the 'purple coalition' of the Social

12. Kamerstukken II 1988-1989, 21163 nr. 3, p. 44.

Democrat PvdA and the conservative VVD ruled the country, together with the social liberals of D66.

By this time, it was known that compliance with the standards laid down in the air-quality Daughter Directive was difficult for the Netherlands, and discussions arose in the Dutch Parliament about the pending implementation. It was still not a hotly debated topic by any means, and certainly was not on the minds of the electorate; it would, however, have severe consequences.

During a general discussion on 14 February 2001, Member of Parliament Klein Molenkamp of the VVD complained about the high costs that correct implementation would incur, and urged the Government to be cautious.[13] In fact, his own preference was to delay executing the Directive until it had been evaluated. The reasons for Molenkamp's suspicion had been generated by statements from Minister Pronk to the effect that implementation of the directive would be exceedingly expensive – the figure mentioned was around 30 billion Euros. Molenkamp wondered how it was possible that implementing the Directive would cost a thousand times more than had initially been thought.

Pronk agreed that even though the matter of air pollution was close to his heart, costs were very high indeed, and he claimed that the EU had provided false estimates. Partial implementation would have a more reasonable price tag, but would still amount to a hefty 2 to 6 billion Euros.

The Minister defended the EU directives, but made an important proviso: the directives contained obligations, but these had to be based on complete information, and they had to manageable. In this case, he considered that a firm result-oriented obligation was not feasible on account of the high costs. He pointed out that he had written a letter to EU Commissioner Margot Wallström (on 14 December 2000, TA), in which he complained about the deadlines and argued for a more lenient interpretation.[14] According to Minister Pronk, the first thing that should be done was to prevent the potential administrative problem from worsening. He asked Brussels to evaluate the Directive before 2003, and he would attempt to change it in order to 'better set priorities' and 'reformulate deadlines'.[15]

Molenkamp remained unconvinced, however, and argued that the air-quality Directive should be implemented not by a General Administrative Order but by a formal law. In Dutch Law, this distinction is important. General Administrative Orders and Ministerial Decrees have a lower status under Dutch law than what is known as a

13. Kamerstukken II 2000-2001, 27 400 XI nr. 56.

14. Letter from Minister Pronk to Commissioner Wallstrom briefnummer MJZ 20001499652.

15. Kamerstukken II 2000-2001, 27 400 XI nr. 56, p. 6.

formal law. According to the Law on Environmental Management, quality standards are stipulated by a General Administrative Order. An important point is that such a procedure is quicker, and Parliament is not required to give its consent to the regulation, but only has to be consulted.[16]

Minister Pronk objected to this procedure, and considered that issuing a General Administrative Order sufficed for the implementation. A 'legislative soap opera in 25 episodes' (Borman 2003, p. 244) started when the AQO was issued in June 2001, because Parliamentarians demanded that the Air Quality Directives be transposed by way of formal law. Their demand was backed up by an article in the Law of Environmental Management, stipulating that if a sizable minority of Members of Parliament demanded a formal law, the issue should be dealt with by using this instrument.

Pronk presented the AQO 2001 to Parliament by way of a letter to the chairmen of the first and second chambers on 7 June 2001.[17] One month later, a total of 38 Parliamentarians from the VVD signed an objection demanding that the implementation take place by means of a formal law.[18] The consequence was that it became necessary for the Government to withdraw the General Administrative Order.

In Pronk's view, the provision in the Law on Environmental Management could not be used to obstruct timely implementation of European law. He felt that Members of Parliament had the right to demand a formal law when the General Administrative Order concept was submitted, but not after submission of the Order itself.[19] He promised to draft a proposal for a formal law but refused to withdraw the AQO 2001. Eventually, Pronk was forced by Parliament to withdraw it, but the Dutch Council of State Advisory Division, the highest administrative advisory body, now objected to the withdrawal.

A year later, Pronk was succeeded by the Christian Democrat Secretary of State, Pieter van Geel, when a new Cabinet took over in July 2002, led by Prime Minister Jan Peter Balkenende (CDA). Van Geel was faced with the opinion of the Council of State that it was impossible to withdraw the General Administrative Order, because a withdrawal would mean the Netherlands would be in breach of the European air quality regulation. The Council of State considered that in the case of withdrawal, the Netherlands would have no air-quality regulation at all, and could be accused of infringement by the European Commission for failing to implement the Directives. He prepared a proposal to arrange the matter through a formal law, and emphasised that he would continue attempts to change the Directive through negotiation at the European level.[20]

16. For more on the different policy instruments see section 3.1.1.
17. Kamerstukken II 2000-2001, 27 793 nr. 1/309.
18. Kamerstukken II 2000-2001, 27 793 nr. 2.
19. Kamerstukken II 2000-2001, 27 793 nr. 3.
20. Kamerstukken II 2002-2003, 27 793 nr. 8.

The discussion about the correct form of implementation demonstrated that worries existed in Parliament regarding air-quality standards. Minister Pronk himself was convinced that the VVD Parliamentarians had only demanded a formal law in order to oppose it in Parliament and bring it down (Jan Pronk, interview). This conflicted with his own strategy of handling the potential problem with respect to air-quality standards. He intended to implement the Directive dutifully, and then renegotiate it when the consequences became dire. He felt that one should first create goodwill and then renegotiate. The Netherlands after all had argued for a more rigorous PM10 policy in Europe, so a withdrawal at that point would not be very convincing. Moreover, Pronk considered that a timely implementation of the Directive would strengthen the Dutch negotiation position where other files were concerned, such as climate change (Jan Pronk, interview).

5.2.3 *Strict implementation of the air-quality Directives into Dutch law*

Minister Pronk was obliged by Parliament in 2002 to withdraw the General Administrative Order, but was unable to comply, which meant that the AQO 2001 remained valid.

Because this regulation became the catalyst for the air quality clash to emerge, some knowledge of its provisions is necessary. In the next sub-section, the contents are interpreted in the light of the accompanying Explanatory Memorandum, but an overview of the articles is presented briefly here.

This AQO contained 33 articles, most of which followed the EU Directives in a fairly straightforward manner. Article 5 to 19 set out the limit values for SO_2, NO_2, PM10, lead, CO, and benzene, respectively.[21] However, only the limit values for NO_2 and PM10 were of importance for the air quality clash. For NO2, a limit value of 40 mug/m^3 as a yearly average was stipulated, and entered into force in 2010. For PM10, a yearly limit value of 40 mug/m^3 was stipulated, as well as a 24-hour average limit value of 50 mug/m^3. This second limit value could be exceeded at 35 days maximum. The PM10 standards entered into force in 2005.

The limit values laid down in the Daughter Directive were stricter than the ones that had already been incorporated into Dutch law. The AQO incorporated the new stricter EU values, as was mandatory under EU law.

Article 1 presented a list of definitions of terms used in the AQO, whose terminology was very similar to that in the Dutch Law on Environmental Management, which became clear from the first definition. The term 'law' in the context of the Air Quality Order 2001 referred to the Law on Environmental Management. The term 'limit value'

21. The substances CO and benzene are not covered by the EU Daughter Directive. They had already been regulated in Dutch law, and the AQO incorporated those older regulations for CO and benzene. They will not be dealt with further within the scope of this thesis.

referred to the term 'limit value' as defined in article 5.1.1 of the Law on Environmental Management. Articles detailing the various limit values for different pollutants used the same terminology as the Law on Environmental Management. Administrative bodies needed to 'observe' the limit values for SO_2, NO_2, PM10, lead, CO, and benzene. Target values were not mentioned in the AQO.

A law concerning quality standards would generally stipulate which administrative bodies were addressed by the standards, and to what kind of decisions the standards would apply. However, the Air Quality Order 2001 did not discriminate between administrative bodies. The articles laying down limit values simply mentioned 'administrative bodies', and did not stipulate to what kind of decisions the standards would apply. The article pertaining to PM10, article 13 for instance stated:

> *'Administrative bodies observe the limit values for Particulate Matter during the exercise of their competencies which may have an effect on air quality regarding PM10, except in the case of a contravening law'.*

The only condition was that the competencies exercised must have consequences for air quality. If competencies used by lower administrative bodies or the Government itself did not qualify in this regard, the limit values would then have to be observed.

As we saw in the section on quality standards in Dutch law, Dutch legal scholars tended to take a hard line when considering the legal consequences of limit values. Decisions regarding permits would need to be refused in situations where the quality standards had been exceeded, and where decisions regarding environmental planning or spatial planning were considered to be under the ambit of the quality standards as well. The law itself did not differentiate according to type of administrative body involved, so in principle these standards needed to be observed in equal measure by all administrative bodies.

The actual implementation of the EU air-quality standards in Dutch law was stricter than the Directive had prescribed. The EU Directive itself did not state that the standards needed to be observed by every administrative body during each decision, which could have some repercussions for air quality. The Directive only mandated that plans were to be drawn up whenever some agglomeration did not comply with the limit values, and that Member States must make sure air quality was sufficiently high on their territory. This left leeway for national Governments with respect to how they would reach the limit values. The Dutch interpretation of linking the limit values to individual administrative decisions was not necessary, and indeed unique according to Fleurke and Koeman (Fleurke & Koeman 2004; Koeman 2006, but contra Backes 2006). In any case, the Dutch interpretation was certainly one of the strictest in Europe (Backes 2006a; Backes et al. 2005). This strict transposition is difficult to explain, because it was already known that the Netherlands would face severe difficulties in meeting the standards.

5.2.4 The Air Quality Order according to the Explanatory Memorandum

Without the promised Memorandum on Air Quality, or the Action Plan Urban Air Quality, the AQO 2001 and the Explanatory Memorandum accompanying it contained the Dutch national policy on air quality. In the AQO, a number of interpretative issues were at stake, and had to be cleared up in the Explanatory Memorandum. The first issue concerned how strictly the limit values had to be interpreted, and what consequences a breach of the standards should have. The second issue concerned provisions in the AQO in the light of the evaluation clause in article 10 of the Directive. The third issue had to do with the question of responsibility. Which administrative bodies would be responsible for meeting the limit values? These three topics all re-surfaced when PM10 became a full-fledged social problem in late 2004.

Consequences of breaching limit values

The wording of the Explanatory Memorandum (E.M.) was ambiguous in terms of how strictly the limit values had to be interpreted, and what consequences a breach would have, but it was strict in certain paragraphs. After first considering that the limit values should not be breached, owing both to health and legal concerns, the E.M. speaks of limit values as 'result-oriented obligations' (Stb. 2001, 269, p. 22). A result-oriented obligation is defined by its result; namely, if a limit value is exceeded when it is in force; the administration is in breach of a hard norm. This wording is reminiscent of the hard line that legal scholars tended to take. Moreover, administrative bodies needed to anticipate the coming limit values, and avoid future situations in which these could be exceeded. That implied that they had to already be aware of the values, and observe them in decisions they would take that potentially had an impact on air quality in the future (Stb. 2001, 269, p. 22).

In regard to those result-oriented obligations, the E.M. stated:

> *'Limit values are binding. They have legal consequences for administrative bodies. They limit the competencies their bodies have in the sense that activities need to be refused or adjusted if they threaten to lead to a transgression of the limit values for quality'* (Stb. 2001, 269, p. 25).

Regarding spatial and transport plans, the memorandum demanded an explicit assessment of compatibility with the air-quality limit values (Stb. 2001, 269, p. 27).

Nevertheless, certain other quotes cast doubts on the intended consequences of breaching a limit value, and on whether every decision that could have some repercussions for air quality was considered to be subject to those limit values. The following quote, for instance, indicated there could be some leeway:

> *'Limit values need to be taken into account when deciding whether intentions can be freely executed or whether additional provisions are necessary regarding sources or spatial planning'* (Stb. 2001, 269, p. 26). On the same page, we read: *'If air quality is burdened to such a degree that a*

> *considerable contribution is delivered to exceeding the limit values, then that particular activity cannot be condoned in this area or in this way'* (Stb. 2001, 269, p. 26).

These statements indicated that limit values should only be seen as a consideration for policy. 'Additional measures' could be necessary, but a full prohibition of the activity in question was not mentioned. Only if it delivered a 'considerable contribution' should an activity be forbidden. Elsewhere, we find a similar consideration. On page 23, it was stated that only *'tasks or competencies by which certain meaningful influence can be exercised on air quality'* (Stb. 2001, 269, p. 23) were considered. This implies that only those competencies that had important consequences for air quality were targeted.

The vagueness of these statements was problematic. Who would decide what comprised a 'meaningful' influence or a 'substantial' contribution? It is clear that the Government did not intend for quality standards to become a rigid instrument that would lock the country in terms of further spatial and infrastructural development.

Evaluation in 2003

As regards the troubling NO_2 standard, the E.M. noted that certain breaches might remain when the standards took effect. However, also in other Member States, some breaches were expected to remain present. According to the E.M., this standard would probably be revised on the occasion of the evaluation in 2003.

For PM10, it was uncertain whether those standards could be met at all in the Netherlands. They would also be evaluated based upon article 10 of the Daughter Directive, however, and the Government was confident that the assessment would lead to their relaxation, as it had noted that many Member States had problems with the PM10 standards: *'During the evaluation this (information TA) will undoubtedly lead to important changes in the norm setting (relaxation)'* (Stb. 2001, 269, p. 17).

The evaluation was a key topic in the E.M. The Government trusted that the standards would be relaxed, and was not concerned about the vagueness of the terminology. As a result, discussion on the legal consequences of the regulation was lacking.

The administrative bodies responsible for PM10 policy

The Explanatory Memorandum explicitly assigned the responsibility for PM policy to the state. Lower administrative bodies could not be held accountable for breaches of the PM10 standards, owing to the many uncertainties about the pollutant, and especially because the standards were exceeded in many parts of the country. However, they did have a responsibility to help reduce the PM10 problem. The imperative to 'observe the limit values' meant that lower administrative bodies should try to reduce PM10 emissions as much as possible. Even at state level, however, a concrete policy could not yet be put into place. According to the E.M., there were still so many scientific uncertainties regarding PM10 that until the results of further inquiry were known, policy would continue to be based on reducing exposure to PM10 as much as possible by applying state-of-the-art technology and ALARA (As Low As Reasonably Achievable). In practice, this meant that the best available technologies had to be used to curb PM10 exposure.

Although it was implied in the E.M. that responsibility for PM policy had been lifted from lower administration bodies to a significant extent, it also specified for what kind of decisions and competencies this memorandum had consequences. These competencies applied to lower administrative bodies as well. Those categories of competencies were listed on page 23 of the E.M. (Stb. 2001, 269, p. 23). The list was long, and consisted of decisions and competencies following from the Air Pollution Law, the Law on Environmental Management, and the Law on Spatial Planning, as well as the Road Infrastructure Law (Tracéwet in Dutch) and the Law on Planning Transport and Traffic ('Planwet Verkeer en Vervoer' in Dutch). The link between quality standards and individual decisions of administrative bodies was clearly kept, and featured prominently in the many debates that followed after 2004/2005.

On 17 April 2001, the Council of State offered its recommendations regarding the proposed AQO. The Council was not overly critical, and its advice was positive, apart from a few minor details. It did not indicate that the AQO might have grave legal consequences for spatial planning in the future, or that the views put forward in the E.M. were somehow deficient. The AQO was published in June of the same year (Stb. 2001, 269).

5.2.5 *Early discussions on the legal character of the Air Quality Order*

As far as legal problems were concerned, there were discussions in Parliament about the possibility of infringement procedures by the European Commission.[22] There was no indication that national legal problems were foreseen at the time the AQO was adopted. Considerations regarding national legal problems started when – owing to findings regarding air quality – the Council of State Administrative Jurisdiction Division in May 2002 annulled a decision to allow the development of a residential area near a highway in Ypenburg. The administration noted that this case made clear that research into the effects of infrastructure development on air quality was an important aspect in decision making.[23] Jan Pronk initiated discussions about the necessity of keeping a safe distance from the highway when building residential areas (Pronk, interview). In the concept involving a new memorandum on spatial planning, Minister Pronk proposed building residential areas at a certain distance from the highway.

22. Kamerstukken II 2002-2003, 28 663 nr. 2, p. 23.
23. Kamerstukken II 2001-2002, 28 089 nr. 3, p. 15.

However, still, the Ypenburg issue did not lead to an immediate awareness that the legal consequences of the standards might be highly damaging to Dutch spatial planning. The implicit understanding among policy makers continued to be that things would turn out fine, and that there would be no grave consequences (Priemus 2006a, p. 6). The Minister stuck to his line of trying to fulfil the terms of the Directives and the consequent AQO as well as possible, even if it meant that new residential areas had to be constructed some distance away from roads. He considered that this situation would be temporary, because he trusted in the possibilities of renegotiating with Commissioner Wallström.

Jan Pronk successor in 2002, Pieter van Geel, was unable to persuade the European Commission to accede to the Dutch point of view. Hence, the Government was still unprepared when the legal ramifications became acute in 2004.

5.3 Concluding remarks

Despite opposition in Parliament the Air Quality Directives were implemented strictly in Dutch law. The main question is why such a strict line has been chosen. The first answer is that these strict consequences were not intended. The resulting legal block on infrastructural development was neither intended nor foreseen. However, even with the knowledge available at the time, I consider the strict implementation to have been risky. In the following section I present two explanations for the strict implementation, one has to do with strategic political choices and the other with a pervading mentality resulting from the eco modernist consensus. To conclude, I present an analysis of the implementation of the Air Quality Directives in terms of the legality of precaution.

5.3.1 *Strategic choices by Jan Pronk*

Initially the Government failed to formulate a consistent stance with respect to the air-quality issue in the 1990s, despite knowledge of the emerging scientific data and the forthcoming European air quality standards. In Europe it supported strict standards thereby legally binding itself, but it was unable to comply with them, as there was no cost-effective national policy that would make these standards reachable. Little attention was paid to this fact, because air quality was not yet a worrying public issue. The few policy documents devoted to the topic endorsed a strategy of 'waiting for Europe'. Because the air pollution problem could not be solved nationally, owing to the lack of a cost-effective strategy, international legislation was considered the only option. The national consequences of this regulation attracted scant attention.

On the issue on air pollution, Minister Pronk chose a tactic of damage control. When he became Minister for VROM in 1999 he was faced with a 'fait accompli', as the standards had already been agreed upon. He was personally committed to the air pollution issue, however, and when a local health scare erupted over severe air pollution in Overschie, he requested further research and made possible a reduction in the speed limit, which was considered to help limit the emission of PM10 and NO_2. The European air quality rules were in any case good news for the people from Overschie. From a health perspective Minister Pronk was pleased with the regulation. The health problems posed by traffic were a matter of concern to him and he supports the application of the precautionary principle. In the absence of further rulings, the implementation by way of a General Administrative Order and the Explanatory Memorandum constituted the whole of Dutch policy relating to air quality.

However budget wise the air quality rules were problematic. Pronk's strategy was to transpose the Directive dutifully into Dutch law, and to seek assistance from the European Commission in order to ensure practical implementation. In the meantime, he tried to curb the impacts of the air-quality standards by demanding that any new infrastructure be implemented some distance from motorways. To this end, he made a division between locations where people spent a considerable amount of time and where they did not. Even if standards were exceeded, construction would be able to take place in areas that people did not frequent. Pronk reasoned that public health was not endangered in those places, and was convinced that his tactic of swaying the Commission might have worked (J.P. Pronk, interview), but in 2002 he was replaced by Secretary of State Pieter van Geel from the Christian Democratic Appeal (CDA) party. Van Geel inherited the potential political problem regarding air quality, but did not manage to convince the Commission to acknowledge the difficulties facing the Netherlands.

The eventual result was that the Netherlands ended up with a strict implementation of the air- quality regulation due to the Dutch system of environmental zoning. In the Dutch legal system, quality standards created 'no-go areas' for further activities when standards were transgressed in an area. Local authorities needed to 'observe' the standards, which meant that they had to cancel permits and adapt spatial development plans when the standards were exceeded locally.

The strictness of the Air Quality Order itself should be contrasted to the wording of the Explanatory Memorandum, which sets out the reasons for and implications of these rules. A discrepancy soon becomes apparent. In many paragraphs in the memorandum, we find indications that this rigour was not intended. In the case of PM10, for instance, it states that due to the magnitude of the problem, local administrative bodies should not be held accountable. Moreover, the evaluation clause in the Daughter Directive is mentioned, and it was strongly considered that upon assessment, the standards would be adjusted downwards; a clamp-down on infrastructural development was never intended. Moreover, scientific uncertainty was used as an argument to demand less from lower administrative bodies, and to blunt the sharp edges of the standards relating to PM.

When one looks at the wording of the AQO itself, it is clear that the subsequent Directive was implemented more rigorously than necessary, however. The text of the law does not include the exceptions and considerations made in the Explanatory Memorandum.

The main reason that Minister Pronk chose a literal and strict interpretation of the Directive was that he considered implementation of the air-quality Directive in the light of a more encompassing Dutch approach to EU environmental policy. A sound implementation of the European rules offered the Dutch Government leeway to negotiate on other terrains, such as climate change and other environmental fields (J.P. Pronk, interview). These considerations explain why the wording of the AQO was strict, but the E.M. allowed for a great deal more leniency. The transposition was never intended to be inflexible, but political concerns led to a rigorous approach.

5.3.2 *The conviction that unworkable standards are not legitimate standards*

The eventual implementation of the air-quality Directives had important legal consequences, but by and large, these were disregarded by the Government. Not enough thought had been given to the way air-quality standards would fit within the national system of environmental law in which quality standards would have potentially far-reaching consequences. I consider that the eco-modernistic tradition of Dutch policy making was conducive to a disregard of legal aspects of the implementation.

In the Netherlands, quality standards played an important role in environmental policy, because in the Dutch system of environmental law, these standards sealed off an area whenever pollution concentrations were exceeded. This system ensured that environmental standards had important consequences for the country's spatial development. However, such standards were to be used prudently, keeping in mind the area under consideration. The Dutch implementation of European air-quality rules did not fit this format, since they did not allow for the required flexibility. The result was that countrywide pollution standards were set that were impossible to achieve within the given time frame. It is odd that policy makers did not take into account much earlier the legal consequences of using the Dutch system of environmental standards to implement the strict European standards. Subsequent Cabinets also considered that air quality was not a real problem and worthy of close examination, because there was little information about it, and the Dutch Government had been unable to formulate a cost-effective policy. The problem was regarded as a European one, until the Overschie issue gave rise to stirrings of alarm. However, even then the legal issues were not foreseen.

This illustrates that in the environmental field, the administration was not highly concerned with the legal consequences relating to air-quality standards. Legal issues arose with the Ypenburg case, but this also did not cause widespread alarm.

Apparently air-pollution policy was still considered reasonably harmless, which is puzzling as well. The Netherlands is a forerunner in regard to lawsuits filed by environmental organisations, and the onset of rigid air-quality standards provided ample opportunity for more of such lawsuits. Noticing this fact amounts of course to hindsight bias, but at the time, certain politicians – for instance, Wijnand Duyvendak – also recognised those opportunities (J. Wijnhoven, interview).

The answer to this puzzle may be found if one takes into account the mentality that years of eco-modernistic policy making had fostered among Dutch politicians. Ecological modernisation was a type of environmental policy that relied on consensus making, flexibility, and sound scientific data. It was a pragmatic way of dealing with environmental problems, and not one that concerned principles. One of the maxims of this pragmatic consensus-oriented policy involved not asking for unreasonable sacrifices from anyone. In the words of Winsemius: 'One cannot demand the impossible'. In the Dutch context, ecological modernisation was achieved through the internalisation of environmental norms and principles, and this internalisation had to be brought about by presenting environmental protection as an opportunity for business as well.

From the Dutch eco-modernistic perspective, the European standards were not reasonable, as they were not based on conclusive scientific data; the feasibility of the standards had not been taken sufficiently into account; and their implementation had turned out to be much more expensive than was previously considered. Therefore, the Dutch Government was convinced that the standards would be revised upon evaluation and concluded such in the Explanatory Memorandum on the AQO. The Government also concluded that the burden of responsibility should not be placed at the level of lower administrative bodies, but at that of the state. However, the wording of the AQO itself did not in any way attempt to circumvent the legal consequences following from its close connection to the terminology used in the Law on Environmental Management.

The Government disregarded the legal dimension because of an unspoken consensus that policy should only be complied with strictly when it does not have undesirable economic consequences. The pragmatic side of eco-modernisation demanded that economic and environmental progress proceed hand in hand. If environmental policy was economically detrimental, it was not sound, and should not be applied rigorously. This eco-modernist principle lies at the heart of the pragmatic way in which the Government tried to deal with consequences of the AQO.2001 Here, however, we witness a clash of rationalities. From a political perspective, this consideration might be reasonable, but from a legal standpoint, the conclusion does not follow. The Netherlands bound itself to the air-quality standards, which had the force of law, irrespective of whether the administration considered them reasonable. The pragmatic rationality of eco-modernisation clashed with the dogmatic rationality of law. At the time the Air Quality Order was implemented in 2001, politicians did not consider this legal dimension, and that oversight would become costly in the years to follow.

5.3.3 *Dutch air quality policy and precautionary legality*

Despite the reputation of the Netherlands as a forerunner in EU environmental policy in the 1990s, the formulation of national air quality policy testified of a different attitude. Even after the epidemiological findings became known in the middle 1990s, Dutch air quality did not change much. A cautious strategy of 'waiting for Europe' was chosen, mainly because no cost-effective policy could be formulated. Bad air quality was not considered a social problem and in the absence of cost-effective solutions, political intervention was considered undesirable. Such considerations remain squarely within the boundaries of the legality of risk and compensation. Health damage should be avoided, but not at all costs and economic development should not be halted.

Only after Jan Pronk became Minister in 1998 and the problems in Overschie became known, did policy change in a more precautionary direction. Pronk visited the area to speak with residents and was sensitive to the comparison between living in Overschie and passive smoking. He proposed measures that can be considered precautionary, such as not building near the sides of the roads.

The most precautionary piece of legislation from this era is undoubtedly the strict AQO2001. However, the Government envisaged that the provisions in this piece of legislation would be dealt with pragmatically. To Pronk as well as his successors it was clear that complying with the Directives in full would not be feasible budget wise and pragmatic administrative practices were devised to deal with the sharp edges of the AQO. His letter to Wallström complaining of the feasibility of the Directives for the Netherlands displayed the same pragmatic concerns.

Nonetheless the relative ease with which these standards could be implemented in the Netherlands and the strictness of the implementation were signs that precautionary legality was to an extent gaining ground. There was no concerted lobby against the provisions of the AQO in any case, at least not one strong enough to force the Government to rescind.

Overall though, Dutch policy concerning air quality fits better within a legality of risk and compensation, rather than precaution, despite the Government formal adherence to the precautionary principle and despite the Dutch reputation as an environmental front runner.

TIMELINE EARLY DUTCH AIR QUALITY POLICY

1983	IMP Air 1984-1988 released
1986	Change in Dutch air pollution law makes setting legally binding air quality standards possible
1986	Implementation Air Quality Directive 80/779/ EC in Dutch Law; air quality standards for SO_2 and Particulates
1989	NMP released
1994	New Criteria Document on PM10 released
1995	Government issued preliminary policy opinion on Particulate Matter and winter smog' (interim beleidsstandpunt fijnstof en wintersmog
1998	Jan Pronk Minister of VROM
1999	Directive 99/30/EC adopted on ambient air quality
2001	Air Quality Order 2001 proclaimed, together with Explanatory Memorandum
2002	Jan Peter Balkenende is installed as Prime Minister, Pieter van Geel Secretary of State for the environment, under Minister Dekker of VROM.

6 THE RISE OF THE PRO-HEALTH CAMP IN THE NETHERLANDS 2001 – 2005

INTRODUCTION

During his time at the Ministry and especially after 'Overschie' in 1999, Jan Pronk demanded that more political attention be paid to air quality. From the point of view of public health, he was keen to reduce concentrations of polluted air, and initiated a policy that forbade the building of new infrastructure in places where the quality standards were being exceeded, and where people might spend a significant amount of time. In Europe though, he tried to gain more time for the Netherlands to achieve the standards, because he knew that full compliance would not be feasible. In 2002, however, Dutch politics was about to take a turn that would greatly affect environmental policy.

The Dutch public was shocked when Pim Fortuyn, one of the main contenders for the Dutch premiership, was assassinated in May 2002, just before the elections. The perpetrator was an environmental activist. In the instantly polarised political landscape, 'the left' was blamed for creating an environment in which the populist politician Fortuyn could be shot. With their criticism of Fortuyn's supposedly extremist ideas, left-wing political parties were accused of having elicited the politically motivated murder.

Conservative and populist parties went on to win a substantial victory in the 2002 elections, and environmental issues were relegated in political importance. The portfolio was granted to Christian Democrat Pieter van Geel as Secretary of State, a junior position in the Dutch Cabinet of Ministers, and Jan Peter Balkenende from the Christian Democrat CDA became Prime Minister. During the Balkenende Cabinets, the air-quality clash erupted in full. I will argue in this chapter that this clash has been a response to the environmental and mobility policies of these Cabinets.

The mobilisation of a pro-health camp – a set of actors that opposed the policies of parties that favoured infrastructure interests – inside and outside of Parliament is the main subject of this chapter.

We will look first at the environmental and mobility policies of the first Balkenende Cabinet, as they laid the groundwork for the environmental policy of his later

Cabinets.[1] Subsequently, two emblematic political issues will be considered: namely, health near highways and the matter of traffic congestion. The first problem caused a debate over the speed limit in the Netherlands, and critics argued that in light of the health problems, the speed limit should be lowered. The problem of congestion meanwhile forced the Government to propose an Emergency Law to expand Dutch highways. In the third section of this chapter, three seminal court cases that became crucial in the eruption of the air-quality clash are discussed, and finally the mobilisation of a pro-health camp inside and outside of Parliament, is recounted. This camp opposed the mobility policies, and used health-based arguments to demand an increase in clean air measures.

6.1 The first Balkenende Cabinets

A bookish Christian Democratic politician, Jan Peter Balkenende became Prime Minister after the resignation in 2002 of the Purple Cabinet, which consisted of the liberal VVD, the socialist PvdA, and the social-liberal D66. The biggest left-wing party, the PvdA, lost heavily during the 2002 elections, and the parties that formed the new Cabinet were all situated on the right of the political spectrum (Hippe et al. 2004). The VVD was the only party from the old Cabinet that retained its position, and it governed together with the CDA, the main Christian Democratic party. The third partner in this three-party coalition was the LPF ('List Pim Fortuyn'), the party founded by Fortuyn before his assassination. After his death the party was strongly favoured by voters. The party can best be described as populist, as it argued for law and order, lower taxes, and less congestion on the roads.

The hapless Cabinet lasted only 86 days in power before it collapsed due to infighting within the LPF. It is an important Cabinet to discuss, however, because in this short time, important shifts in environmental policy and transport policy took place.

Box 5: Cabinets, Ministers, and Secretaries of State in the 2000s

Between 2000 and 2008, five different Cabinets ruled the county. Until July 2002, the Purple Cabinet between the PvdA, D66, and the VVD was in power. Jan Pronk (PvdA) was the VROM Minister, from 1998 until 2002 and dealt with the environment. Tineke Neetelenbos (PvdA) was Minister of Transport and Water Management. The Kok Cabinets ruled during the 1990s, but weariness set in during the first years of the new decade. Moreover, turbulent political and historical occurrences, such as Pim Fortuyn's rise and assassination, along with events of '9-11', destabilised the coalition as well.

1. Jan Peter Balkenende would go on to lead four Cabinets in total, and was Prime Minister from July 2002 until October 2010.

The political landscape changed in 2002 when Balkenende took over as Prime Minister. A conservative Cabinet was formed between the CDA, the VVD, and the LPF. The environment was relegated to a State Secretariat, and was granted to Pieter van Geel (CDA). Roelf de Boer (LPF) became Minister for Transport and Water Management (V&W). In 2003, this Cabinet resigned, and was succeeded by the second Balkenende Cabinet, which consisted of the CDA, the VVD, and D66. Pieter van Geel retained his post, and Carla Peijs (CDA) became Minister for Transport and Water Management. The Minister of VROM during this Cabinet was Sybilla Dekker from the VVD, but she did not have the environment in her portfolio. Balkenende 2 lasted until the middle of 2006, when it was forced to resign. Cabinet Balkenende 3 was quickly formed, but was in fact a minority Cabinet mandated to propose new elections and to prepare a budget for 2007. Carla Peijs retained her post. Dekker was forced to resign and from September 2006 to February 2007, Pieter Winsemius acted again as interim Minister of VROM. Van Geel, however, remained the Secretary of State, responsible for environmental matters.
The elections in 2007 brought the PvdA back into power, to govern the country together with Christian Democrats of the CDA. The environment was reinstated as a Ministerial topic, and Jacqueline Cramer headed the VROM Ministry. Camiel Eurlings of the CDA became the new Minister of V&W.

6.1.1 *Environmental policies of the Balkenende 1 Cabinet*

The formation of Balkenende's first Cabinet ran smoothly, and in July 2002 the new Cabinet took the helm, laying down its plans for the forthcoming period in what was called the 'Strategic Accord'. In this relatively short document, the paragraph regarding the environment did not bring much new to the table, but its tone revealed that environmental rules and regulations were not looked upon favourably. The Cabinet sought to decentralise environmental policy, and argued that economic development should not be constrained. A sentence like '*To keep the countryside alive, one should not lock it down (to economic development, TA)*'[2] indicated that the emphasis was on development and not on environmental protection. The Cabinet also stated: '*The ambitions and instruments from the Fourth National Environmental Policy Plan will be adapted to financial possibilities*'.[3] The fourth national environmental action plan was a document drafted by the previous Minister, Jan Pronk, and contained plans for a transition towards a green economy. The sentence quoted above indicated that budget cuts would be considered, and that those ambitions would be shelved for the time being.

2. Kamerstukken II 2001-2002, 28 375 nr. 5, p. 21. Because of the large amount of citations of Parliamentary documents in this chapter, footnotes have been chosen rather than in text referencing to refer to these documents, the same applies to court cases.
3. Kamerstukken II 2001-2002, 28 375 nr. 5, p. 22

The new Cabinet indeed amended the fourth National Environmental Policy Plan, in the form of a memorandum entitled 'Vaste Waarden Nieuwe Vormen' ('Steady Values, New Forms'). The memorandum intended to *'set different priorities'*, *'use existing means in a more focused way'*, and *'indicate where short-term goals need to be "temporised"* (VROM 2002, p. 3). 'Temporisation' was a euphemism for postponement, and many environmental targets were postponed in this way, including targets for air quality. The Cabinet sought to delay until 2015 the deadlines for standards, and it also abolished subsidies for cleaner cars (VROM 2002, p. 35).

The consequences of this postponement were considered in the memorandum, but only infringement procedures of the Commission against the Netherlands for breaching the Air Quality Directives were taken into account (VROM 2002, p. 27). In Europe, the Cabinet intended to lobby in Brussels to have the targets for air quality deferred from 2010 to 2015.[4]

The new line of the Balkenende 1 Cabinet regarding environmental policy constituted the breach of a trend. In the years prior to 2002, the funding for environmental protection had risen steadily, but cuts were now being considered. A calculation made in

Figure 2: Yearly environmental spending by the Government, 1985-2006. Dotted lines indicate the budget of the second Purple Cabinet (2005 is considerd the baseline year), Balkenende 1 and Balkenende 2.

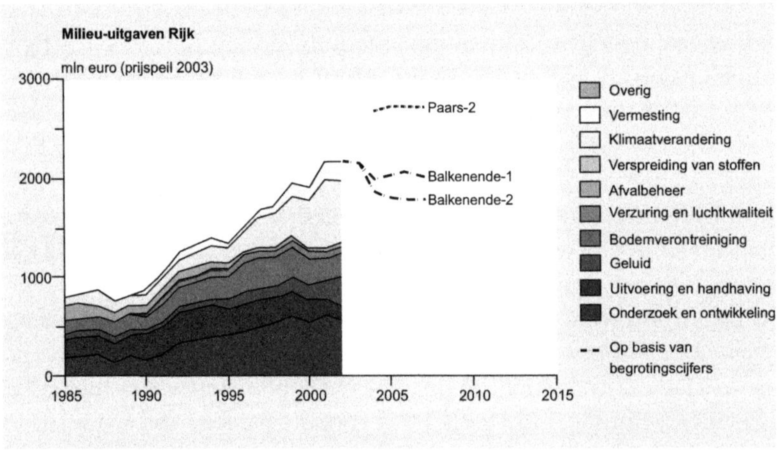

Figure taken from RIVM, Milieu en natuureffecten hoofdlijnenakkoord Balkende II, 2003, p. 8

4. Interestingly, the PM10 standards would enter into force in 2005, but that date is rarely mentioned. The Cabinet seemed most worried by the standard for NO_2 which would indeed enter into force in 2010. The reason why 2005 is seldom mentioned is not clear to me, but I speculate that it is because the Cabinet considered that PM10 policy was defined as a national responsibility in the explanatory memorandum to the AQO and not a burden on lower administrative bodies.

2003 displayed graphically the differences between the environmental expenditures of the Kok Cabinets and Balkenende's first and second Cabinets (RIVM 2003, p. 8). The dotted line in the upper right-hand corner indicates the projected expenditure of the Kok Cabinet, while the lines below represent the foreseen spending of the Balkenende 1 Cabinet.

6.1.2 *Mobility policies of the Balkenende 1 Cabinet*

In contrast to its cutting back of environmental expenditures, the first Balkenende Cabinet planned to accommodate the growth of auto-mobility. Just before the elections, the pro-mobility party VVD torpedoed the new transport and mobility plan – the National Traffic and Transport Plan (NVVP) – as was detailed in chapter 3. In particular, the VVD opposed road pricing, because it was not popular with the party's motorised electorate.

In this National Traffic and Transport Plan, the Kok Cabinet was already adopting a more 'laissez faire' approach towards accommodating the growth of auto-mobility than in its predecessor, the SVV2. Large investments in new roads were scheduled, and in exchange for road expansion, the motorist would be charged for the use of the roads instead of paying a fixed price in road taxes.

From the political perspective, road pricing was a highly controversial dossier in Dutch politics, and in his dissertation at Tilburg University, the debates were documented extensively by Maarten Smaal (Smaal 2012). Concern for the environment was the main reason the Government included road pricing in the SVV2. The VVD and the motorists' association ANWB were avowed opponents of the scheme when the idea was first launched in the late 1980s.

The conservative Cabinet of 2002 aimed squarely at appeasing the transport sector and the individual motorist. A levy of 25 cents per litre on fuel introduced by the Kok coalition would be abolished and various high-profile road-building projects were proposed, such as construction of the A4 Midden Delfland. Cabinet also planned a second 'Coen' tunnel, the 'Westrand weg', and the A9, according to the Transport Ministry's budget for infrastructure.[5] The road-pricing scheme by way of a kilometre tariff was dismantled (VROM 2002, p. 35), though debates about road pricing reappeared in 2003 and after 2005, when the air quality clash erupted.

However, the new Cabinet could not base its proposals for road expansion on a transport and mobility plan, and therefore there was no solid basis for expansion and construction. The Cabinet devised a solution to make road expansion possible by way of an Emergency Law on Road Expansion, which would make it possible to expand roads and shorten procedures. Together with discussions on the speed limit, the Emergency Law on Road Expansion was one of the two salient political issues in

5. Kamerstukken II 2002-2003, 28 600 A nr. 2.

which the interests of mobility clashed with those of the environment and public health, as discussed below.

The new Cabinet's policies were the culmination of a trend in mobility policy from the 1990s onwards. With slogans like 'gateway to Europe' and 'Netherlands, distribution country', the Dutch Cabinet sought to position the Netherlands as a transport country in the 1990s. The new Cabinet favoured policies of growth over environmental policies, and the emphasis on transport and mobility was a natural consequence. In the Netherlands, the transport sector wielded considerable economic power, and mobility was considered the lifeblood of the economy (Brokking 2001).

The environmental and mobility policies of Balkenende1 laid the foundations for policy in these areas that remained largely unaltered during at least the first three Cabinets led by Balkenende. Certain changes did occur, but, as I will argue, they were due mainly to the eruption of the air-quality clash.

6.2 Two salient political issues: health near highways and traffic congestion

The Government's mobility and environmental policies had significant repercussions for two problems which predated the Balkenende Cabinet, but for which it sought a solution, namely the emerging awareness that air quality near roads was bad and road congestion. It is necessary to dwell on these two problems and the Government's solutions for them at length because they have led to the formation of a pro-health camp in Parliament, a set of political parties voicing concerns over Balkenende's mobility and infrastructure policies.

6.2.1 *Highways and health: the debates on speed limits*

As mentioned previously, discussions about the conflict between mobility and health in the context of PM10 in the Netherlands occurred first in the area of Overschie. Epidemiologists pointed out that people living near busy roads could be vulnerable to negative effects on their health. In Overschie, the local health service (GGD) and residents managed to have their voices heard, and to involve influential politicians in their struggle for acceptable air quality.

Following Minister Pronk's visit to the Overschie area in February 1999, the ecologically minded party GroenLinks was the first to pick up on the issue, almost a year later. After publication of the 'Milieubalans 2000' (State of the Environment 2000), (MNP 2000) in September 2000, GroenLinks Parliamentarians complained during a Parliamentary discussion with the Minister of V&W in October stating that no concrete measure had yet been taken to address the issue of poor air quality in the Overschie area. They arranged an excursion to Overschie, assuring the Minister that they had seen with their own eyes how cars drove straight through that residential area at a speed of 120 km/h.[6] On 16 November 2000, the Transport Minister Netelenbos responded in writing to questions posed by the GroenLinks faction about lowering the speed limit, stating that only in exceptional situations would the Ministry consider lowering the speed limit.[7] From then on, the speed limit became a bone of contention.

In April 2001, the Minister acceded to the request for a speed limit of 80 kilometres for a tract of highway near Overschie, because of the involvement of her colleague Jan Pronk. She promised a package of local measures, and announced that the first calculations had indicated that reducing the speed limit might have positive effects.[8] She agreed to lower the speed limit in 2002 in the vicinity of Overschie, and presented the reduction as an experiment.

Parliamentary discussion led to increased awareness of potential air-quality bottlenecks, referred to as 'Overschie situations'.[9] PvdA Parliamentarian Jeroen Dijsselbloem introduced a motion that led to a report being submitted in May 2002, entitled 'Knelpunten leefomgeving op het Rijkswegennet' (Liveability Bottlenecks on the Dutch Road Network).[10] The report provided an overview of the number of road trajectories where the air quality was poor.

The experiment to lower the speed limit was unique in the context of Dutch mobility policy. The Transport Ministry generally insisted on at least 100 km/h on highways (Ecorys kolpron 2002, p. 57), but the test involving speed reduction commenced on 11 May 2002.

In July 2002, Balkenende took over as Prime Minister, and he inherited this potentially polarising discussion on highways and health. Within his own Cabinet, two governing parties, the VVD and the LPF, were not keen on speed reduction, and in fact wanted to raise the speed limit. The LPF in particular was not sensitive to environmental issues; in Parliament, it was the left-wing parties that had been the most vocal in that regard.

As could be expected, the successors of Netelenbos, the new Ministers first Roelf De Boer (LPF) and later Carla Peijs (CDA), respectively, were more cautious than she was on the issue of lowering speed limits. Carla Peijs took over in 2003, after the collapse of the short-lived Balkenende 1 Cabinet. In principle, Peijs was not against reducing the limits on other roads, but insisted that they deliver substantial contributions to air quality, but also that they should not create any further bottlenecks for traffic. She

6. Handelingen II 2000-2001 nr. 13, p. 889.
7. Handelingen II 2000-2001 nr. 24, p. 1980.
8. Kamerstukken II 2000-2001, 27 408 nr. 26.
9. Kamerstukken II 2000-2001, 27 408 nr. 21, p. 48.
10. Kamerstukken II 2001-2002, 28 000 A, nr. 29.

planned to lower speed limits only if they did not have a negative effect on traffic congestion, and if they contributed to an even flow of traffic.[11]

Meanwhile, epidemiological studies that pointed to a connection between premature mortality and living close to highways continued to appear.[12] The newspaper Trouw combined the figures presented in the research, together with the number of houses situated close to highways, and asserted that 100,000 of these were in unsafe areas, and that the 230,000 occupants of those houses ran serious health risks (Trouw 05-08 2003). Secretary of State Pieter van Geel had to inform Parliament about air-quality bottlenecks, and in a letter dated 21 October 2003, he made a point of referring to the aforementioned numbers.[13] Van Geel saw speed reductions combined with strict enforcement of these limits as only a temporary solution regarding the worst air-quality hotspots.

During the discussions on highways and health, a combination of political parties voiced basically the same opinions, and argued for speed reduction. Generally, the PvdA, GroenLinks, D66, and the left-wing Socialist Party (SP) demanded speed reductions. Also in subsequent debates on air quality, public health would be the main concern of this camp. In contrast, the VVD, the right-wing LPF, and – to a lesser extent – the CDA favoured fewer reductions, and insisted on 'compensation' for reductions in the form of raising speed limits on roads elsewhere.

Even though the parties forming the ruling Cabinet were not keen on the lowering of speed limits, they could not avoid them. In the end, Peijs and Van Geel both embraced such measures, because it was clear that the European air quality standards could not be met. Speed reductions were regarded by the Government as a cheap measure that could be taken to ease the burden on air quality in bottleneck areas. Speed reductions were also considered a means to alleviate the consequences of road expansions. Overschie was the first location in which the speed limit was lowered, in May 2002, in the last months of the Purple Cabinet's reign. However, during the air-quality clash, a similar measure was taken in four additional locations, with the speed limit being reduced from 100 to 80 km/h on tracts of tangential roads near the cities of Amsterdam, Utrecht, Rotterdam, and The Hague.

6.2.2 *The symbolic importance of the speed limits*

The speed limit discussion went on during the period of the air-quality clash, and it remained a contentious issue. However, it must be kept in mind that the issue was mostly of symbolic significance. At least three reports appeared regarding lowering

11. Kamerstukken II 2003-2004, 28 663 nr. 11, pp. 1/2.
12. For instance, Gerard Hoek, Bert Brunekreef, and others in *The Lancet*, October 2002 (Hoek et al. 2002).
13. Kamerstukken II 2003-2004, 28 663 nr. 8.

of the speed limits, and in 2006 a colloquium was held to discuss its effects, which included reductions in travel time and congestion. These effects were varied, as in some cases and on some trajectories, congestion became less, while on others, the measures led to a significant increase in travel time. Congestion was often displaced, because a lower speed limit on the highways led to more use of the inner roads (AVV 2006). As regards air quality, the results were mixed as well. In Overschie, the measure led to a decrease of around 25% in transport-related PM10 emissions. In itself, that is an impressive reduction, but emissions from cars are not the only cause of poor air quality. To a large extent, quality is determined by regional background concentrations, unspecified sources, and foreign, non-local sources. Therefore the measure resulted in a fall of only 4% (TNO 2003). The same limited effects on air quality were found on the other tracts of road (AVV 2006).

Even though the political attention to this subject suggested otherwise, the effects of speed reductions were small, both in regard to health benefits and to losses of time. When we use the same frame of comparing air quality to the passive smoking of cigarettes, we can conclude that Overschie residents passively smoked half a cigarette less per day because of the 80 kilometre measure. Similarly, on most trajectories, any loss of time due to the speed reduction was mostly limited. Parliamentarians and Minister Peijs mentioned losses of time in the scope of half a minute to one minute.[14] In some areas, tests involving the speed-limit reduction were considered successful, while on others they were not. However, political and media attention, along with public concern over this debate, was 'exceptionally high' (van Beek et al. 2006).

These comparisons illustrate that the whole debate was a symbolic issue, albeit an important one, judging from its political and social saliency. In discourse analytic terms, Overschie became an emblematic case, and by referring to Overschie, actors on the political stage could refer instantly to an entire host of problems involving highways close to residential areas. It was not the extra cigarette or the time lost on the road that was the issue, but whether auto-mobility and transport should be given a free reign. Lowering the speed limit was a symbolic recognition that the expansion of transport and mobility had its limits, and that it had to be toned down in order not to jeopardise environmental interests. Parties that felt that economic development should have priority saw nothing in such limits, even if the time loss was small.

In summary, the pro-health parties considered it important to lower speed limits on as much of the road as possible, while conservative parties in the Government objected. Minister Peijs feared the measure would have negative consequences for the subjective appreciation of the motorist.[15] The struggle over the speed limits is important

14. Kamerstukken II 2003-2004, 28 663 nr. 9; Kamerstukken II 2004-2005, 28 663 nr. 26.
15. Kamerstukken II 2004-2005, 28 663 nr. 26, p. 4.

because it sensitised discussion participants to the problems and possibilities relating to road transport before the air quality clash began. The desirability of the expansion of road transport was called into question through the symbolic measure taken at Overschie, and thus became the prism through which further mobility measures and the air-quality problem would be viewed.

6.2.3 *Congestion and the proposal for an Emergency Law on Road Expansion*

The second significant issue that helped give rise to the eventual clash was the problem of traffic congestion. Congestion is a mainstay on the Dutch political agenda, because the Netherlands is a densely populated country, where car density is high, and the transport sector is economically important. The most common occupation among Dutch males is that of a trucker (Website Loonwijzer.nl last accessed 15 July 2014).

The issue of road congestion carried for pro-mobility and pro infrastructure parties the same weight as the speed limit did for the pro-health and pro environmental factions. Though political parties on both the left and the right of the political spectrum promised to reduce the problem of congestion, their proposals differed. The VVD was generally in favour of road construction, while GroenLinks was in favour of more environmentally friendly solutions. Already in the old SVV2 transport plan from 1989, reducing congestion had been the main aim, together with environmental improvement.

It is important to note that the VVD, one of the main political parties that shaped Dutch eco-modernistic environmental policies, still framed its solutions in terms of win-win scenarios. A new road around Overschie in addition to the one through the area for instance would unburden the area in terms of air quality and traffic congestion. However, in the 'old' transport plans from the late 1980s, road construction was still considered an option in last resort, an 'ultimum remedium'. The VVD that took part in the Balkenende Cabinets appeared to have lost its faith in behaviour change and relied on road construction. During one debate a GroenLinks Parliamentarian mentioned that he longed for the days of Pieter Winsemius again.[16]

Mobility was a spearhead of the Balkenende 1 Cabinet, and its solution was based squarely on road construction. However, the procedures to construct new roads were long and winding. A short-cut was proposed in the form of an Emergency Law on road expansion that would curtail procedures.

On the whole, the first Balkenende Cabinet was a spectacular failure, but on 18 November 2002, just prior to the new elections, Minister Roelf de Boer (LPF) managed

16. Handelingen II 2001-2002 nr. 25, pp. 1802-1844.

to propose his Emergency Law on Road Expansion. The Law was the icing on the cake of that Cabinet's pro-mobility agenda, because it made the expansion of existing roads possible in various ways: for instance, by allowing motorists to use the emergency lane during heavy traffic. This law was designed to shorten and simplify decision making procedures. The expansion of roads could normally only be achieved after a long procedure involving an environmental impact assessment, consultation with interested parties, and an investigation into the consequences of the expansion for environmental quality. Among the consequences taken into account were air quality, noise nuisance, and the consequences for plants and animals in the area. In addition, road expansion required various different administrative decisions based on diverse laws, such as the Law on Spatial Planning and the Road Law (Tracéwet).

The Emergency Law sought to shorten these procedures. In addition, it was designed to cut appeal proceedings, in the event that administrative decisions were challenged in court. Lower administrative bodies, as well as interest groups, would have shorter periods in which to object to decisions. The Emergency Law limited appeal proceedings to only one instance. All in all, the Government hoped to shorten by two years the procedure to expand roads.[17]

The proposal presented by LPF Minister Roelf de Boer was strongly worded. According to De Boer, ease of accessibility and the opening up of areas to road transport were essential for further economic growth and well-being, and in this regard *'The Netherlands is all too often halted on the road'*.[18] The Minister considered that unnecessarily broad decision making mechanisms had hampered finding a solution to the problem of congestion in the past. He proposed introducing a decision making culture with less negotiation – or 'polderen', as it is known in Dutch – in favour of a more hierarchical approach:

> *'In an administrative culture with less negotiation and more decision making, substantial gains in time can be made. Elaborate consultation can be limited, exceedances of terms can be pushed back, coordination mechanisms can be better used, and unnecessarily broad forms of decision making may be cut back.'*[19]

Before being sent to Parliament, the proposal was submitted to the Council of State Advisory Division, the highest administrative advisory council in the Netherlands. The Council of State is an important institution in Dutch policy making, because it acts as the highest advisory body, advising on all proposed legislation, as well as being the highest administrative judge.[20]

17. Kamerstukken II. 2002-2003, 28 679 nr. 3, p. 13.
18. Kamerstukken II. 2002-2003, 28 679 nr. 3, p. 1.
19. Kamerstukken II. 2002-2003, 28 679 nr. 3, p. 2.
20. For a full overview of the activities of the Council of State see box 6.

The Council of State Advisory Division recommended sending the proposal to Parliament, but only after it had reconsidered a number of critical issues. The Council was critical concerning the proposal's place in the light of current mobility policies, because it felt that this law had the potential to generate further mobility growth, which ran counter to official Dutch national and international policy commitments. Moreover, the Council felt that the law was not in accordance with ideas proposed in the Kyoto protocol on climate change and in the European 6th Environmental Action Programme (Council of State 2002).

The Council of State Advisory Division was also not happy with the shortening of the procedures, and considered that rules for environmental impact assessment should still be applicable, and that those regarding air quality and noise nuisance needed to be observed as well. Therefore, time gains could only be reached through shortening the periods in which various decisions could be taken. The Council also advised that – in order to maximise legal protection – the designated court of appeal be the Council of State itself, in its capacity as highest administrative court.

The advice provided by the Council of State Advisory Division was significant, because it demonstrated the Council's interest in the direction that policy was taking with regard to the environment and to legal protection.

Box 6: The Council of State

The Council of State is a historical institution within the Dutch administrative system. Nowadays it has two functions: that of the highest administrative court, and that of the highest advisory council regarding new legislation. It was instituted by emperor Charles V in 1531 in order to aid his regent, for the Netherlands, Mary of Hungary (Website Council of State, last accessed 19-06 2015). In it the most notable gentry was represented alongside the highest clergymen and a number of lawyers. The Council of State should provide advice in regard to matters of the gravest importance for the Dutch territories. Since 1861, civilian were able in some instances to appeal to 'the Crown' in conflicts between administrative bodies and citizens. The Crown consisted of the reigning monarch and current Ministers. In these cases, the Council of State needed to be heard before the Crown reached its decision.
Over time, it became standard practice for the Crown to adopt the advice provided by the special division of the Council of State that dealt with such administrative appeal cases, the Council of State Administrative Jurisdiction Division. Nonetheless the European Court ordained in the Van Benthem case that appeal to the Crown could not be considered access to an independent court in the sense of the European Convention of Human Rights. Since 1994, the Council of State Administrative Jurisdiction Division has acted as an independent Court. However, the Council of State also retained its political function of advisory council, The Council of State Advisory Division. The Advisory Division mostly consists of

former politicians and elder statesmen and its task is to advice the current generation of politicians on political and legal issues pertaining to legislative proposals. Discussion arose about the separation of these two functions when the European Court ordered in the Procola case that the combination of juridical and administrative duties undertaken by the Council of State also compromised the independency of the judges, and constituted an infringement of Article 6 of the European Convention for Human Rights.
This case led to discussion about the separation of powers of the Council of State. According to the Council's yearly report on 2005, the functions were executed separately, but the two divisions exchanged knowledge and experience (Council of State 2006, p. 47). In 2010, a new law on the organisation of the Council was enacted, which restructured the Council, and separated both functions more clearly (Council of State Restructuring Act 2010 [Wet herstructurering Raad van State], Stb. 2010, 175).

6.2.4 *Parliamentary discussions on the Emergency Law on Road Expansion*

The Emergency Law on Road Expansion was proposed in autumn 2002, at a time when air-quality concerns were on the rise, at least in Parliament. In October 2002, the RIVM issued a report indicating the damage that bad air quality could cause. The institute concluded that premature mortality afflicted possibly 10-15,000 people per year in the Netherlands (Buringh & Opperhuizen 2002, section 0.2), owing to their chronic exposure to PM 10. This report coincided with the aforementioned publication in October 2002 by Dutch epidemiologists Hoek and Brunekreef (Hoek et al. 2002). On 6 November, some two weeks before the road expansion proposal, the Transport Ministry sent a letter in which he explained the possible health effects.[21] On 25 November 2002, this letter was discussed between Secretary of State Pieter van Geel and the delegates of the Second Chamber of Parliament dealing with environmental issues.[22]

During the debate, air quality was heatedly discussed, and the Government's strategy of trying to postpone implementation of the standards until 2015 was lambasted by spokespersons for the PvdA and GroenLinks. Wijnand Duyvendak, Parliamentarian for GroenLinks, and Ferd Crone, spokesperson for the PvdA on environmental matters, connected the emblematic issues of Overschie and passive smoking to the forthcoming Emergency Law on Road Expansion. According to Ferd Crone, the Government was gambling irresponsibly on postponement and on the lowering of standards in Europe. He used the familiar trope that living near the highway was as unhealthy as the passive smoking of 17 cigarettes a day. In terms of public health, he doubted that these standards would be lowered in Europe, despite the

21. This letter is mentioned in the general discussion, but was not traceable.
22. Kamerstukken II 2002-2003, 28 600 XI nr. 73.

Government's expectation, and considered that the Emergency Law would only worsen the situation.[23]

The proposal for the Emergency Law itself was discussed in a Parliamentary consultation on 16 December 2002, during which the Parliamentary Commission for the Environment and the Commission for Transport were present, as well as both Ministers.[24] It appeared that in all probability the Law would be adopted, since the conservative parties had an important majority, but it granted the opposition the opportunity to point out any weaknesses in the proposal.

One of the more forceful participants in this discussion was GroenLinks Parliamentarian Duyvendak, who, as former chairman of the environmental pressure group Milieudefensie, had good contacts with that organisation. Duyvendak argued for a withdrawal of the Emergency Law, and for combating traffic congestion through the use of road pricing.[25]

According to Milieudefensie campaign leader Joris Wijnhoven, Duyvendak noticed early on that the European limit values for air quality could have important legal ramifications for infrastructure projects (interview J. Wijnhoven, interview W. Duyvendak). In the debate on the Emergency Law on Road Expansion, Duyvendak tightened the screws on Secretary of State Pieter van Geel. Firstly, Duyvendak wondered how the Emergency Law would bring air-quality targets closer, and in his argument he mentioned the victims of health problems, premature mortality, and asthma.[26]

Secondly, Duyvendak forced Van Geel to concede explicitly that the standards for air quality would be 'hard conditions', and that the Air Quality Order would remain in full force, despite the Emergency Law. When the meeting drew to a close, Duyvendak demanded to hear from Van Geel precisely whether provisions in the Air Quality Order applied. In response to Duyvendaks' insistent questioning, Van Geel affirmed explicitly that the standards needed to be observed in the context of the Emergency Law on Road Expansion. Van Geel informed Parliament that the Air Quality Order would be applicable in full. He put it as follows: '*There is no word of French in what I am saying; regarding air quality, the standards need to be met*'.[27]

During the debate, there was again no indication that anyone, except perhaps Duyvendak, had taken into account that the standard could be enforced by the Dutch Court in proceedings against administrative decisions. Only the possibility of fines by the Commission was mentioned. Parliamentarians tended to concentrate on

23. Kamerstukken II 2002-2003, 28 600 XI nr. 73, p. 6.
24. Kamerstukken II 2002-2003, 28 679 nr. 40.
25. Kamerstukken II 2002-2003, 28 679 nr. 40, p. 12.
26. Kamerstukken II 2002-2003, 28 679 nr. 40, p. 14.
27. Kamerstukken II 2002-2003, 28 679 nr. 40, p. 56.

2010, when the NO_2 standards would become applicable, but the standards for PM10 were already due in 2005. An explanation for this is perhaps that the Explanatory Memorandum to the AQO had stated that in regard to PM10 the state instead of lower administrative bodies would be responsible for policy, because the limit values were being exceeded everywhere.

In the end, the Emergency Law was accepted with a large majority. In the Second Chamber of Parliament, even the socialist PvdA supported the law, while left-liberal D66, the hard-left SP, and the ecological party GroenLinks rejected the proposal. After adoption in the second and first chambers of Parliament, the Emergency Law entered into force in 2003. During the debates, the opposition did manage to score an important political point, even though the significance was not yet clear. Secretary of State Van Geel was forced to proclaim that air-quality standards took precedence over the Emergency Law, as the significance of the Air Quality Order had been affirmed by the Government itself.

6.2.5 *The emergence of a pro-health camp in response to the speed limit and road expansion*

During debates on the Emergency Law and speed limits a number of political parties made similar claims criticising Government policy. In general left leaning parties such as the SP, GroenLinks, D66, and the PvdA argued that liveability concerns should trump the concerns of mobility. The air quality bottlenecks, such as the Overschie region featured prominently in the demands to take the environment and public health into account. GroenLinks representatives in particular posed critical questions and warned of potential health threats.[28]

The issue of the speed limits also piqued the interest of the environmental movement. In autumn 2002, the first posters of environmental pressure group Milieudefensie appeared in Voorburg, alongside the busy A12 highway, at the time that the Emergency Law was under discussion. The pressure group demanded a speed reduction in order to mitigate the effects of road expansion on noise nuisance. In subsequent years Milieudefensie also tried to influence Parliamentary debates on these issues.[29] However, during the years before the air quality clash, this coalition of parties and the pressure group lacked the political clout to actually force abatement measures beyond the most modest ones.

However, the foundation has been laid for what I refer to as the pro-health camp, a set of actors making similar claims about the social significance of air quality. They used the claims of scientific institutions such as the RIVM on health impacts of air quality to

28. Among others, Handelingen II 2002-2003 nr. 62, p. 3709; Kamerstukken II 2003-2004, 28 663 nr. 9; Handelingen II 2003-2004 nr. 24, p. 1683.
29. Letter from campaign leader J. Wijnhoven to the Parliamentary Commissions of VROM and V&W, Oct. 21 2003, regarding the upcoming debates on the speed limits, letter on file with the author.

underscore the need for protective measures. During these early years the pro-health camp did not have a very strong hand. The Parliamentary majority clearly favoured road expansion and air quality was not considered a pressing public issue.

It must be noted that at that point the pro-health camp consisted of parties in the opposition (except for D66, the most moderate one, it entered the Cabinet in 2003), and had ulterior motives for obstructing mobility policies. It was also an opportunity to oppose the reigning Government. Nonetheless, they did commit themselves discursively to an anti-mobility storyline, much as the Labour party had done in England when childhood asthma became a social problem. The tide for Labour turned when grass roots opposition against the road expansions soared and asthma became a widely discussed phenomenon. In the Netherlands the legal intervention of the Council of State Administrative Jurisdiction Division was instrumental when it started to annul road expansions because of conflict with the Air Quality Order.

6.3 Emergence of the air-quality clash

In the period between mid-2004 and April 2005, concerns regarding air quality rose to an unprecedented level in the Netherlands. The pro-health camp in Parliament received a major impetus when the Council of State Administrative Jurisdiction Division started to terminate high-profile road expansions and other infrastructure works in 2004. Pro-environmental parties and movements suddenly found themselves on the offensive.

In the following, I recount how this offensive played out within and outside of Parliament, and how court decisions contributed to a change in the balance of power between the pro-health camp and the governing parties, especially the CDA and the VVD. Before the actual clash can be discussed, it is necessary to look at air-quality policies in the years 2003 and 2004. The policies initiated in those two years indicated that the Government was aware that air quality in the Netherlands was potentially problematic, but it was not considered high enough a priority to spend a great deal of money on. According to documents from that period, the solution regarding the improvement in Dutch air quality was thought to lie in technological progress and in the European adoption of clean technology.

6.3.1 *Pieter van Geel's air-quality policy in 2003 and 2004*

After the fall of the first Balkenende Cabinet, a new Cabinet was instated in May 2003. It still contained the CDA and the VVD, but was now completed with the more environment-friendly D66 instead of the LPF. The Secretary of State for the Environment, Pieter van Geel, retained his post. Roelf de Boer, the Minister of Transport and Water Management responsible for the Emergency Law, was replaced by Karla Peijs, already mentioned in the context of the speed limits. Van Geel promised to come up with a Memorandum on Traffic Emissions in response to the situation in Overschie, and to the continuing lack of a consistent policy on air quality.

Before this Memorandum appeared, Van Geel explained the main lines of his policy in a letter dated 21 October 2003.[30] This letter is of interest, because it revealed how the Government regarded the situation before serious legal trouble became manifest. Van Geel felt that the best the Government could do was minimise PM10 and NO_2 emissions in the Netherlands, and he considered it more effective to fight emissions through European policy. In the meantime, he continued the policy line of not building residential areas and 'sensitive destinations' – such as schools or playgrounds – near highways. He noted, however, that this policy was causing problems for municipalities, because a great deal of building space was being lost. In regard to PM10, lower municipalities were required to use the ALARA principle to limit exposure to PM10 as much as possible. His commitment to mobility was clear from his consideration that a situation like the one in Overschie could only be resolved by taking the pressure off the main road by way of another one leading around the area: the controversial A4 Midden Delfland road. He announced further measures to be taken in the forthcoming Memorandum on Traffic Emissions.

The Memorandum was issued in June 2004, and contained several goals, principles, and concrete actions regarding emissions of pollution, noise nuisance, and CO_2, which influenced climate change. The aim was to reduce harmful environmental effects of emissions, but the document asserted explicitly that mobility as such was not a problem for the environment, as mobility fulfilled an 'indispensable societal function'. From an environmental perspective, there was no objection to mobility per se (VROM 2004, p. 7).

The Government considered that the transport and mobility sector should start to contribute more to environmental well-being. Industry had borne the brunt of previous measures, and now it was up to transport to become cleaner. The Government, however, would keep in mind that a part of the transport sector was internationally competitive (VROM 2004, p. 13). This turn of phrase indicated that, if possible, the important transport sector would be spared tough measures, as it had been in the past.

The document recommended a strategy combining green technology and tax-based solutions, together with international initiatives for a more stringent source policy. National measures were only to be taken if they were demanded by the EU:

> *'National measures for polluting substances and greenhouse gasses will be taken if they are necessary to fulfil international agreements, or if international agreements are no longer possible'* (VROM 2004, p. 7).

It became clear from the Memorandum that waiting for Europe to act and to influence decision making at that level would remain important components of Dutch

30. Kamerstukken II 2003-2004, 28 663 nr. 8.

air-quality policy. The Netherlands did not consider any infrastructure or volume measures. Instead, though not reducing mobility per se, it chose to implement taxes and subsidies to motivate people to opt for cleaner cars. The Government did not consider any measures that could hurt financially. For example, a modest stimulation package aimed at encouraging a choice for cleaner lorries in the Netherlands was considered, but only if the necessary money (23 million Euro) could be found within the VROM budget. In hindsight, a sum of 23 million Euros was a small amount compared to what would be spent on air quality later on. The fitting of cars with soot filters was to be encouraged via a budget-neutral change in taxation. In order to alleviate local bottlenecks, speed limits were to be lowered, and municipalities would be informed about possibilities to designate areas where only clean cars and lorries were allowed.

The same strategy of waiting for solutions to come out of Europe and for cautious budget-neutral policies was proposed in the National Air Quality Plan 2004 (Nationaal Luchtkwaliteitsplan 2004), which was submitted in February 2005. It outlined further policy in order to reach the air-quality standards imposed by the EU Air Quality Directives. This plan resulted from an EU obligation, because the Netherlands had failed to meet the NO_2 and PM10 targets, and in that case the Directive obliged the Government to draft a plan detailing reduction policies. Therefore, it was just as much a plan to prescribe policies in the Netherlands, as it was a document intended to convince the EU of the validity of the Dutch strategy (VROM 2005, p. 5).

It did not bring much new to the table, but offered a number of arguments aimed at convincing the EU that the Netherlands needed more time to comply with the standards. The main argument was that the effective reduction policies for PM10 were far too expensive, as well as being unfeasible (VROM 2005, p. 8). The plan conveyed the message that it would not be reasonable to demand that the Netherlands meet the standards on time.

This 'reasonability criterion' appeared numerous times in the document (VROM 2005, pp. 10, 21, 39, 41), and the appeal to a shared idea of what was reasonable could be heard often in Dutch discussions about environmental policy. However, at that time there was no indication from the European Commission that the Netherlands would be granted a postponement or a relaxation of the standards.

In the European arena, the Dutch Government continued its export of ecological modernistic ideas and it made use of the familiar storyline presenting win-win scenarios and 'the environment as an opportunity'. The Dutch Government tried to push the notion of 'sustainable mobility' on the EU agenda (VROM 2004, p. 27). When the Netherlands became President of the European Council in 2004, the Government organised an informal Environmental Council meeting in Maastricht from 16 to 18 July 2004, and it unveiled the Dutch initiative 'Clean, Clever, and

Competitive', a measure to highlight the importance of technological innovation to counter environmental problems and to become more competitive on the world stage.[31]

6.3.2 Three seminal verdicts in 2004

The political situation around air quality changed drastically in September 2004, when environmental and health interests switched from being a minor issue to a major problem for the realisation of the Balkenende Cabinet's ambitious road expansion programme. The reason for this turn of events was a string of verdicts given by the Council of State Administrative Jurisdiction Division.

Before 2004, the Council of State Administrative Jurisdiction Division was asked to judge in cases that related to air quality standards and to Dutch compliance with them. However, the verdicts up until May 2004 had not demonstrated a clear and consistent line of reasoning,[32] and so they attracted little attention. For instance, air quality was not even mentioned in the Council of State's annual report in 2003. In 2004, the Council of State formulated a consistent and strict guideline regarding interpretation of the AQO 2001. In its yearly report in 2004, the Council predominantly discussed three verdicts: one regarding a decision to expand a road between the Dutch towns of Barneveld and Hoevelaken on 12 May 2004;[33] one regarding another road expansion on the trajectory Vught Ekkerswijer on 15 September;[34] and one on the decision to halt development of a commercial zone in Hendrik-IdoAmbacht, also on 22 September.[35] The Council mentioned the three verdicts in its yearly report in 2005 as well, and reported that they constituted the approach the Council had taken on air quality. Therefore, we will examine in more detail these verdicts and their impact.

12 May 2004, Road expansion Hoevelaken - Barneveld

According to Loes Schutte Postma, air quality became a topic of discussion after 12 May 2004 (Schutte Postma 2006, p. 11), the day the Council of State Administrative Jurisdiction Division passed its judgement regarding the complaints of residents in the small village of Ter Schuren in the east of the Netherlands. Residents protested against a planned road expansion between the towns of Hoevelaken and Barneveld, invoking a number of arguments, including the increase in noise nuisance and the additional burden on air quality in the vicinity. The administration argued that air-quality standards would not be exceeded near people's houses, and concluded that

31. Kamerstukken II 2004-2005, 21 501-08 nr. 187, p. 5.
32. Van der Feltz mentions ABRvS 12 Nov. 2003 LJN AN7847 200202139 (Overslag Europoort) and ABRvS 17 Dec. 2003, 200301366/1 (Zwijndrecht) as verdicts in which the Council seems to accept the policy of the Government to observe only the norms in places where people live – referred to as 'sensitive destinations' (Van der Feltz 2006, p. 28).
33. ABRvS 200308160. Barneveld Hoevelaken.
34. ABRvS 200401178/1. Vught Ekkerswijer.
35. ABRvS 200307780/1. Hendrik-Ido-Ambacht.

therefore the air quality standards had been observed. The Council agreed with the residents, and judged that the standards needed to be observed everywhere in the country.

The verdict was a setback for the Dutch administration, because it meant that the administrative practice of observing the air-quality standards strictly only in places where people lived – the sensitive destinations – was considered a faulty interpretation (De Hoop 2006, p. 63). The Council did not intend to terminate the expansion altogether, however. Instead, it judged that the road could be expanded if the speed limit on the road were lowered to 80 km/h. It is interesting to note that the Council chose sides in the debates on the speed limit, because it would allow this road expansion if the limit were lowered.

September 2004, Road expansion Vught - Ekkerswijer
In September of the same year, the Council reached two more crucial verdicts. The first was passed on 15 September in another road expansion case, this time involving the junction between Vught and Ekkerswijer. A number of different residents' associations and interest groups appealed against the expansion, and again the plaintiffs claimed that the administration had not taken sufficiently into account the impact of road expansion on air quality.

This case is a highly interesting one, as PM10 concentrations near the junction exceeded the standards, and the administration therefore needed to take measures to reduce pollution concentrations at that location. The data presented showed that the expansion actually had benefits for air quality in the area, because it alleviated traffic on other roads. The Council of State Administrative Jurisdiction Section did not contradict those arguments, but stated that the municipal administration was responsible for reducing the concentrations as much as possible. It took this responsibility to mean that the most adequate measure for reducing air quality had to be taken. However, the administration could not prove that in terms of air quality this specific measure was the best solution, and as a result, its decision was not sufficiently motivated. The Council therefore annulled the decision to allow expansion of the road.

This decision placed a virtual bomb under the whole road expansion programme and the Emergency Law. Throughout the country, air quality standards were being exceeded near roads, so it would be very difficult to show that expanding a road anywhere would be the most effective solution to meet the standards. According to the Ministry of V&W the Vught Ekkerswijer decision *'was the reason for all the efforts the Cabinet has made to reduce the air-quality problem, international efforts, financial efforts, policy efforts, as well as an amendment of regulations'*.[36]

September 2004, Commercial zone Hendrik-Ido-Ambacht
The third crucial verdict was reached one week later, on 22 September 2004. In this case, residents challenged a provincial decision to approve a plan to allow the

36. Kamerstukken II 2006-2007, 30 646 nr. 2, p. 2.

realisation of an industrial and commercial zone near Hendrik-Ido-Ambacht, a town in the South of Holland. Again they argued that the provincial Government had not sufficiently observed the air-quality standards. Construction of the industrial zone led to an increase in traffic in the region, and hence to an increase in pollution concentrations. The provincial administration put forward a number of counter arguments. Firstly, it stated that air quality had not worsened in the area where residents lived, the by now familiar argument of sensitive destinations. Secondly, the administration argued that in the case of PM10, the responsibility lay with the Central and not the Provincial Government, an argument based on the Explanatory Memorandum as discussed in the previous chapter. Thirdly, the provincial Government contended that the increase in air-quality concentrations was insignificant, and did not need to be taken into consideration.

The Council of State judged all three of those arguments to be invalid. Firstly, the administration was wrong to presume that the standards only needed to be observed in residential areas. Secondly, even though the Central Government was responsible for PM10 policies, it was the responsibility of lower administrative bodies to show that their policies would lead to a reduction in pollution as well. Thirdly, the Council of State concluded that neither the Air Quality Order nor the Explanatory Memorandum made a distinction between significant or insignificant increases in concentrations. That last part of the verdict meant that every increase, however minimal, could be a reason to terminate a decision in a situation in which air-quality norms were exceeded.

The Hendrik-Ido-Ambacht decision was in line with the earlier decisions relating to road expansion. The verdict implied that not only were expansions possible targets of termination, but other construction projects ran the risk of annulment as well. Moreover, lower administrations were also considered responsible for air-quality policy, and the sensitive destination doctrine was proven twice in a row to be untenable. Road expansion and spatial development were now under threat of being halted because of the air-quality standards.

Those last two verdicts attracted large scale attention. Newspapers ran headlines like 'Highway expansion in danger' and 'Stuck between car and health' (de Volkskrant 25-09, 2004). After Hendrik-Ido-Ambacht, Secretary of State Pieter van Geel warned that those verdicts would cause the Netherlands to become 'blocked'[37] from further infrastructure-related developments (Volkskrant 01-10 2004). The image of the

37. The Dutch phrase is 'Nederland op slot', which translates to 'The Netherlands locked down'. I use the phrase 'blocked', which I think is a faithful translation of the Dutch phrase. Priemus and Schutte Postma use the translation 'off limits', but I feel it captures the meaning less well than' blocked'. The translation is important because 'Nederland op slot' was a phrase heard frequently in the PM10 discourse. It is an important turn of phrase, because it would become emblematic of the damage that air-quality standards could do, according to some political parties who opposed the strict regulation.

Netherlands being blocked in this manner would later become an emblem in a storyline arguing for relaxation of environmental standards.

The May verdict was the first to deal with the Emergency Law on Road Expansion. Termination of the administrative decision on the basis of the AQO 2001 was unexpected.[38] The court's order to lower the speed limit was also considered conspicuous, because speed reductions in the context of the Emergency Law would fall within the competency of the Minister of V&W.[39] In regard to the speed limit, the Council of State Administrative Jurisdiction Division positioned itself on the side of the pro-health camp with this verdict.

The September verdict on road expansion near the Vught Ekkerswijer junction made clear that additional demands were to be made on administrative bodies to motivate their decisions. They had to take into account the impacts a decision had on air quality in an area beyond the junction in question.[40] The Council also demanded this integral approach to air-quality research in its Hendrik-Ido-Ambacht decision.[41]

VROM Secretary of State Pieter van Geel was asked to remedy the situation immediately, as parties on both the right and the left of the political spectrum called for a solution.[42] The Government first reacted in disbelief, and reiterated that mobility itself was not a problematic issue. The Ministry of V&W in fact intended to go ahead with the planned road expansions, since, according to the Minister, it was impossible to legally bar the Netherlands from pursuing further development of the infrastructure based on the EU Directive.[43]

At the end of 2005, projects that should have come to fruition under the Emergency Law on Road Expansion were hit especially hard. Road expansions incurred hefty delays because of the Council of State's strict interpretation of the AQO 2001, and some projects were cancelled.

The judgements implied that a great deal of additional research had to be done when it came to air quality, but Parliamentary documents mentioned that there was only one research institute in the Netherlands capable of handling those kinds of detailed investigations.[44] As Struiksma noted in his annotation of the Hendrik-Ido-Ambacht case, the Government had to adapt the AQO 2001, but then it remained to be seen

38. ABRvS 200308160, Annotation A.G.A. Nijmeijer.
39. ABRvS 200308160, Annotation A.G.A. Nijmeijer.
40. ABRvS 200401178/1, Annotation J.M.H.F. Teunissen, J.W.A. Fleuren.
41. ABRvS 200307780/1, Annotation J. Struiksma.
42. Handelingen II aanhangsel 2004-2005 nr. 148, pp. 313, 314.
43. Idem p. 313.
44. Kamerstukken II 2005-2006, 28 679 nr. 50, p. 2.

whether it would remain in compliance with the European Directives.[45] For the Government, this represented a conundrum.

For environmentally inclined parties, the verdicts were proof that the Emergency Law on Road Expansion needed to be shelved. The Council effectively put a halt to road expansion because of the possible harm regarding air quality, and as a result, the verdicts provided the perfect opportunity to launch an offensive against mobility from an environmental perspective.

6.3.3 *The letter of 30 September: Van Geel's struggle with the Council of State*

The Government was well aware of possible consequences of the September verdicts. The Secretary of State tried to remedy the situation quickly by sending a letter on 30 September to provincial and municipal administrations, in which he defended his interpretation of the Air Quality Order 2001. He defended explicitly the distinction between sensitive destinations where people spend time and non-sensitive destinations where they do not (Van Geel 2004). In this '30 September letter', Van Geel restated the opinion that when applying the air-quality standards, priority should be given to sensitive destinations, such as houses, schools, hospitals, and sports fields. Lower administrative bodies should see to it that projects designated as sensitive destinations would not be planned in areas in which the quality of air was bad.

In this letter, Van Geel still claimed that the judgements by the Council of State displayed a mixed pattern. In some instances, the Council of State condoned the distinction, and only opposed it in two judgements regarding the Emergency Law on Road Expansion. Van Geel claimed that in Europe the idea of making a distinction between sensitive and non-sensitive locations was gaining ground. The Secretary of State reiterated that the Central Government was responsible for PM10, because breaches of the PM10 standards were widespread. Lower administrative bodies should only see to it that the best options in terms of air quality is chosen, and that the best available techniques be prescribed when granting permits. He claimed that the Council of State's verdicts on the issue were mixed as well.

According to Van Geel, all administrative bodies should do everything possible to reach the standards, but non-compliance with the NO_2 and PM10 standards would still occur. The Government did still not opt for expensive infrastructure measures like vaulting roads, but for lowering the speed limit or restructuring traffic, as well as for making changes in European regulation. Van Geel ended his letter with the hope that he had granted lower administrative bodies the necessary clarity and tools for the appropriate application of the Air Quality Order.

45. ABRvS 200307780/1, Annotation J. Struiksma.

With his letter of 30 September, Van Geel went against the line taken by the Council of State in its recent verdicts. This is a crucial episode because all through the air quality clash Van Geel would keep struggling with the Advisory Section. In two follow-up letters to Parliament, dated 28 October[46] and 26 November 2004,[47] Van Geel elaborated upon his position, basically reiterating the policy plans already mentioned in the Memorandum on Traffic Emissions. His aim was to meet the standards in 2015, instead of in 2005 and 2010. However, he now promised that extra money would be reserved for air quality. In the forthcoming Memorandum on Mobility, 300 million Euro would become available to remedy bottlenecks in the period 2011-2014, and the use of soot filters would be encouraged by way of a package totalling 100 million Euros. This package was a first success for the pro-health camp, because in the strategic accord of 2001, the subsidies on cleaner cars had been scrapped.

Van Geel's letter of 26 November was less elaborate but more interesting, because in it he set course for a direct confrontation with the Council of State. He considered that the Council was blocking necessary projects, against the intentions of the European Directives.[48] Van Geel insisted that giving priority to sensitive destinations was defendable, and he announced that a Ministerial Decree[49] would be drafted that would closely match the wording of the Directive. According to Van Geel, this meant that a distinction would be made between norms that protected public health and norms to protect eco-systems.

The concept version of the Ministerial Decree was ready in January 2005, and was far more lenient than the current AQO 2001. Article 2 of the concept version of the Decree contained the stipulation that those limit values that protected public health were only valid at the places mentioned in the letter of 30 September, such as schools, houses, and sports fields.[50] With regard to PM10, a separate article was inserted, stipulating that lower administrative bodies must try to minimise exceedances (Art. 3 sub 4) when making decisions, unless this were to have unreasonable financial consequences (Art. 3 sub 5). Moreover, article 3 sub 1 stated that when using their competencies, the Ministries of Transport and VROM would strive to comply as much as possible with the limit values for PM10. Those articles would take the sting out of the standards, because now it would be enough if the administration did everything reasonably possible to comply. In this proposal, the result-oriented character of the limit values was dropped. This article reduced the PM10 limit values in practice to target values, because it took away their obligatory character. The pro-health camp objected that the decree was tailor-made for the Emergency Law on Road Expansion and for other infrastructure projects.

46. Kamerstukken II 2004-2005, 28 663 nr. 27.
47. Kamerstukken II 2004-2005, 28 663 nr. 30.
48. Kamerstukken II 2004-2005, 28 663 nr. 30, p. 1.
49. A Ministerial Decree is a piece of legislation that can be enacted without the approval of the two Chambers of Parliament. It is quicker to enact, but is lower in rank than a General Administrative Order.
50. Concept Ministerial Decree Art. 2, sub 1, 2.

Before issuing the Decree, Van Geel asked the advice of the Council of State in its capacity as advisory council. The Council submitted its recommendation on 30 March 2005 (Council of State 2005a), in which it did not approve Van Geel's intentions at all. It reasoned that although some standards aimed to protect human health, it did not mean they should not be observed in places where people generally did not spend time. According to the wording of the Directive, the standards needed to be observed everywhere in the Netherlands. The Council advised negatively as well with regard to the weakening of obligations following from the limit value for PM10.

According to the Council, the Government as well as the lower administrative bodies should refrain from taking decisions that jeopardised obligations following from European law, especially in situations when those obligations were already being breached. The Council advised the Ministries of VROM and V&W to confer with their colleagues in Brussels as to whether compliance was impossible, and demonstrate that they had done everything possible to comply with the obligations (Council of State 2005a, p. 7).

The Council's final remark contained certain cautionary words that read like a threat:

> *'Trying to find a solution on the basis of one's own interpretation is risky, because it may elicit litigation that leads to prejudicial questions that will have to be answered by the European Court of Justice. Experience teaches that this results in a delay of a number of years with an uncertain outcome'* (Council of State 2005a, p. 8).

According to the Council, the Government should provide a plan illustrating that all possible measures had been taken.

The Council of State acted in this case in an advisory capacity, but the same organisation also housed the highest administrative court. With advice like this, it is hard to conceive of the administrative court accepting the Decree, especially because the Advisory Division warned that prejudicial questions needed to be posed to the European Court. There was nothing left for Van Geel to do but come up with a new plan and revoke his earlier letter of 30 September, since in view of the Council's advice, the letter would be unworkable.

6.4 Mobilisation of the pro-health camp inside and outside of Parliament

In this section, the further mobilisation of the pro-health camp in and outside of Parliament takes centre stage. After Van Geel's letter of 30 September, opposition became more forceful. The period between September 2004 and March 2005 saw a great deal of Parliamentary discussion on the issue. In previous years, a camp of left progressive parties that favoured lowering the speed limits had already been formed. That camp was now bolstered by the Council of State's verdicts, and later on by the Council's advice on the concept of the Ministerial Decree, and it saw a political opportunity to attack the Government's policy on mobility and the environment.

Outside of Parliament, environmental groups began to intensify their campaigns against transport and air pollution. The campaign acquired a specific legal dimension when the environmental movement started using the court verdicts in order to mobilise public support to initiate more cases and to bring Government policy to the attention of the court.

6.4.1 *Pressure by the pro-health camp in Parliament at the end of 2004*

Directly after the 2004 verdicts, the pro-health camp scored further victories. In October 2004, The Minister of V&W decided to lower the speed limit to 80 km/h on another four sections of highways, and on additional sections she lowered the speed limit from 120 km/h to 100 km/h. She explained that she was in a dilemma, because there was no support from automobile users regarding lowering the speed limit.[51]

The VVD's desire to determine whether the speed limit might also be raised could not be taken into consideration. The verdicts of the Council of State prevented this, because raising the speed limit would cause air quality to deteriorate, and such a measure would certainly be annulled.

After the initial test in Overschie, this was a further victory for left-leaning parties in their struggle to lower the speed limits. As we saw in section 6.2.4, in Parliament this pro-health camp comprised the left-wing parties in opposition, such as Groen-Links and the PvdA, but was also supported by the governing party, D66. This camp of political parties met Van Geel and Peijs during a general discussion on 4 November 2004,[52] during which all significant recent events regarding air quality were on the agenda, such as the Memorandum on traffic emissions, questions regarding the Emergency Law on Road Expansion, speed limits, consequences of the Vught Ekkerswijer judgement relating to road expansions, and Van Geel's new remedial policy efforts. Parties in the pro-health camp demanded that the speed limits be lowered on further tracts of road, but compensation in the form of higher speed limits elsewhere, as the VVD demanded, was out of the question. Both the left-wing SP and GroenLinks offered the Secretary of State alternative strategies regarding air quality, consisting of a number of taxation plans; for instance, the GL wanted to raise taxes on diesel and diesel cars, while both the SP and the GL wanted to lower the maximum speed to 80 km/h on significant tracts of highway and inner city roads. The GL was keen to see the Ministry designate inner city areas where only clean cars would be allowed, but the Ministry informed the faction that this was a municipal prerogative. The

51. Kamerstukken II 2004-2005, 28 663 nr. 26, p. 1.
52. Kamerstukken II 2004-2005, 29 667 nr. 8.

Socialist Party demanded a halt to construction in places within 150 metres of a highway. Van Geel agreed that this was a desirable situation in principle, but declined to name any concrete distance.[53] These demands are highlighted here, because in fact some of them would later be introduced.

At the end of the debate, Duyvendak introduced motions that were supported by a significant number of the opposition, and in one instance by D66 as well. The motions were eventually rejected, but by a narrow margin,[54] and those made by the Socialist Party were supported by the left-wing parties. The most broadly supported motion contained the desire to reduce speed limits on ring roads around the large cities in the Netherlands, and the SP asked for a minimum distance between a highway and surrounding buildings. This motion was also rejected, but the measures were now firmly on the agenda.

6.4.2 *Mobilisation of pro-health groups outside of Parliament*

The Council of State's verdicts and its advice concerning the Ministerial Decree created opportunities for the pro-health camp in Parliament to take the offensive, but at the same time, extra Parliamentary opposition was organising. Campaigns for clean air were started by different pressure groups, but the most important one was the campaign by Milieudefensie, the Dutch branch of Friends of the Earth. This campaign involved the mobilisation of people to initiate proceedings against the Dutch state.

Milieudefensie belonged to the more radical of the environmental action groups, and it campaigned against the mega projects, such as the expansion of Schiphol Airport, the Betuwe rail route, and expansion of the harbour at Rotterdam. In the beginning of the 2000s, Milieudefensie decided to forego it's up till then more cooperative strategy and enter once more into a conflict-minded style of campaigning, taking detrimental effects of transport as a campaigning target. As previously mentioned, this was a controversial choice, because the transport sector was – and remains – economically important, and campaigns against auto-mobility targeted a large number of ordinary car owners.

However, road expansions and the new road-building projects triggered Milieudefensie to add its weight to the drive to lower the speed limit after the Overschie test. In autumn 2002, the first posters already appeared in Voorburg, alongside the busy A12 highway, at the time that the Emergency Law was under discussion. The campaign gained momentum after the verdicts of September 2004. In October 2004, Mili eudefensie released a booklet called 'Snelheid geboden, 80 de limiet op de snelweg' (Speed is a necessity, 80 is the limit on the highway'), in which it urged lowering the speed limit on Dutch highways. In this campaign booklet, a number of citizens, politicians, and scientists argued in favour of reducing the speed limit on important tracts of road, mainly around the big cities.

53. Kamerstukken II 2004-2005, 29 667 nr. 2.
54. Handelingen II 2004-2005 nr. 34, p. 2248.

The brochure had been compiled at an earlier date, however, and the most recent court decisions on the Emergency Law had not been taken into account. Nevertheless, the verdicts created an opportunity to intensify the campaign. Campaign leader Joris Wijnhoven and others at Milieudefensie realised that *'this could become dynamite'* (Joris Wijnhoven, interview). The road expansion verdicts provided Milieudefensie with a powerful argument against auto-mobility. As campaign leader Joris Wijnhoven put it:

> *'Residents found out that they could claim their rights. Then we thought, well, clean air is not something abstract, but it is simply a right, it is in the law. There cannot be more dirt in the air then is stipulated in the Directive. So we thought, well we will claim this'* (Wijnhoven, interview).

At the end of 2004, Milieudefensie initiated the campaign 'Nederland in Ademnood' (the Netherlands gasping for breath). This campaign was aided by satellite images that became available in October 2004, and which showed that air above the Netherlands contained high concentrations of NO_2.

Image of satellite photo showing NO_2 pollution above the Netherlands.

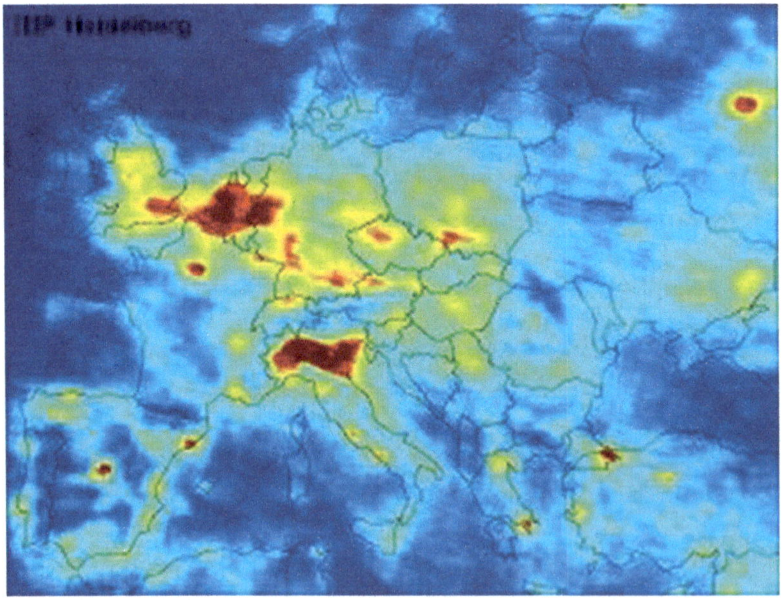

http://www.kennislink.nl/publicaties/satelliet-brengt-luchtvervuiling-in-kaart

The images displayed NO_2 concentrations in the atmosphere, however, and not at street level. Moreover, NO_2 is far less harmful than PM10, but those nuances were lost in the imagery, as well as in the media attention that began to gain momentum. The image was all over the newspapers. On 15 October, the Dutch daily newspaper Telegraaf opened with the headline 'Our country is the dirtiest in Europe' (Telegraaf, 15 October 2004).

The campaign was broad in its scope. Milieudefensie used the disturbing satellite images as displayed in the picture below, and also drafted a list of the 50 dirtiest streets in the Netherlands. According to Wijnhoven, the campaign became an instant media hit.

6.4.3 *Milieudefensie's legal campaign*

Perhaps the most striking feature of the air-quality clash was its rapid juridification. The clash gained decisive momentum with the Council of State's verdicts, and the judiciary's heavy involvement – provoked by the actions of the environmental movement, especially Milieudefensie – was also instrumental in the social problem's further evolution.

Milieudefensie added a legal dimension to its campaign after the road expansion verdicts and the debacle involving the Ministerial Decree. The group decided to send a letter to Minister Peijs in March 2005, in which it demanded that Peijs lower the speed limit on roads through and around cities, totalling some 200 km of highway (Website Milieuactueel.nl, last accessed 18-07 2014) Wijnhoven, interview). In addition, Milieudefensie demanded that the municipality of The Hague close two notorious air-quality bottlenecks in The Hague to traffic, the 'Stille Veerkade' and the 'Amsterdamse Veerkade'. According to Milieudefensie, those two streets were among the dirtiest in Europe. In a press release, Mileudefensie stated that its eventual goal would be to force road pricing for lorries, to raise diesel fuel prices, to lower the speed limit, and to curtail road expansion as well as the expansion of Rotterdam Harbour and Schiphol Airport (Website Milieuloket, last accessed 18-07 2014).

https://www.indymedia.nl/en/2004/11/23461.shtml / Michiel Wijnbergh

If the municipality and Minister did not comply, Milieudefensie threatened to initiate court proceedings together with residents' interest groups. According to an interview in the daily BN De Stem with campaign leader Wijnhoven, Milieudefensie would unleash an 'avalanche' of proceedings (BN De Stem, 07-04 2005).

Minister Peijs refused to agree to the demands made in Milieudefensie's letter, but the municipality itself made certain concessions, though it refused to block the roads leading to the two streets, or to block the streets to lorry transport. Not accepting the demands made by Milieudefensie constituted a decision in the sense of the Dutch Administrative Code, and therefore it was open to appeal. The pressure group promptly appealed the decision, and summoned the municipality of The Hague before the Administrative Court, where it demanded that the municipality take a number of measures, because the air in those parts of the city was not in accordance with the air-quality standards.

Milieudefensie lost this case, first in the court of first instance and later in appeal before the Council of State Administrative Jurisdiction Division.[55] The judge decided that although air quality was indeed poor, the administrative body had considerable leeway in deciding how to achieve compliance. The court sanctioned the municipality's defence that these measures could lead to a worsening of air quality elsewhere in the city. However, Milieudefensie was unable to force the municipality to take certain decisions.

Nevertheless, the attention that such legal proceedings garnered was considerable. According to Breeman and Timmermans, this strategy created a boost in media attention regarding air quality (Breeman & Timmermans 2008, p. 38). After the first court case, more followed, with Milieudefensie losing a substantial number of them (Wijnhoven, interview).[56] However with the aid of these court cases, Milieudefensie was able to present itself as a champion of the struggle against air pollution and its effects on liveability. Moreover, it was able to use its expertise to help others, and it advised and mobilised pressure groups of residents living in 'dirty' areas, as well as people living alongside roads, who subsequently acted as litigants (Wijnhoven, interview).

During the actual proceedings, Milieudefensie frequently played a role in the background. On its website, Milieudefensie published a 'do-it-yourself' kit detailing what steps to take in order to bring a clean-air case before the Council of State (website Milieudefensie). When the proceedings started to become more technical and elaborate calculations were necessary, Milieudefensie assisted residents, using their own experts who could do the necessary math (Wijnhoven, interview). According to

55. ABRvS 200604266/1, Veerkaden Den Haag.

56. The Council of State website lists 14 verdicts containing both the terms 'air quality' and 'Milieudefensie' between 01-01 2005 and 31-12 2008. In 13 of them, Milieudefensie was a party, and of those cases Milieudefensie lost eight and won five. The number is small because Milieudefensie was often not a direct participant, but merely acted in a supporting role for other pressure groups. However, this gives an indication of the success rate of appeals lodged by pressure groups.

Wijnhoven, Milieudefensie members spent days answering phone calls from residents' groups, and were staffed to maximum capacity.

In April 2005, those people residing on 'dirty streets' as assigned by Milieudefensie received another campaign folder, entitled 'Air alarm', which featured all the emblematic aspects of the air-quality problem, including the Overschie situation and the '17 cigarettes'. The folder also mentioned research indicating that planning schools close to the highway was a bad idea, and the populace was called to start proceedings against the administration if air quality exceeded the standards; in this regard, the 'do-it-yourself kit' was also pointed out.

Milieudefensie used the court cases as a crowbar to force opportunities to influence air-quality policy, but also added an explicitly legal storyline by discussing the issue in legal terms. In their campaigns, the group made explicit references to the law and to human rights, and proclaimed clean air to be 'a right' (Joris Wijnhoven, interview; Van Klink 2005; Milieudefensie 2005). The movement urged local pressure groups to use this argument as well (Website Leefmilieu, last accessed 25-06 2015) and it also managed to attract media attention for it (de Volkskrant 07-05 2005) The focus of the arguments was still on the public health aspects of PM10, but a legal storyline was added to its repertoire. The storyline portrayed the Government as hypocritical because it was not prepared to stick to its own laws, which argued that the Government was in violation of citizens' rights by exposing people to polluted air. Residents were summoned to fight back and claim their rights to clean air. Milieudefensie took its claim that clean air was a right to court in 2005. Although the Council of State judged in 2008 that clean air is not an enforceable right,[57] the court cases granted the environmental pressure group the opportunity to drive its point home.

6.4.4 *The presentation of worrying scientific findings by the MNP/RIVM*

The claims of the pro-health camp gained in credence when the MNP and RIVM published another set of worrying scientific findings in 2005. The MNP issued a report indicating that the health problem posed by PM10 was worse than expected. The MNP publishes an annual 'state of the environment' report, The Environmental Balance (Milieubalans), which always contains a section on air pollution. In the report of 2004, an estimate of 5000 deaths linked to PM10 was given (MNP/RIVM 2004, p. 15), and in 2005, the RIVM arrived at a new estimation, which painted a much more serious picture. At the presentation of the report in May 2005, Klaas van Egmond, director of the MNP, stated that as many as 18,000 people could die prematurely each year due to PM10. The figures were taken from a report pending at the time (Knol & Staatsen 2005), in which it was stated that 18,000 people were estimated to die 10 years earlier than might have been expected (Knol & Staatsen, 2005, p. 56). The estimate was also mentioned in a report on PM10, 'Fijnstof nader bekeken' (MNP 2005a, p. 56). This

57. ABRvS 200704973/1, 80 Kilometre.

Figure 3

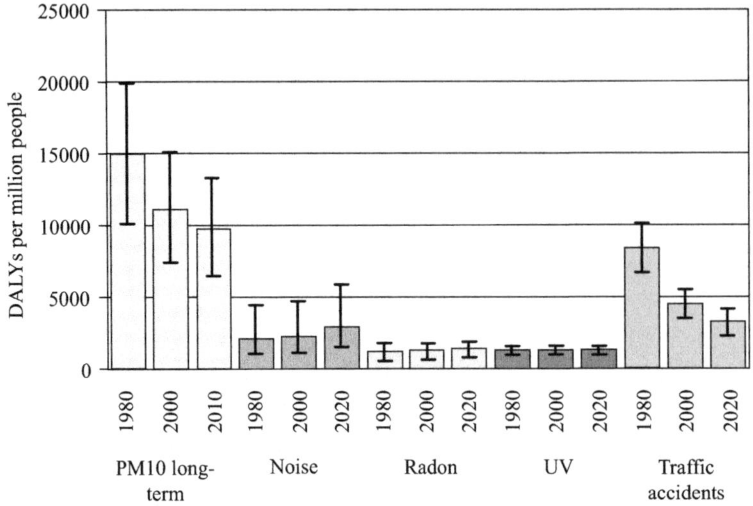

meant that, at a stroke, PM10 became the most serious environmental problem in the Netherlands, far worse even than traffic accidents (Knol & Staatsen 2005, p. 72), see figure 4.

At the presentation of The Environmental Balance in May, the media seized upon the issue, and discussion about the figures mounted. The Dutch national daily newspapers Volkskrant, Telegraaf, and Trouw reported on the large number of expected deaths resulting from high levels of PM10 (de Volkskrant, 11-5 2005; Telegraaf, 11-05 2005). Media attention for PM10 was high in 2005 in general. During the spring of 2005, a television programme called Zembla, which dealt in investigative journalism, turned its attention to the deleterious effects of air pollution, and to the role of mobility in this regard. It contained the emblematic issue of Overschie and noted epidemiologist Bert Brunekreef revealed their finding about the harmful effects of air pollution on children. In addition, the debate in Parliament in April concerning air quality was a prominent item on a political news show called Den Haag Vandaag (Website Den Haag Vandaag, last accessed 25-06 2015).

The large increase in the number of prospective PM10-related deaths in May 2005 was remarkable, especially considering that the two American studies upon which the data were based dated from the 1990s. These were the Harvard Six Cities Study and the ACS Study discussed in chapter 2. All the information was already present in 2004, when 5000 premature deaths were considered to be plausible, as indicated in The Environmental Balance. The scientists Leendert van Bree and Erik Lebret defended their findings in the newspaper Trouw of 10 June 2005, but also pointed to uncertainties and assumptions in the numbers. These uncertainties were recounted in a later RIVM report that mentioned numerous doubts regarding every aspect of PM10 (Mathijssen & Visser 2006, p. 2).

In the media, the figure of 18,000 premature deaths was reported often to indicate the harm done by PM10, but it is striking that this large increase did not greatly influence the political debates. As far as I could determine, in 2005 only GroenLinks leader Halsema mentioned that specific number of deaths,[58] and in the documents, we generally encounter the number 5000, also in debates as late as 2006 and 2007.[59] Although public health was referred to by the environmentally conscious factions, apparently the exact number of deaths was of secondary importance in the debates. In an interview, Joris Wijnhoven, campaign leader for Milieudefensie, put it as follows: *'They are all epidemiological studies; well, whether it is 100,000 deaths in Europe, or 300,000, or perhaps 50,000, it is simply crystal clear that breathing polluted air is bad for you'.*

However, the media-attention strengthened the case of the pro-health camp. The claims of the political parties belonging to this camp, the environmental movement and these scientific institutes overlap; bad air quality due to air pollution is a significant health threat and should be mitigated as much as possible. Therefore the scientific institutes are situated within the pro-health camp.

6.4.5 The Council of State caught in the middle

Milieudefensie tried to provoke the Council of State Administrative Jurisdiction Division into demanding concrete measures from administrative bodies, In general though the Council did not go as far as to directly condemn Government policy. Milieudefensie lost the proceedings it had started against The Hague municipality, in which it demanded that the court order the administration to take concrete measures. The Council of State itself documented its role as advisory body and administrative judge in this period in two yearly reports: 2004 and 2005. In its annual report for 2004, the Council pointed to the mismatch between the national system of environmental law and the European Directives. European regulation had its own logic and was not automatically compatible with national legislation. In this regard it was up to the national legislator to weld European and national law into a coherent whole, especially because the national legislator was also co-legislator at the European level (Council of State 2005b, p. 25).

That line of reasoning was also present in the 2005 annual report. The Council of State devoted considerable space to the air-quality clash, reiterating that the implementation of measures had been guided too much by national concepts and considerations. The national legal ramifications of the air-quality Directives had been underestimated. Van Geel's letter of 30 September as well as the Ministerial Decree had been rejected by the Council of State, because they were based on mistaken assumptions regarding obligations stemming from the Directives. The proposal

58. Handelingen II 2004-2005 nr. 82, p. 4918; Handelingen II 2004-2005 nr. 103, p. 6264.
59. For instance, Kamerstukken II 2004-2005, 29 667 and 28 663 nr. 18; Kamerstukken II 2007-2008, 30 175 nr. 46, p. 5.

by the Secretary of State would take away the result-oriented nature of the air-quality standards. In the yearly report, the Council of State offered two recommendations: firstly, the Government should redress the situation where it had begun – in the European arena, and it would have to seek a solution together with other Member States and the European Commission; secondly, the Government would need to take all possible measures to meet its obligations at home (Council of State 2006a, p. 134).

As to the court cases, the Council of State indicated that the three seminal verdicts discussed above displayed the Council's line. The report stated that non-compliance with the standards directly was a reason for the Court to annul administrative decisions, but more often thorough research was lacking and this lack of research was the primary ground from annulment. Administrative bodies should make a credible case that their decisions would not lead to deterioration in air quality in situations where the standards were exceeded. The brunt of the annulments occurred because administrative bodies did not comply with the courts' more rigorous demands for research.

6.4.6 *Further victories for the pro-health camp in the spring of 2005*

In the spring of 2005, Milieudefensie was waging its legal and societal campaigns, and the Administrative Jurisdiction Division of the Council of State kept terminating projects, some of them with a very high profile. Between September 2004 and April 2005, a number of prestigious projects were annulled, such as development of the 'Southern Axis' in Amsterdam, an ambitious commercial zone,[60] development of a football stadium for ADO Den Haag,[61] development of Amsterdam Central Station, and expansion of the A27 highway.[62] Hugo Priemus mentioned in an article in 2006 (Priemus 2006a) that further expansions of the A27, A1, and A2 were under threat. Residential areas were also not immune, and Priemus listed residential areas such as IJburg, Ypenburg, Almere Poort, Haarlemmermeer, and Leidsche Rijn, which had also been mentioned by Van Geel in April 2005. In that same month, the clash reached a preliminary climax when Minister Peijs (V&W) lamented: *'Money is not my biggest worry; air quality is my biggest worry! I have money enough, but I cannot get one spade to dig into the ground'* (Laverman 2005, p. 7).

In this unenviable position, Secretary of State Van Geel drafted another letter dated 20 April 2005 to outline his next moves.[63] He still aimed at obtaining a postponement from Europe, and he also sought to implement a distinction between sensitive and non-sensitive destinations into a future EU regulation. He also promised that 300

60. ABRvS 200406190/2, 22 November 2004.
61. ABRvS 200400465/1, 26 January 2005.
62. ABRvS 200407748/1, 13 April 2005.
63. Kamerstukken II 2004-2005, 29 667 nr. 12.

million Euros would be designated for air quality after 2010, in addition to which he pledged an extra package on top of that. He proposed additional measures, such as encouraging the use of soot filters and raising taxes for diesel fuels. Van Geel insisted that those measures would make the Netherlands a forerunner in Europe, and he hoped that they would lead the European Commission to view the Dutch situation in a favourable light. Van Geel felt that with this package, the Dutch Government was doing everything that was reasonably possible. He also outlined his intention to attempt to amend the Air Quality Order once again.

The letter was under discussion on 26 April 2005, together with the other air-quality policies.[64] In the April 2005 debate, Van Geel defended the package as the maximum that could reasonably be undertaken, but he came under heavy fire from PvdA MP Diederik Samsom for that statement. Samsom insisted that it was unclear what 'doing all that is reasonable' meant in the context of the threat to public health. The Government, he insisted, was too focused on meeting the standards, and disregarded the people that needed to be protected. The Secretary of State was seen as having lost valuable time by trying to 'wiggle out' of the standards by way of a Ministerial Decree.[65]

Together with Wijnand Duyvendak, Samsom came up with more measures, such as extra speed reductions, banning old cars from using the ring roads around Amsterdam and Rotterdam, and encouraging the use of soot filters on all diesel cars. Together with D66 MP Van der Ham, Duyvendak insisted on the need to look at road pricing.

Samsom's thrust that Van Geel was not doing all that reasonably could be done, and his challenge to make explicit what 'reasonable' meant in this context, were a particularly venomous, because the pragmatic notion of reasonableness was a central feature of the tacit Dutch consensus on environmental issues. Policies had to make sense from an economic and an environmental point of view, and should not demand the impossible either way. However, Samsom now forced Van Geel to explain for whom those measures were reasonable, and to what extent the Government was doing all that could reasonably be required.[66]

Van Geel had obvious difficulty meeting this charge in the debate. He repeated that he was doing everything that could be considered reasonable, without being able to answer the question as to why he was not doing more. Van Geel accused the PvdA and GroenLinks of 'cheap rhetoric',[67] and insisted that air quality had improved consistently over the years. The PM standards could not be met 'with all reasonable measures', and he alluded to the considerable scientific uncertainty that existed when the

64. Kamerstukken II 2004-2005, 29 667 nr. 18.
65. Kamerstukken II 2004-2005, 29 667 nr. 18, p. 2.
66. Kamerstukken II 2004-2005, 29 667 nr. 18.
67. Kamerstukken II 2004-2005, 29 667 nr. 18, p. 11.

standards had first been agreed upon. He promised that the Government would look into more far-reaching measures, and that he would try to find funds for extra measures within the budgets for 2006.

At the end of the debate, Duyvendak stated that he was pleased. A general speed reduction was not in the cards, but all the measures GroenLinks had been demanding were back on the political agenda. In this debate, the environment-friendly opposition scored important victories. Van Geel was given the green light to draft a new law to come up with a solution regarding the deadlock on infrastructure development, but he and Minister Peijs were forced to consider environment-friendly measures to appease the opposition, 'Europe', and the Council of State.

The debate in April 2005 proved to be a turning point. The pro-infrastructure parties that had hitherto been put firmly on the defensive began to organise their counter arguments, and to claim forcefully that the air-quality rules were misguided and would lead to economic ruin. In April 2005, however, the country's infrastructure projects appeared to have been brought effectively to a halt.

6.4.7 *The pro-health camp and its discursive strategies*

The pro-health camp and especially GroenLinks and the PvdA saw air quality standards as an opportunity to oppose the policies of the Balkenende Cabinet. A number of arguments were made that illustrated the storylines the pro-health camp was using to criticise the Government. GroenLinks politician Duyvendak was one of the most critical of the Parliamentarians, and in general he was supported by members of the PvdA and the SP. Duyvendak used the large mortality rate foreseen by epidemiologists to convey his criticism of Van Geel's policies. For instance, he criticised the Ministerial Decree that Van Geel planned to issue as *'merely a stop gap, while in the meantime 5000 people die yearly because of bad air*.[68]

Another politician who opposed Pieter van Geel vehemently in Parliament was Diederik Samsom, who, like Duyvendak, had a background in the environmental movement. A former campaign leader for Greenpeace, Samsom applied pressure to Van Geel as well by referring to the damage to people's health that his policies would cause:

> *'No actual measures to make cars drive slower, no policy to reduce the number of cars on the roads or to keep off the most polluting lorries, no, the Secretary of State drops all these options and makes his most talented lawyers devise a trick! As a kind of Cato I repeat: every year 5000 people die of the consequences of air pollution by transport. That is more than 13 a day! How many human lives will the Secretary of State save with his Ministerial Decree full of legal tricks? Is he prepared to answer that? Under the mantra that the country cannot be blocked, human lives are put at risk.'*[69]

---

68. Handelingen II 2004-2005 nr. 27, p. 1756.
69. Kamerstukken II 2004-2005, 29 800 XI nr. 93, p. 6.

Samsom was critical of the way Van Geel was handling the classical ecological modernistic assumption that economy and ecology could progress side by side. He appreciated Van Geel when he argued for environmental protection as a possible economic opportunity in Europe, but felt that on the national front, too few of these opinions resulted in policy.

All through 2004 and 2005, the environmental camp attacked the measures taken by the Government as insufficient. Discursively, Duyvendak took the strongest pro-environmental position. In a debate with the Minister of Transport, he criticised the Government's policies as being a 'sickening recipe' of more asphalt and less attention to public transport. He called the politicians belonging to the Governmental parties a 'bunch of ostriches'.[70] Duyvendak's intentions went beyond PM10 abatement and towards limiting auto-mobility per se, which became even clearer in an argument he presented late in 2005, at the high point of the crisis:

> *'This new generation of environmental problems, the difficult problems can only be solved by setting strict standards and observing strict compliance. There are too many ostriches in Parliament, the Cabinet and in society, who stick their head in the sand instead of facing this reality. We hear it echoing around that further development in the Netherlands has been blocked, but the question is in what way we intend to unblock it.'*[71]

Other parties in the pro-health camp generally did not go as far. Instead, they presented other packages of measures, and criticised the Secretary of State. The death toll resulting from concentrations of PM10 was a leading argument, as well as the charge that the Secretary of State was trying to 'wiggle out' of implementing European standards.

Diederik Samsom, for instance, stated in one of the last debates of 2004 on the issue:

> *'Parliament and Government should consider the health of these people and the tens of thousands of other people who live in areas polluted by exhaust fumes as leading. They should do everything possible to reduce air pollution. With this attitude, the Secretary of State would not have lost valuable time with an attempt to wiggle himself out of implementing European standards by Ministerial Decree'.*[72]

And:

> *'If people were truly central in the debate and not just the standards, then the Government would have presented a package of measures with everything that is reasonably possible to solve the problem of air pollution, apart from the question of whether the standards are being met'.*[73]

This charge undermined the tacit Dutch environmental consensus based around the tacit agreement about 'doing what is reasonable'. The PvdA, however, did not

70. Kamerstukken II 2004-2005, 29 800 A nr. 43, p. 20.
71. Handelingen II 2005-2006 nr. 19, p. 1162.
72. Kamerstukken II 2004-2005, 29 667 nr. 18, p. 2.
73. Kamerstukken II 2004-2005, 29 667 nr. 18, p. 2.

question the reigning eco-modernist approach of relying on clean technology and reconciling the interests of mobility and environment.

The political party D66 was in a difficult position. It was known as being pro-environmental, but from 2003 to 2006 it was in a Government with the right-wing VVD and CDA, both of which were not very environmental friendly during those years. As such, D66 was responsible for Government policy, even though it was the junior partner in the coalition. During debates, the party operated cautiously, but often sided with the pro-health camp. It especially urged the introduction of road pricing in order to combat both congestion and emissions.[74] For obvious political reasons, it could not be persuaded to support fully the pro-health camp on breaking-point issues (Wijnand Duyvendak, interview).

The last party belonging to the pro-health camp was the socialist SP, and it did not have a truly independent profile, as its objections conformed to those of GroenLinks and the PvdA. There were differences, however. The SP was not in favour of a road pricing, for instance, but intended to make motorists pay to use the roads by raising the prices of fuel.

The pro-health camp organised itself first around the speed limit and later around the issue of air quality. These parties opposed the reigning mobility policies of the Balkenende Cabinet, and the air-quality issue provided an ideal public issue from which to launch their criticism. In 2005, this strategy led to policy changes in a pro-health direction.

6.4.8 *Pro-health policies at the end of 2005*

Combined with the strict interpretation of the Council of State and the findings of the environment and health institutes, the efforts of the pro-health camp in 2004 and 2005 brought about a number of pro-health policy developments. Some of these involved concrete measures, while others consisted of reformulations of policy.

The most conspicuous development was the publication of a package of measures to encourage cleaner transportation. These measures were backed up financially to a significant extent, and the package was presented in September 2005 when the Government revealed its budget for 2006.[75] Hence, it became known as the 'Prinsjesdag package', because on 'Prinsjesdag'[76] the Government traditionally presents the budget for the coming Parliamentary year.

74. Kamerstukken II 2004-2005, 29 667 nr. 8, p. 8, Kamerstukken II 2004-2005, 29 800 A nr. 43, p. 20; Kamerstukken II 2004-2005, 29 667 nr. 18.
75. Kamerstukken II 2005-2006, 30 175 nr. 10.
76. Prinsjesdag refers to the third Tuesday in September, when the Dutch budget is released. The budget for 2006 contained a significant financially supported package, the goal of which was to reduce concentrations of air pollution.

Among other things, the package included – by way of a fiscal deduction – the incentive to produce and to drive cleaner lorries and trucks: for instance, subsidies for diesel taxis and delivery vans that contained soot filters; subsidies for retro-fitted soot filters; tax differentiation favouring low-sulphur diesel; campaigns to encourage driving in ways that saved fuel, lowering the speed limit in a number of trajectories; pushing for the use of air scrubbers in intensive cattle farming; and concluding covenants with the industry to reduce PM emission from production processes.[77]

This package aimed to *'improve public health and to prevent the Netherlands from becoming "barred" from further economic development'*.[78]

This package was supported financially by the budgets of VROM, the Ministry of Transport and Water Management, and the Fund for Strengthening the Economic Structure (FES). An amount of 900 million Euros was reserved for air quality, which was a significant development, because the measures outlined in the Memorandum on Traffic Emissions, for instance, lacked financial back up. Moreover, 900 million Euro was a considerable sum if one considers that – as late as 2004 – budget cuts had been foreseen in the field of the environment.

However, The Prinsjesdag package would not significantly reduce those areas where air- quality standards were being exceeded; it would reduce total emissions by only 1%. (Hammingh et al. 2005, p. 9). It was still considered cost-effective, however, because it would reduce by 10% the most health-relevant emissions – those of traffic and transport. The RIVM noted that the package would have been more cost-effective if a kilometre tariff had been included.

The package was a major success for the pro-health camp (Wijnhoven interview), as it meant that the Government had to let go of its tactic of waiting until 2015 before implementing any measures. However, the omission of a kilometre tariff also indicated that this success had its limits, as the inclusion of such a tariff was politically too unsavoury for the governing parties.

However, the Prinsjesdag package was not the only significant policy change. Policy plans in general became more sensitive to environmental and health concerns as well. In a number of documents, concerns for environmental aspects of mobility, especially for air quality, came prominently and more often to the fore. During the period of the air-quality clash, the new transport plan – named Memorandum on Mobility – was under discussion. This memorandum would replace the old SVV2, and had to do with the future of mobility in the Netherlands. The first draft was published in September 2004, and a second one was published in August 2005. In comparison to the first draft,

77. Kamerstukken II 2005-2006, 30 175 nr. 10, pp. 5/6.
78. Kamerstukken II 2005-2006, 30 175 nr. 10, p. 1.

the sections on air quality, public transport, and innovation, were worded more strongly in the second one.[79]

Sustainable mobility became a priority for the Ministries of VROM and V&W.[80] In the Explanatory Memorandum for the VROM 2006 budget, considerable attention was paid to a transition to sustainable mobility, brought about because of air quality and climate concerns. In the budget of the Ministry for Transport and Water Management, a separate paragraph was devoted to the theme 'Mobility and Society'. Air- quality improvement was noted as a top priority because of health concerns and the legal blocking of further activities relating to the country's infrastructure. Innovative strategies were considered necessary because *'More of the same no longer works'*.[81] The blockade of infrastructural development apparently forced the admission that new directions needed to be explored.

In Europe, the Netherlands increased its efforts to push the subject of sustainable mobility onto the EU agenda. This effort resulted in strengthened cooperation at home and abroad and the Netherlands considered itself a front-runner in many transport files.[82]

A further development of note is the reappearance of road pricing on the political agenda. The Platform 'Paying Differently for Mobility' (Anders Betalen voor Mobiliteit 2005) issued a report at the end of 2005 that endorsed road pricing. This Commission was instituted to investigate some form of road pricing, though the topic was still met with little enthusiasm by two out of three the political parties in Cabinet at the time. The CDA and especially the VVD had never warmed to the idea, but the third coalition partner, D66 was keen on it.

The Commission to investigate road pricing was chaired by Paul Nouwen, former front man for the Dutch Association of Motorists (ANWB). The ANWB had always opposed road pricing, but the Nouwen Commission presented a report indicating that such pricing was a laudable idea, with positive environmental effects being cited as one good reason. In its advice, the Nouwen Commission used the same storyline as the pro-health camp; it referred to the hazards that highways presented to children living or going to school near them, and to the air quality in Overschie (Platform Anders Betalen voor Mobiliteit 2005, p. 26). The Nouwen Commission's recommendations were embraced in the Mobility Memorandum.

Even the Council of State made cautious remarks that indicate it felt that a shift in policy from economy to environment and health was necessary. In its advice on

79. Kamerstukken II 2005-2006, 29 644 nr. 12.
80. Kamerstukken II 2005-2006, 30 300 XI nr. 2, p. 13 and Kamerstukken II 2005-2006, 30 300 XII, nr. 2, p. 10; Kamerstukken II 2006-2007, 31 031 XI nr. 1.
81. Kamerstukken II 2005-2006, 30 300 XII nr. 2, p. 10.
82. Kamerstukken II 2006-2007, 31 031 XI nr. 1, p. 108.

the budget for 2006, we may find some markedly environmental concerns, especially regarding air quality. The Council of State Advisory Division remarked that both the Dutch Government and the EU had a one-sided focus on economic well-being, and in addition, it considered that Western European economies, including the Dutch one, presupposed high economic growth as a given; it also noted that structural weaknesses could make it necessary to forego these growth ambitions. One of those factors was environmental:

> *'There is pressure on the physical living environment, space is scarce and there is a need for a sustainable economic growth in which there is no cutting back on a healthy living environment, but on the contrary improvements are necessary, such as regarding CO2 and particulate emissions'*[83] (Council of State 2005c).

6.5 Concluding remarks

By April 2005, air quality and PM10 in particular had become a social problem in the Netherlands, and it became the subject of political debates, media attention, public concern, and campaigns by pressure groups. In the following sections, I will analyse key issues in the construction process from 2000 up to 2005: firstly, the emergence of a pro-health camp; secondly, the rupture of the eco-modernist consensus over mobility policy; thirdly the emergence of a storyline in which clean air is portrayed as a right; and fourthly, the Council of State's involvement in this political discussion.

6.5.1 *The emergence of a pro-health camp as a reaction to marginalisation*

In Europe and in the Netherlands, the construction of a social problem gained momentum as soon as medical experts became involved. The participation of experts such as the WHO made a European air-quality policy possible, and in the Netherlands the opposition also coalesced around a medical storyline. Political parties used numbers of premature deaths cited in reports to up-end Government policy, and to accuse the Government of complacency in the face of the imminent danger posed by air pollution.

During discussions on the speed limit, political parties belonging to this camp were already sharing similar concerns, and the deaths mentioned in reports and in epidemiological research added ammunition. A discourse coalition was formed between political parties, the environmental movement and the health experts of scientific and health institutes like the RIVM and GGD. The pro-health camp, however, began to succeed in affecting Government policy significantly only after the court cases that blocked the Emergency Law on Road Expansion, and other infrastructural projects. Not only could it accuse the Government of complacency, it could also argue that it

83. Kamerstukken II 2005-2006, 30 300 nr. 3, p. 3.

was incompetent, because it had failed on both fronts: air quality and infrastructural development. In the table below, political parties belonging to the pro-health camp are summarised, together with their storylines and political position.

It is not coincidental that a broad pro-health camp included political parties, but also the environmental movement. The camp emerged against the threatened marginalisation of environmental interests. After the assassination of Pim Fortuyn the tide was not with the environmentalists, or other groups that were associated with 'the left'. When, in this climate, the conservative parties took over they did away with the tightrope walking exercise between environmental and economic interests. Construction and mobility would be the interests catered to and environmental targets were shelved or postponed. Subsidies for clean cars for instance were abolished by this first Balkenende 1 Cabinet.

Table 7

Party	Main spokesperson	Position	Storyline
GroenLinks	W. Duyvendak	Against further growth of auto-mobility	Health problems display the ongoing environmental degradation caused by road transport. The health effects associated with air quality are a symptom of misguided emphasis on development.
PvdA	D. Samsom	In favour of more environmental/health measures and adherence to EU standards, speed reductions, road pricing, tax measures.	The disregard for health problem displays the complacency of the current Government.
SP	A. Gerkens	In favour of more environmental/health measures and adherence to standards, speed reductions, raising fuel prices, tax measures; building should take place away from roads.	Public health should have precedence over infrastructure development.
D66	B. van der Ham	No restriction of mobility per se. In favour of speed reductions and a kilometre tariff to make motorists pay for use of the road.	Sound infrastructure and good air quality can both be realised. The current impasse should be resolved by structural solutions that reduce PM10 and NO_2, such as road pricing.
Milieudefensie	J. Wijnhoven	In favour of speed reductions and arguing for closing off roads in polluted residential areas. In general against road expansion.	The people have a right to clean air, the Government is not caring for the health of its citizens.
Scientific institutes (RIVM / MNP / GGD)	RIVM / MNP scientists and prominent epidemiologists	Bad air quality due to high concentrations of PM is a health threat	Scientific findings indicate 18,000 people may die prematurely because of bad air quality.

The emergence of air quality concerns in Overschie as a minor environmental issue was an opportunity to at least take a stand against the impending road expansions. There was no way to stop the Emergency Law, but at least the pro-health camp could control the damage done from an environmental perspective by making sure pledges were made regarding the air quality standards. In a sense Van Geel's cornered himself with his comments during the Emergency Law debates that the air quality standards would be applicable in full. In this regard it is understandable why the road expansions were terminated. The Council of State was not positive about the Emergency Law and Van Geel himself had admitted that the Dutch Government had legally bound itself to the European air quality standards.

After the annulments of the road expansions, air quality was no minor issue anymore and the hitherto marginal pro-health camp could take the offensive. In the already polarised environment it did so and forced the Government to take many costly air quality measures that the Balkenende 1 Cabinet had been very unwilling to take.

6.5.2 *Cracks in the eco-modernist consensus*

The threatened marginalisation of environmental interests led to cracks in the hitherto rather firm ecological modernistic consensus. The strain in this discourse caused by the relinquishment of the ideals of ecological modernisation is illustrated by the debates over the speed limits. The speed limits became the locus of a fierce battle between the pro-health camp and parties that favoured auto-mobility. The issue was highly symbolic, because although the lower speed limits led to only a marginal loss of time for motorists, – and did not resolve the air-quality situation – they remained a thorny issue. In a nutshell, the speed limit symbolised what was at stake in the air quality clash: namely, freedom for motorists or protection for residents along highways, and there seemed to be no middle ground. Lowering of the speed limit came down to a recognition that the expansion of mobility could not continue indefinitely, and such a curtailment on development ran counter to the core assumptions of the Dutch eco-modernist environmental policy. According to this discourse, development and protection needed to flourish together, and even to reinforce each other.

After the verdicts the pro-health camp hardened its stance and demanded measures, but Van Geel continued the administrative practice of demanding compliance with air-quality standards only in places where people spent time. Moreover, he did not treat the standards as strictly result-oriented obligations, but contended that lower administrative bodies should apply the ALARA principle, which meant that exposure to PM10 should be as low as could reasonably be achieved. Van Geel stuck to this interpretation of the air-quality standards, even when it became clear he was not only opposed by the pro-health camp, but also by the Council of State.

Those practices complied with the pragmatic policy outlook of ecological modernisation: namely, one should not be ordered to do the impossible, and solutions should be 'smart', meaning geared optimally to serve the environment as well as the economy. However, something fundamental had changed; the verdicts made short work of this pragmatic solution, and triggered accusations of incompetence and complacency. Initially the Government tried to patch things up by reiterating its interpretation of the AQO in a letter of 30 September 2004 and subsequently by proposing a new Ministerial Decree. Both approaches were turned down by the Council of State Advisory Section as contrary to the Government's legal obligations stemming from European Law. After this debacle it gave in to a number of pro-health demands by putting aside extra funds for air quality, by lowering speed limits at a number of locations, by promising to encourage the use of soot filters, and by raising the tax on diesel fuels. Van Geel stated that with these measures, the Netherlands had done all that could reasonably be expected.

Van Geel implicitly invoked the Dutch eco-modernist consensus of not asking for the unreasonable, but the spokespersons for the pro-health camp, Duyvendak and Samsom rejected this appeal. In fact, the definition of what could be considered reasonable in light of the potential damage to health was now up for grabs. The pro-health camp managed to present the air-quality clash as an either/or choice between mobility and public health, with members accusing the Government of trying to circumvent European obligations by means of a Ministerial Decree that – from a legal perspective – was shoddy, implicitly portraying the Government as sacrificing public health on the altar of unbridled mobility. At the same time, they argued that if they were in power, they would take new and speedy measures to protect the public.

D66 was the only party in the pro-health camp to argue that both efficient mobility and health protection could be possible. The other parties however, portrayed the issue as a clash of interests, and the Government pro-mobility parties had no convincing solution. They argued that they would look for intelligent solutions, but in practice those turned out to be the taking of piecemeal measures and trying to get road expansion back on track. Such one-sided solutions were incompatible with ecological modernisation, and if such choices needed to be made, the eco-modernist dream of win-win solutions would turn out to be a pipe dream.

6.5.3 *Emergence of the 'clean air is a right' storyline*

The political battle raging inside Parliament intensified in the spring of 2005, but the cause of the clash lay in the courtroom. After the Council of State had annulled three infrastructural projects, and it became clear that it would stick to its line of reasoning, the Government was in political trouble.

Outside Parliament, however, the verdicts inspired the environmental movement to innovate its strategy. It added a legal dimension to its clean air campaigns, and demanded measures be taken by the municipality of The Hague and by the Ministry of Transport and Water Management. Milieudefensie lost those cases, but was able to profile itself as a champion in the fight for clean air, and it urged groups of residents to take their complaints to court as well.

The courts ruled that one could appeal against administrative decisions, but could not force administrative bodies to take specific measures. Nonetheless, the court cases attracted considerable attention, and the media was quick to pick up on it. Subsequently, residential groups turned to Milieudefensie for advice on how to undertake such a procedure against the administration.

Involving the courts was in fact a clever move, both practically and discursively. The Council of State annulled many high-profile projects, and due to this pressure the Government was forced to take extra measures to clean the air. Moreover, the legal battle kept air quality firmly in the public eye. The court cases led the environmental movement to argue that clean air was a right that was legally protected by the air-quality standards, and Milieudefensie interpreted the involvement of the courts as a sign that fundamental rights were at stake.

The campaign of legal mobilisation waged by Milieudefensie confirmed findings by Lisa Vanhala, who had concluded that NGOs used the court to demonstrate the failings of the existing system (Vanhala 2012). In this case, the legal campaign revealed the shortcomings of the Dutch policy on air quality and raised environmental awareness with the citizenry.

Discursively, the storyline that clean air is a right was an innovative one, though the idea of environmental human rights, or a 'right to clean air' was not new (Hiskes 2012, p. 399) – it had been used in human rights treaties and discussed by scholars in the 1990s. However, Milieudefensie's approach in the case, and the Council of State's verdicts, gave the notion a new dimension. Milieudefensie used the storyline in order to mobilise the public to initiate court cases. The environmental movements' actions turned the debate on environmental rights from a theoretical issue into concrete social action involving the courts of law. The storyline ceased to be an academic exercise, and entered the legal practice. I consider it to be one of the most important discursive innovations to come out of the air quality clash.

6.5.4 *The Council of State: its politicisation and its storyline*

The Council of State itself did not discuss the issue of air quality in terms of it being a problem regarding enforceable rights. In fact, in 2008 it would specifically rule out such a view. During 2004 and 2005, it developed its own storyline, based on the primacy of European law vis a vis national regulation. The Council of State noted that the implementation of air-quality standards was indeed strict, but that the Dutch

Government itself had agreed to the standards, and had chosen to implement them in a rigorous manner by relating them to the terminology of the law on environmental management. The Council stated that European law could not be circumvented by the Government own interpretation of it, but should seek support for its position in Europe. This storyline placed European law squarely above national law and national administrative practices. In practice, the Council of State's line of reasoning came down to an increased obligation to research the effects of administrative decisions on air quality, and to display convincingly that those effects had no deleterious consequences.

The Council of State's arguments were legal ones, and, on the face of it, not very political. The Council ignored the social consequences of its verdicts, and appeared concerned mainly with legal implications. However, in the advice provided by the Council on the Emergency Law on Road Expansion, and on the proposal for a Ministerial Decree in 2005, certain political elements could be detected. Even the seminal court cases themselves contained certain elements that revealed a particular political position.

As to the Emergency Law on Road Expansion, the Council of State remarked that it seemed to be incompatible with objectives of national environmental and mobility policy, and that it was also not consistent with international obligations, such as those on climate change. In its advice on the Ministerial Decree, the Council not only warned the Government not to interpret the standards in ways that were contrary to European law, but it also demanded that the Government draft a plan containing all measures that could be taken, and present it to Europe in order to renegotiate. This advice contained a demand that the Government take earnest measures to protect the quality of air in the Netherlands, and such a requirement exceeded a merely legal point of view.

The Council of State Administrative Jurisdiction Division mostly used strictly legal reasoning based on the approach that administrative practices were not laid down in the Air Quality Order itself, and were not consistent with the wording of the Directive and the Air Quality Order. Such argumentation appeared legalistic as well, but even the Court displayed occasionally a more political point of view in its decisions. In the first road expansion case, the Court ordered that expansion could continue, but that the speed limit on the trajectory be lowered to 80 km/h. In this instance, it took a position in the discussion on the speed limit, and forced the Government to take a concrete measure if it wanted road expansion to continue.

These Council of State concerns indicated that more was at stake than merely a legal debate. Especially the Advisory Division was concerned with the way politics handled its environmental obligations, and pushed through an agenda of road expansions with little regard for participatory and legal procedures. The Council of State may well have been concerned with the way the Emergency Law on Road Expansion curtailed the power of the judiciary. However, by taking the side of the concerned residents, and by defending the Air Quality Order against being amended, the Council of State came to be regarded by the pro-health camp as a champion of environmental standards. It became caught up in a highly controversial polarising political discussion, even though it tried to appear concerned solely with a sound legal interpretation of the standards.

Even though the Council of State at times endorsed position taken by the pro-health camp, its different storyline prevents me from situating it within that camp. The position of the Council of State is a different one. It acts as a kind of political legal consciousness for policy. The Advisory Division is mainly populated by ex-politicians of a high calibre. It makes sure the Dutch political ship does not deviate from its course too much. In that regard its advice concerning the Emergency Law on Road Expansion should be read. This Emergency Law was a short cut to make road expansions possible, without their being a solid base for them in a policy plan. Transport policy was in principle still governed by the rather environmental SVV2. The Council frowned upon such 'rash' swings of policy. However, because of its two pronged nature as a high advisory body of state and the highest administrative court, the power base of this 'political consciousness' is potentially very strong. In theory, the Council's advice does not bind the Government, but if the advice comes from the same organisation that provides binding legal rulings as well, the situation is different. Even though the two divisions of the Council of State are separate, policy makes such as Pieter van Geel did not want to take the risk.

The desirability of this influence is the subject of much debate. In principle the two divisions are strictly separate, but how insurmountable this 'Chinese wall' between the Administrative Jurisdiction Division and Advisory Division is, is unknown. A negative advice for the Advisory Division may have dire consequences if it translates directly to a negative judgement of the Administrative Jurisdiction Division. Moreover, from the point of view of political legitimacy it is questionable whether elder politicians should be allowed to govern in this way after retiring from open politics. On the other hand, much can be said for an institution that makes sure political transitions transpire relatively smoothly. At the beginning of the 2000s the Dutch political landscape was in turmoil. It befits the Dutch consensus structure of policy to have an institution guarding against major upsets. In the end, consensus was restored and that may partially be attributed to the tenacity of the Council of State. I consider the consensus structures such as the Council of State to be bound up by the Dutch political tradition. In the next chapter we will see how the Council itself became a target of criticism though because of its position in the air quality clash.

TIMELINE DUTCH POLITICAL DEVELOPMENTS 2000 – 2005: EMERGENCE OF AIR QUALITY CLASH

2000-2002	Gradual mobilisation of a pro-health camp around air quality and the issue of the speed limits
2001 July	Air Quality Order 2001 entered into force
2002 May	Test with reduction of speed limit to 80 km/h in Overschie commenced
2002 July	Balkenende took over as Prime Minister, Pieter van Geel Secretary of State for the Environment
2002 Nov.	Minister of V&W R. De Boer proposed Emergency Law on Road Expansion
2002 Dec.	Debate on Emergency Law, Van Geel conceded Air Quality Order would be applicable in full
2003 May	2nd Balkenende Cabinet installed, Van Geel retained post, Karla Peijs took over as Minister for V&W
2003 October	Emergency Law on Road Expansion entered into force
2004 May	RIVM issued 'Milieubalans 2004 and mentioned 5000 premature deaths due to PM
2004 May	Council of State Administrative Jurisdiction Division gave 'Barneveld' verdict and annulled first road expansion ABRvS 200308160
2004 June	Government issued Memorandum on Traffic Emissions
2004 Sept.	Council of State gave 'Vught Ekkerswijer' and 'Hendrik-Ido-Ambacht verdicts, annulled second road expansion and decision to construct commercial zone ABRvS200401178/1 and ABRvS 200307780/1
2004 Sept.	Van Geel presented 'Letter of 30 September' defending the Government's interpretation of AQO 2001
2005 Jan.	Van Geel proposed new Ministerial Decree on Air Quality
2005 March	Council of State Advisory Division advised negative on Decree.
2005 April	Van Geel promised 300 million extra for air quality
2005 Sept.	Prinsjesdag package unveiled, 900 million extra for air quality

7 The rise of the pro-infrastructure camp and the resolution of the clash 2006 – 2010

Introduction

The Netherlands was in danger of becoming 'blocked' with regard to infrastructure development after the successive 2004 verdicts by the Council of State Administrative Jurisdiction Division. Moreover, the Advisory Division's heavy criticism of Van Geel's letter of 30 September and of his proposed Ministerial Decree in March of the same year prevented any quick fix. The pro-health camp was quick to take the initiative, and demanded more environmental and health measures.

In the spring of 2005, the resulting blockade of infrastructure development sparked a counter reaction. Parties that favoured infrastructure development were starting to put forward a diverging storyline, featuring a different definition of the problem and a different solution.

The theme of this chapter concerns the emergence of the pro-health camp, along with the subsequent redefinition of the problem of air quality and its eventual resolution. The first section discusses the emergence of a pro-infrastructure camp of parties and pressure groups, while their opposing scenario is the main focus of section 7.1.2.

In section 7.2, legislative initiatives occurring at the end of 2005 and in 2006 are described. First a renewed Air Quality Order was issued, and afterwards a new law was introduced, aimed at mitigating the air quality clash by serving both the interests of infrastructure development as well as the demands for clean air. This Air Quality Law contained a novel idea involving a 'programmatic approach' by which the Government sought to break the legal deadlock and enlist multiple actors to help resolve the air quality clash. During debates on this law, the pro-infrastructure camp intensified its political offensive.

In the third section, the adoption of the new air quality law and the annulment of another important road expansion by the Council of State Administrative Jurisdiction Division are discussed. During these events, the pro-infrastructure camp managed to redefine the problem in Parliament, taking it from being a public health problem to a regulatory one.

In the fourth section, we examine the resolution of the air quality clash by way of the programmatic approach that was made possible under the new Air Quality Law adopted in 2007. The implications of the approach itself are under review as well as further legal and political developments following directly from the air quality clash. At the end of this section, an overview of the various participants in the conflict and their positions vis-a-vis each other are presented. In the chapter's final section, I conclude by discussing four salient points that came to the fore: namely, the eventual re-emergence of an ecological modernistic consensus; the discursive changes brought about by the air quality conflict on the subject of infrastructure and mobility; the political position of the Council of State; and the problematic relationship between scientific and legal rationality.

7.1 The mobilisation of the pro-infrastructure camp

In 2005, the topic of air quality was at the zenith of its political importance. The Council of State continued to terminate projects, and the pressure of the pro-health camp yielded pro-environmental regulatory results. Van Geel proposed a package of measures, a number of which the pro-health camp in Parliament had already been demanding.

However, in opposition to the terminations and annulments, a lobby emerged from a variety of representatives of economic sectors, demanding a solution to the problem by adapting the Dutch regulatory structure in order to accommodate infrastructure development. The emergence of a countervailing camp that challenged the pro-health camp's presentation of the situation is the main topic of this section.

7.1.1 The emergence of a new discourse coalition in 2005

In the spring of 2005, the pro-health camp had everything going for it. The Government was under fire because of the termination of projects, and in his letter of 20 April 2005, Van Geel forced to promise new measures to improve air quality.[1] Media attention to air quality was increasing (Breeman & Timmermans 2008), and scientific institutes were raising the alarm over the possible impact of PM10 on public health.

Nonetheless, the debate in April 2005 regarding Van Geel's letter was one of the first in which groups and parties that represented mobility, infrastructure development, and other economic interests took an offensive course. Lobby groups representing many important economic sectors in the Netherlands raised their voices as well, urging a quick solution to the problem, and this call was taken up by political parties on the conservative side of the political spectrum.

1. Kamerstukken II 2004-2005, 29 667 nr. 12.

Just before the April debate, the employers association VNO-NCW sent a letter to the commission of Parliamentarians that discussed environmental and transport issues.[2] The VNO-NCW is the largest organisation of Dutch entrepreneurs, and in its letter it asked Parliamentarians to take into account that the Council of State had thus far terminated 16 highly important construction projects such as roads and commercial zones. The VNO-NCW also pointed out that air quality was steadily improving, and it therefore asked members of Parliament to accept Van Geel's proposal to amend the Air Quality Order and make it more lenient. The VNO-NCW endorsed the distinction Van Geel had made between places where people lived and places where they did not, which, as we have seen, was contrary to the line taken by the Council of State.

In the April 2005 debate, the representative of the populist party LPF, Joost Eerdmans, was the first in Parliament to state directly that the problem had been caused by the Dutch system of linking administrative decisions to environmental quality standards. He stated that this link was not required by the EU Directives, and pointed out that a prominent professor of environmental law, Niels Koeman, had written an article in the daily newspaper Het Financieele Dagblad, in which he argued that this link was not necessary (Koeman, 2005a).[3] The notion that the Netherlands had blocked further development itself due to its own legally strict approach to the air quality standards would become a persistent theme in further debates.

As previously discussed, this link entailed that lower administrative bodies were obliged to observe quality standards, which meant that in zones where environmental standards were not met, further harmful activities could not be undertaken. Quality standards led to metaphorical 'picket lines' (Koeman 2010) being formed around zones that were experiencing sub-standard environmental quality, in an effort to seal them off from further contamination. For instance, certain activities that created an odour nuisance were forbidden within a perimeter around a waste incineration facility whose emissions already exceeded the odour limits. This inflexible system was already unpopular with lower administrative bodies (De Roo 2003), but the air quality standards had far greater effects on spatial planning. Whole swathes of the country became blocked with regard to further development because the air quality standards were being exceeded.

However, Van Geel did not intend to release the link at that time. He explained that the Netherlands had an integral review of spatial planning decisions, and this included testing whether they complied with air quality standards. Other countries did not have as integral a review, and this allowed them to 'sweep environmental

2. Letter from J.H. Schraven to the Chairman and members of the Regular Commission of VROM and the Regular Commission of Transport and Water Management 21-04 2005. The letter was CCed to the Ministers of VROM and Transport and to the VROM Secretary of State Van Geel. On file with the author.
3. Kamerstukken II 2004-2005, 29 667 nr. 18, p. 9.

problems under the carpet', as he phrased it.[4] He reasoned that this could lead the EU to impose fines on those European countries. He promised to look into the matter, but warned that Koeman's opinion was not representative of all legal specialists.

From that time onwards, however, the link became the prime target of pro-infrastructure factions and pressure groups. A new discourse coalition coalesced around the idea that the main culprit of the air quality clash was 'the link'. In Parliament, this camp comprised the CDA, the VVD, and the LPF, the parties that had supported road expansions in 2002 and 2003, and that were sceptical about lowering the speed limit. Of the three, the CDA took the most centrist position, and supported its Secretary of State Pieter van Geel.

7.1.2 *Discursive strategies of the pro-infrastructure camp in 2005*

In the course of subsequent debates, the pro-infrastructure camp strengthened into a discourse coalition of political parties and pro-infrastructure pressure groups that developed its own storyline, containing a definition of the problem and the favoured solution. This happened most conspicuously in June 2005 when a number of important Parliamentary meetings were scheduled such as the discussion on the spring budget. During this discussion the powerful Minister of Finance would be present. A day later, a general discussion was planned between Secretary of State Pieter van Geel and the Parliamentary commissions for VROM and Transport and Water Management.[5]

Just before the debate with the Minister of Finance, a number of interest groups representing construction, transport, and small and medium enterprises issued a joint press release on the 28th of June 2005, in which they portrayed the air quality clash as a regulatory problem that the Netherlands had brought upon itself. The camp consisted of interest groups in the construction and service sectors, such as Aedes, Bouwend Nederland, MKB-Nederland, NEPROM, NVB, NVM, VBO, and the VNO-NCW (Bouwend Nederland et al. 2005, Press release 28-06 2005). It was supported later by the transport organisations EVO and TLN (EVO 2005 press release 22-11 2005).[6] Apart from claiming that the link was the main culprit in terms of restrictions on infrastructure development, the lobby groups announced the results of a quick scan made by the organisation of Dutch municipalities. The scan indicated that some 7.7 billion Euro were at stake, because construction projects were being kept on hold throughout the country.[7] Some 100,000 employees in the construction sector

4. Kamerstukken II 2004-2005, 29 667 nr. 18, p. 16.
5. Handelingen II 2004-2005 nr. 96, pp. 5776- 5804; Kamerstukken II 2005-2006, 30 175 nr. 9, respectively.
6. For the full names of these organisations, I refer to the list of acronyms found at the end of the book.
7. Hugo Priemus, professor of spatial planning at Delft Technical University, provides the same figures in Priemus (2006b) 'Wat ging er mis met het fijnstofbeleid?', Internationale Spectator nr. 60 jaargang 6, 2006.

were in danger of losing their jobs, and half of the 80,000 houses planned might not be built. In its press release, the coalition of interest groups wrote:

> *'The association of Dutch municipalities has quantified the problem in a quick scan at 7.7 billion euros. It is expected that the real costs will be even higher. For every billion euros in construction turnover, 12,500 people may be employed for a year, according to the Economic Institute of Construction (EIB). This means that now 96,250 man years of employment are at stake. Moreover, enterprises can only invest limited amounts in the Netherlands because the construction of some 4,500 hectares of commercial zones is delayed. This harms our international competitive position'* (Bouwend Nederland et al. Press release 28-06 2005).

CDA and VVD Parliamentarians referred to this 'emergency call'[8] during the subsequent debates, especially prominent spokespersons were Liesbeth Spies of the CDA and Paul de Krom of the VVD. The alliance of interest groups did not state that literally 100,000 people were about to lose their jobs, but the distinction between man years of labour and jobs was lost quickly in the debates. CDA Parliamentarian De Neree tot Babberich stated for instance the following during the debate on the spring budget:

> *'The reason I bring up this subject again is that there is a problem with the link between spatial planning and the environment in regard to air quality. The Netherlands is one of the few countries in the European Union that has established such a link in its regulation, and we know what consequences it has. The Netherlands is under threat of becoming blocked. Thanks to this link, municipalities and other lower administrative bodies are in dire straits. It is said that projects totalling 7.5 billion euros, which cannot or almost cannot be executed, are delayed'.*[9]

Later in the debate he stated: *'The CDA faction will not have it that thanks to this strangling link between spatial planning and the environment in relation to air quality, 100,000 people will become unemployed in the construction sector'.*[10]

The lobby of the construction and transport sectors first attracted media attention in the financial and business daily Het Financieele Dagblad. In a series of articles, the newspaper examined the economic problems relating to the air quality regulation (Het Financieele Dagblad, 31 May 2005; 2 June 2005; 3 June 3 2005; 28 June 28 2005), and after the press release, the daily newspapers Parool (28 June 2005) Telegraaf (28 June 2005), and NRC (28 June 2005) also adopted the story of the job losses and the foreseen financial damages.

After these two debates, the pro-infrastructure camp's storyline unfolded as follows: 'Brussels prescribed unworkable standards that have been implemented much too strictly because of "the link", a particular feature of Dutch environmental policy. This

8. Kamerstukken II 2005-2006, 30 175 nr. 9, p. 2.
9. Handelingen II 2004- 2005 nr. 96, p. 5781.
10. Idem p. 5781.

legal and administrative problem will cost 100.000 jobs and billions of euros. The solution is to separate administrative decisions from quality standards'.[11] Moreover, legal professionals such as Niels Koeman en Chris Backes lent their weight to the notion that the Council of State interpreted the AQO too strictly, or that uncoupling the link was a viable solution. The opinions of these two professionals have been cited in the media outlets pressure groups such as VNO-NCW's magazine 'Forum' (Van den Broek & Smit 2006) and the pressure groups have made use of their arguments, especially concerning the release of the link. Therefore I include these legal professionals in the pro-infrastructure camp, the claims they make are similar to the claims of the pressure groups: the Netherlands does not need to be blocked due to air quality.

The definition of the problem in this storyline was different from that of the health-based version, because it presented the air quality clash as an administrative issue that was bad for business. The pro-infrastructure camp pushed for recognition of the administrative economic problem posed by air quality regulation, and wanted to uncouple the link. This was in conflict with the pro-health camp's narrative, in which the health aspects were emphasised. This storyline called for a defence of the link.

The Government found itself stuck between these conflicting storylines. In 2005 Van Geel argued that the European directives seemed to demand such a link, because spatial planning projects had considerable effects on air quality, and the Government was obliged to do everything possible to prevent exceedances. He claimed that it was doubtful whether releasing the link was even possible under European law, and speculated that the link was 'essential' because spatial planning dealt with the weighing of different interests that all demanded public space.[12] The link ensured that this deliberation would take place early on in the process. Van Geel considered that the only real solution would be an improvement in air quality,[13] and in 2006 he added the argument that dropping the link was not condoned by the Council of State.[14]

It is likely that fear of the Council of State's opinion influenced Van Geel's own position. The Council of State Advisory Division harshly rejected both Van Geel's Ministerial Decree to repair the Air Quality Order 2001 as well as his letter of 30 September 2004 on the doctrine that the standards were only valid in places where people lived. He would not easily risk another collision with the Advisory Division.

---

11. This is my condensed version of a storyline that resurfaced in many debates, although the exact wording differed from time to time.
12. Kamerstukken II 2004-2005, 30 175 nr. 2, p. 4.
13. Kamerstukken II 2004-2005, 30 175 nr. 2, p. 4.
14. Kamerstukken II 2005-2006, 30 489 nr. 12.

7.2 End 2005 – 2006: legislative initiatives

At the end of 2005, when the Prinsjesdag package was proposed, the budgets for the Ministry of V&W and VROM were gloomy; they warned about delays in important projects,[15] and the VROM budget mentioned that tension had become manifest between the interests of public health and economic development.[16] Projects that would have been realised under the Emergency Law on Road Expansion were hit especially hard. These expansions incurred hefty delays because of the strict interpretation of the Air Quality Order by the Council of State. The administrative Court demanded additional research as to whether projects conformed to air quality standards, but only one research institute in the Netherlands was equipped to handle these detailed investigations. As a result, more delays were foreseen.[17]

A new law was needed to relieve the delays confronting infrastructure projects and to make sure that the standards were met, but the debates only deepened and became more heated. The regulatory initiatives and debates from 2005 and 2006 are discussed in this section.

7.2.1 *The Air Quality Order 2005*

After the failure of the Ministerial Decree in March 2005, the onus was on Van Geel to quickly repair the current Air Quality Order 2001. He needed to find a regulation that made infrastructure development feasible again but that did not lower the standards or entirely uncouple the link.

During the debate on 26 April, he was given the green light by Parliament to devise a new amended Air Quality Order. Already on 10 May 2005 he offered a concept of it (henceforth the AQO 2005) to Parliament. The AQO 2005 was basically the old Air Quality Order 2001 but with two changes. In the first adjustment, it became possible to deduct the sea salt element in Particulate Matter from the total. Since sea salt forms a part of the mixture of particulates in the air above the Netherlands, this meant the measured PM concentration could be reduced by a couple of micrograms. This would at least help a little to reach the standards, though it was not a structural solution.

The second adjustment had more structural legal consequences, as it became possible to balance a project's negative effects on air quality with other possible positive effects. For instance, the construction of a road in a certain area might create extra exceedances on the curb side, but it could also unburden other roads in the vicinity.

15. Kamerstukken II 2005-2006, 30 300 A nr. 2, p. 14.
16. Kamerstukken II 2005-2006, 30 300 XI nr. 2, p. 6.
17. Kamerstukken II 2005-2006, 28 679 nr. 50, p. 2.

The construction would be admissible if the balance on the whole were to be positive: that is, if it were to lead ultimately to improvements in air quality.

The link between air quality standards and concrete administrative decisions was kept, but the Government hoped the balancing approach would make it more flexible. The Council of State was sceptical, however, and indicated that balancing was restricted in 'time, space, and scope' (Council of State 2005d). This restricted the possibility of balancing positive and negative effects between different projects in a different area. The administration had to demonstrate that a single project, independent of other projects or other measures, would improve air quality more than damage it. This demand was a tall order, however, considering that the projects under threat of cancellation were generally harmful.

In its advice, the Council of State urged the Government to think beyond the traditional approach of implementing European environmental directives in Dutch law by closely conforming to the terminology of the Law on Environmental Management. According to the Council of State, this reflex led to unnecessarily complex regulation, as the recommendation was basically an admonition to come up with a new legal format to integrate European directives.

Van Geel carried the Air Quality Order 2005 through Parliament though, and the regulation was adopted on 23 June (Stb. 2005, 316). It was known that the AQO 2005 would be simply a stopgap. Parliament still insisted on regulating air quality by means of a formal law, which was a heavier regulatory instrument than a General Administrative Order. Van Geel began to work on a new law immediately after the AQO 2005 was adopted.

7.2.2 *The programmatic approach and the role of legal experts in 2005/2006*

Even during the time the AQO 2005 was being conceived, Van Geel and the VROM Ministry were looking for ways to end the blockade on a more permanent basis. Civil servants from VROM invited legal scholars – among whom were Niels Koeman and Chris Backes – well known specialists on administrative law, to a brainstorm session in an unassuming office space in the Dutch city of Utrecht.[18] A couple of more of such meetings would follow.

The main idea that emerged from the session was to use the balancing approach from the AQO 2005, but to extend it to larger areas and to different projects and measures. Instead of balancing on a project-by-project basis, projects would be combined and fitted together in a broader programme. Projects that were harmful because they worsened air quality would be listed on a kind of balance sheet, and the negative

18. Niels Koeman, interview.

effects of these projects would then need to be balanced by a package of measures that would improve air quality. The sum of all the projects and measures should result in air quality that met the standards. In essence, the approach worked like a kind of score sheet, and as long as the balance remained positive for air quality, new projects could be realised. In August 2005, Van Geel informed Parliament by letter of this new strategy regarding the problem of air quality. He termed it the 'programmatic approach'.[19]

In addition to the balancing of projects and measures, another characteristic of the approach was that projects that did not harm air quality 'to a significant extent' did not have to be listed separately. The effects of such small projects would be included when the background concentrations of pollutants were determined. This possibility of a programmatic approach would form the core of a forthcoming new proposal for a formal law, which would give the Minister the competency to draft such a comprehensive programme.

This approach amounted to a regulatory innovation. The exact way the regulation would work was not explained in Van Geel's short letter, but he did inform members of Parliament that he had sent the concept of the new law to the Council of State for advice. If all went smoothly, the law could be adopted in spring 2006. This timetable turned out to be overly optimistic, as we will see in the following sections, because the negotiation process lasted all through 2006 and was finalised as late as autumn 2007. Following the advice of the Advisory Division, the proposal would be sent to Parliament.

Even though the proposal was partly the work of legal professionals, the Advisory Division was critical again in its advice of 11 November 2005 (Council of State 2005e). The Council of State Advisory Division mostly criticised the complexity of the proposal. Firstly, it contained a *'complicated system of programmes and plans which seem to overlap partially'* (Council of State 2005e, p. 1/2). Secondly, the Council was also not certain whether this programmatic approach complete with a balancing sheet of projects and measures constituted an adequate transposition of the European Directives. In the end, the European Court would have to judge whether the Council of State was referring again to the possibility of asking prejudicial questions. Thirdly, the Council considered that the legal protection civilians had against the programmatic approach itself and against individual decisions by lower administrative bodies was unclear. The proposal allowed plaintiffs to appeal against the whole programme of measures and projects, but the Council considered that this was unwelcome. The programme contained general measures that applied to the Netherlands as a whole, and such measures were not open to appeal. The Government had intended to limit appeal to the whole programme instead of individual projects, but the Advisory Division indicated

19. Kamerstukken II 2004-2005, 29 667 nr. 21, p. 1.

that this route was also not desirable, as projects should still be open to appeal because the measures might not have the estimated beneficial effects. The Council of State therefore considered that:

> *'In such cases, the air quality aspect cannot be ruled out in an appeal procedure considering the obligation the Netherlands is under in the scope of the European Directives'* (Council of State 2005e. p. 13).

The Council of State once again took up a position as the critical guardian of citizens' access to the court, and warned that this proposal might not do justice to the EU Directives and to the Dutch system of decentralisation. Nonetheless, the opinions of legal professionals played an important role in shaping the debates and outcomes of the air quality clash. The Council of State was critical, but the direction proposed in the programmatic approach would determine the subsequent developments. It started when the links was questioned in the articles in het Financieele Dagblad by Niels Koeman, but also the opinions of Chris Backes featured during numerous Parliamentary debates[20]. Moreover, Parliament had a plenary session with these professionals in which they explained their ideas.[21]

7.2.3 *A revised proposal for the Air Quality Law*

In spring 2006, Van Geel submitted a revised proposal for a new Air Quality Law to Parliament. Liesbeth Spies, prominent spokesperson of the CDA in this file, insisted that this proposal was 'Council of State proof'.[22] The proposal consisted of a change in the Law of Environmental Management that made the programmatic approach possible. The Minister would also gain the competency to declare that some projects would not influence air quality to a significant extent.

In a letter of 10 March, Van Geel explained that the proposal had been accepted in the Ministerial Council,[23] and, as was customary, the proposal was sent to Parliament together with the Explanatory Memorandum, the advice of the Council of State, and Van Geel's comments on the advice.[24] In his elaborate discussion of the comments made by the Council of State, Van Geel highlighted that the programmatic approach was a co-operative one between the state and lower administrative bodies, and together they would prepare the lists of projects and measures. Van Geel considered that the approach was necessary to facilitate urgent housing, infrastructure, and other spatial planning projects.[25] The programme would be named the National Cooperation Plan Air Quality (NSL).

20. Kamerstukken II 2005-2006, 30 489 nr. 7, p. 6, 9, 21, Kamerstukken II 2005–2006, 30 175 nr. 15, p. 6.
21. Lisa Vermeer, e-mail correspondence, 1-7, 2013.
22. Kamerstukken II 2005-2006, 30 175 nr. 17.
23. Kamerstukken II 2005-2006, 30 175 nr. 16.
24. Kamerstukken II 2005-2006, 30 489 nr. 2, nr. 3 and nr. 4.
25. Kamerstukken II 2005-2006, 30 489 nr. 4, p. 6.

In Parliament the proposal proved to be controversial and Parliamentarians wished to know what they could expect from it, as it was a complex and new type of regulation about which they required advice. The pro-infrastructure parties in particular feared that this new law would fail again before the Council of State (Interview Wijnand Duyvendak). On April 20, the Second Chamber of Parliament convened for a round-table discussion with legal experts, health professionals, pressure groups, and lower administrative bodies regarding the contents and implications of the Air Quality Law and its main component, the NSL.[26] Participants included, among others, the legal scholar Chris Backes, the environmental campaigner Joris Wijnhoven, and the epidemiologist Bert Brunekreef.

On the basis of the report drafted by the Parliamentary Commission of VROM and the response by the Secretary of State,[27] it became apparent that the general idea of the law was considered to be viable. However, much in the exact practical application needed to be fleshed out, as the literature indicated that legal scholars remained divided on the issue. The solution was supported by legal scholar Niels Koeman, who considered it was not necessary to link decisions directly to the standards, but that it was enough to list projects in a plan or programme (Koeman 2005b, p. 504). Chris Backes favoured a different solution to loosen the hold that the air quality standards had on infrastructure development (Backes, 2006b, 88).

With the NSL, Van Geel intended to appease critics of the link without disconnecting it entirely, but by making it 'flexible'.[28] Projects would no longer be linked directly to air quality norms, but would be taken up in a broad programme designed to make sure that air quality standards were complied with everywhere.

The programmatic approach constituted a new legal feature in spatial planning policy and law, as air quality standards would no longer directly affect lower administrative decisions. This entailed a break with the old regime in which environmental quality standards had direct spatial consequences: namely, polluting activities could no longer be condoned in an area where standards were exceeded.

7.2.4 *The struggle over the programmatic approach in 2006*

The NSL would have important consequences for the traditional relations between lower administrative bodies and the central Government. It broke with the practice of linking spatial planning directly to environmental quality, and it influenced the decision-making procedures around infrastructure projects. These projects had previously been assessed on a project-by-project basis, but now they would have to fit within an overarching

26. Lisa Vermeer, e-mail correspondence, 1-7, 2013.
27. Kamerstukken II 2005-2006, 30 489 nr. 7, and nr. 8.
28. Kamerstukken II 2005-2006, 30 489 nr. 8, p. 42.

long-term plan. The plan would be drafted centrally, and therefore this new approach would deviate from the trend towards decentralisation in the Netherlands. It is understandable that this change of policy led to heated discussions in Parliament.

The policy documents indicate that the opposition between the various actors concerned especially the position of the Council of State, the European Union, and the desirability of the NSL. There are many debates, questions, and other Parliamentary documents in which air quality was mentioned in 2006, but here I will concentrate on five key documents dealing with the air quality law: the report from the regular Commission of VROM;[29] Secretary of State Van Geel's response to this report;[30] a set of Parliamentary questions and their answers supplied by Van Geel on 14 June;[31] and the minutes of two debates in September and October 2006, one with the regular commissions of VROM and V&W;[32] and a plenary session in October 2006.[33]

The regular Parliamentary commission of VROM debated environmental questions, but also issued reports about legislative proposals such as the Air Quality Law. In the report on the proposal for the Air Quality Law,[34] various political parties brought forth their considerations about the proposal. As could be expected, parties that formed the pro-health camp were critical. They would have preferred a more ambitious package of measures in order to improve air quality instead of a proposal aimed at reaching the standards by way of balancing. The pro-health camp stuck to its argument that the Government was not doing anything to improve air quality, but was concerned only with devising tricks to 'wiggle out' from under the standards. The PvdA disliked the bureaucratic character of the proposal, and feared it would be used to show that the standards had been met simply on paper rather than that it was a strategy to actually improve air quality.[35]

In the September debate, Both Duyvendak and Samsom spoke about the approach as a 'Bureaucratic Monster',[36] and Van Velzen (SP) called it a 'Legislative Abomination'.[37] The pro-health camp warned that this complex piece of legislation would only make consultancy agencies happy, and urged for more measures to achieve good air quality. They especially feared that the NSL would be used to 'smuggle in' dirty projects. Samsom stated in this regard:

29. Kamerstukken II 2005-2006, 30 489 nr. 7.
30. Kamerstukken II 2005-2006, 30 489 nr. 8.
31. Kamerstukken II 2005-2006, 30 489 nr. 12.
32. Kamerstukken II 2006-2007, 30 489 nr. 25.
33. Handelingen II 2006-2007 nr. 14, pp. 884-897.
34. Kamerstukken II 2005-2006, 30 489 nr. 7.
35. Kamerstukken II 2005-2006, 30 489 nr. 7, p. 9.
36. Kamerstukken II 2006-2007, 30 489 nr. 25, p. 9.
37. Kamerstukken II 2006-2007, 30 489 nr. 25, p. 18.

'Let me be clear about this, the NSL is in the eyes of my faction the ultimate smuggle strategy. It is not meant to clean the air, but to gain approval for construction projects with the given quality of the air, approval which it would not get otherwise'.[38]

The CDA was the most supportive of the new approach, but the VVD was less so. The CDA faction emphasised the broad consensus on which the new law could draw, and that referred mostly to pro-development groups such as the VNO-NCW, NEPROM, and the transport branch. Milieudefensie remained an avowed opponent of the NSL (J. Wijnhoven, interview).

From the written questions put to Van Geel, it became clear that the pro-infrastructure faction VVD was not happy, as it preferred a total release of the link.[39] Before Van Geel could answer these questions, he received further advice from the Council of State Advisory Division (Council of State 2006b). On June the first 2006, the Council of State Advisory Division issued another warning that even though, strictly speaking, the European Directives did not prescribe such a link between projects and standards, it would still be necessary to determine whether individual projects had negative effects. From the standpoint of 'effective air quality management', the Council considered that some sort of test was unavoidable (Council of State 2006b). The link was not considered *absolutely* necessary, but a complete release of the link was also not justified. The Council considered it undesirable to scrap altogether the testing of projects against air quality standards.

Van Geel had to inform the Parliamentary commission, and especially the VVD faction, that he was afraid the Council of State would not accept doing away completely with the link. In fact, he stated this opinion multiple times in his responses to questions in 2006.[40] During the debates in September and October concerning the proposal for the Air Quality Law, the pro-infrastructure camp complained about the bureaucratic nature of the proposal as well. In the words of Liesbeth Spies:

'After an enormous number of polder negotiations, juridical procedures, and one consultation after the other in Brussels and with the Council of State, we have quite a complex legislative proposal [...] it all looks dirigistic and centralistic, and is not really a showcase for deregulation'.[41]

Spies was especially angry with the Council of State and the Europe Union, which she blamed for the Netherlands having to swallow this solution. The Council of State blocked attempts to 'de-link', and 'Brussels' burdened the Netherlands with unrealistic standards. In her opinion, therefore, there was nothing better than this 'ugly

38. Kamerstukken II 2006-2007, 30 489 nr. 25, p. 9.
39. Kamerstukken II 2005-2006, 30 489 nr. 12.
40. Kamerstukken II 2005-2006, 30 489 nr. 12, pp. 3, 4, 10, 11.
41. Kamerstukken II 2006-2007, 30 489 nr. 25, p. 4.

duckling', as she called the proposal. According to the VVD, the only good thing about the NSL was that the connection between administrative decisions and air quality standards was made less stringent. De Krom (VVD) felt that the whole programmatic approach with its expansive balance sheet of measures and projects would create a 'phantom reality' of numbers and measures. He claimed that there was a great deal of scientific uncertainty regarding PM10, and that harsh measures were not called for, especially because Brussels did not cooperate.[42] He would still favour the law, however, because it did provide a solution for a mess he felt was created in Brussels.

After being discussed by the Commission, the proposal was dealt with in a plenary session,[43] during which Parliamentarians proposed amendments to policy proposals; De Krom proposed one, for instance, that would bar the Council of State from testing projects directly against the air quality standards. This amendment contradicted squarely the line of the Council, but for De Krom it was a matter of principle. He insisted that Parliament rather than the Council of State made the laws. The introduction of this amendment caused the debate to revolve largely around the role of the Council of State in the air quality clash. De Krom stated that his amendment was not supported, only because the Secretary of State feared the Council's opinion. According to De Krom, this influence on the part of the Council of State was politically undesirable and constitutionally incorrect.

Liesbeth Spies (CDA) agreed with the proposed piece of legislation, but also felt that the Council of State rather than the legislator seemed to be pulling the strings. Together with De Krom, she took aim at both the Council of State and the European Commission: '*[...] my faction has the unnerving feeling that the European Commission and the Council of State are pulling the strings and that the legislator seems to be offside*'.[44] In the end, however, she went along with the proposal because she did not want to jeopardise a solution in favour of a principled fight with the Council of State.[45]

7.2.5 *The redefinition of the problem by the pro-infrastructure camp*

The programmatic approach was seen as necessary by the Government and by the pro-infrastructure camp, both of whom considered it the only solution albeit a poor one. The parties accepted it partially because the interest groups with whom they were allied also agreed that the programmatic approach had to be accepted.

The pro-health camp was put on the defensive during these debates. Its public health storyline was not directly refuted, but began to play a more marginal role, and the

---

42. Kamerstukken II 2006-2007, 30 489 nr. 25, p. 23.
43. Handelingen II 2006-2007 nr. 14, pp. 884-897.
44. Handelingen II 2006-2007 nr. 14, p. 889.
45. Handelingen II 2006-2007 nr. 14, p. 889.

managerial problem of how to de-link air quality standards from spatial planning decisions took centre stage. This discursive turn caused Duyvendak to lament that the 'true' nature of the problem – human health – was increasingly relegated to the background.[46]

The anger towards the European Union and especially the Council of State is of importance. The issue became more profound than a simple political difference of opinion, but concerned the institutional structure of the Netherlands itself, the role played by environmental law in the Dutch legal system, and the relationship between the administration and the judiciary.

The pro-infrastructure camp managed gradually to redefine the problem by hammering on the release of the link and emphasising the economic consequences of the air quality regulation. From a problem about public health, the issue evolved into an administrative problem regarding changes in the national system of environmental spatial law.

In table 8 below the positions of the actors that comprise the pro infrastructure camp have been listed

7.3 The adoption of the Air Quality Law and the Burgerveen-Leiden verdict

In February 2007, a new Cabinet, the fourth under PM Jan Peter Balkenende, was inaugurated. However, there was a change of parties, because instead of the VVD, which had lost the election, the PvdA and the small Christian faction ChristenUnie (CU) took seats in this Cabinet.

The PvdA was a member of the pro-health camp, and this change in the Cabinet could lead in theory to a stronger position for environmental concerns. The environment was now also no longer represented by a Secretary of State but by a fully-fledged Minister. Jacqueline Cramer (PvdA), a noted environmentalist, took over as Minister of VROM, and her portfolio included the environment and hence the air quality clash. The credo of this Cabinet was 'Work together Live together', and in the agreement between the three parties outlining their plans for the future, we may read that they sought to cooperate for growth, sustainability, respect, and solidarity.

The formation of this new Cabinet led to delays in the decision-making procedure pertaining to the Air Quality Law though. The law had been accepted in the Second Chamber of Parliament after the plenary debate in October 2006, but not yet by the First Chamber of Parliament, the Dutch Senate. By that time, however, the tone

46. Kamerstukken II 2006-2007, 30 489 nr. 25, p. 4.

Table 8

Actor	Main spokesperson	Position	Storyline
CDA	L. Spies	In favour of the proposal for a new air quality law because it would make infrastructure development possible again and the representatives of economic sectors endorse it.	Pressure groups have shown worrying economic figures, the block on infrastructure development should be removed without entering into open conflict with the Council of State.
VVD	P. De Krom	Sceptical of the proposal for a new law because it is unnecessarily complicated, in favour of releasing the link totally.	The link is causing large economic damage and should be released as soon as possible.
LPF	J. Eerdmans	Argues for a release of the link.	'Europe' and the Council of State have unnecessarily blocked the Netherlands and the way to unblock it is to release the link.
Pressure Groups (Aedes, Bouwend Nederland, MKB-Nederland, NEPROM, NVB, NVM, VBO, and the VNO-NCW)	Various	A solution needs to be found to make infrastructure development possible again a release of the link would be welcome	The air quality regulation threatens delays costing 7.7 billion Euro and threatening 100.000 jobs. The Netherlands has caused this blockade itself because of linking air quality standards and administrative decisions.

regarding air quality had become far more upbeat because the RIVM and MNP had released new figures indicating that air quality had suddenly improved. This unexpected turn of events is the subject of section 7.3.1.

The optimism was short-lived, however, because before the procedure to adopt the Air Quality Law could be finalised, the Council of State – on the instigation of the environmental group Milieudefensie – terminated one more high-profile road expansion. This was an important event because it thrust air quality back into the social and political spotlight. In 2007, air quality was no longer making the same headlines, but the annulment of the A4 road expansion showed that it still posed a significant threat to infrastructure development. The verdict itself is recounted in section 7.3.2, while the Parliamentary consequences of this verdict are under consideration in 7.3.3. In 7.3.4, the eventual adoption of the Air Quality Law is considered.

7.3.1 *The MNP's new figures: an improvement in air quality?*

In the midst of the debates in 2006, word had it that an unexpected solution might be in sight. The RIVM published new scientific figures, and stated publicly that the problem of air quality had been overestimated. After recalibrating the measuring and

Figure 4

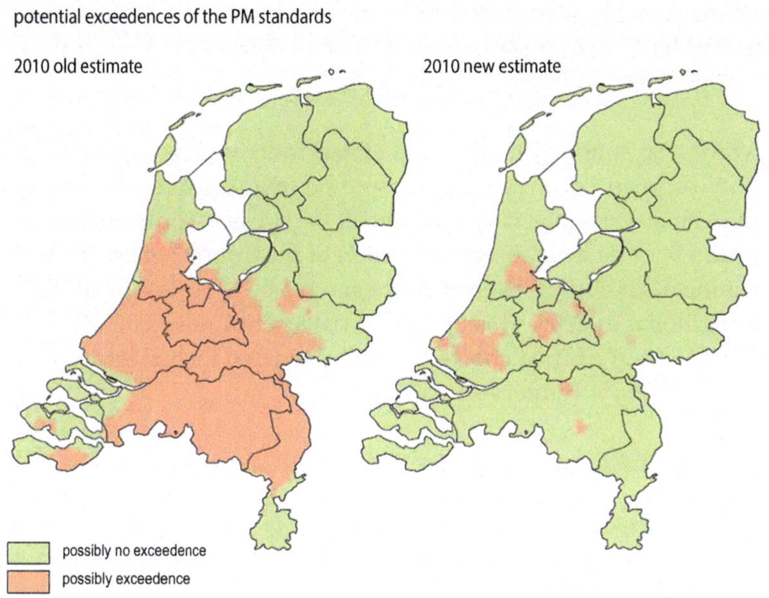

modelling tools, they found that the concentration levels on PM10 had been overestimated by 10 to 15%. Released in March 2006, these new figures caused the Government to view the issue of air quality in a less gloomy light than it had in 2005.

New measurements indicated that actual concentrations in rural areas especially had fallen unexpectedly between 2003 and 2005, and the most likely explanation for this fall was a change made to the measuring network (MNP 2006a, pp. 6/7). Researchers provided a map of exceedances expected in 2010 that included the new estimates both in a report in March and in the Environmental Balance 2006 (MNP 2006a, p. 12, MNP 2006b, p. 14). It is clear that the red spots had considerably diminished.

These new figures resulted in optimism within the Government, as many areas now suddenly complied with the standards. In several areas, concentrations were just a little above the standards, and with this 10 to 15% reduction, many areas passed the test. The Ministry of VROM's yearly report was drafted in May 2007, and in its overview of the past year it was much more upbeat on the issue of air quality.[47]

The Government and governing parties like the CDA tried to gain politically from the changing situation. On the basis of national measures and inventive solutions by municipalities, the VROM budget announced that predictions that air quality would

47. Kamerstukken II 2006-2007, 31 031 XI nr. 1.

lead to a further stagnation of construction work would not materialise.[48] Liesbeth Spies complimented the Government because its measures had resulted in cleaner air, and she hoped that citizens would notice the improvement.[49] She did not mention, however, that the change was due to a recalibration of the measuring and modelling rather than to any objective changes in the quality of the air.

The Ministry of V&W also sounded happy, and stated in its yearly report that the impact of air quality on road construction had become 'manageable'.[50] Under the new AQO 2005, the Ministry estimated that 80% of all the projects could now be realised,[51] and it hoped to limit delays caused by verdicts of the Council of State. On the whole, at the beginning of 2007, the Government seemed to be in far less trouble than it had been at the beginning of 2006. This optimism lasted until July 2007, when the Council of State Administrative Jurisdiction Division terminated the Leiderdorp road expansion in the case of the A4 Burgerveen-Leiden.

7.3.2 *The case of the A4 Burgerveen-Leiden*

On 13 February 2006, the administration took a 'trajectory decision' to expand the A4 highway between Burgerveen and Leiden, a noted traffic bottleneck. It had been in the top ten congestion-prone trajectories for years, and was also a bottleneck for air quality because PM10 and NO_2 levels around this highway were exceeded. Because under the Air Quality Order 2005 a project could be realised if it led to an improvement in air quality as a whole, the administration took a number of flanking measures to make sure the expansion did not worsen air quality or lead to an increase in noise pollution. The road would be broadened from four to six lanes, but it would also be dug so that it lay deeper than the surrounding landscape. In addition, the speed limit on the Burgerveen-Leiden trajectory would be lowered from 120 km/h to 80 km/h, and large screens would be planned along the side of the road to abate noise nuisance. The measures were outlined in a report drafted by the Ministry of V&W (V&W 2006).

On 19 April, the environmental pressure group Milieudefensie appealed against the decision to broaden the trajectory. As we saw, instigating legal procedures was a key ingredient in this environmental group's clean air campaign. In its appeal, Milieudefensie listed five reasons why the Council of State should annul the decision of the administration, with its core argument being that expanding the road would generate more traffic. This extra traffic in turn would offset the improvements proposed by the administration. There was no guarantee that the measures the Minister was proposing would ensure that standards were met (Milieudefensie 2006).

---

48. Kamerstukken II 2006-2007, 31 031 XI nr. 1, p. 13.
49. Kamerstukken II 2006-2007, 30 489 nr. 25.
50. Kamerstukken II 2006-2007, 31 031 XII nr. 1, p. 16.
51. Kamerstukken II 2006-2007, 31 031 XII nr. 1, p. 16.

The Council of State considered that the Milieudefensie objections were by and large justified, as the administration had not performed the necessary 'integral inquiry' into the consequences of the road expansion for the network of secondary and provincial roads.[52] The administration had also not made sufficiently credible that with the speed limit reduction to 80 km/h the air quality standards could be met. In addition, the research should have taken into account the possible air quality situation extending beyond 2012, the year foreseen for the opening of the roadworks. The verdict of the Council of State in July 2007[53] was a blow, especially because the Government thought that the air quality crisis was finally becoming manageable. Again the Council of State suspended the expansion of a road because the administration had not conducted a sufficiently thorough investigation.

7.3.3 *The effects of 'Burgerveen-Leiden' in Parliamentary discussions*

The cancellation of another high-profile road expansion could be regarded as a victory for Milieudefensie, but it turned out to be a Pyrrhic one. The A4 Burgerveen-Leiden court case was catalytic (VROMraad 2009, p. 15, Borgers 2012, p. 24), because after this verdict Parliamentarians started to criticise the environmental movement as well over its strategy of using the courts to further its clean air campaign. Expansion of the road was considered urgent in light of the Government's efforts to combat congestion.

In Parliament, the discussion about de-linking flared up again, but this time the environmental movement itself drew flack as well.[54] In a letter to Parliament, Minister Jacqueline Cramer stated that the issue of air quality standards was being used by pressure groups to put a halt to spatial developments. I consider this statement highly significant, because Cramer belonged to the pro-health camp; moreover, she had once been chairman of Milieudefensie. Such an argument could be expected from parties in the pro infrastructure camp, but apparently her position as Cabinet Minister made her more sensitive to infrastructure concerns. Minister Cramer stated that the discussion of the link had been 'reopened', thus suggesting a full uncoupling might be possible after all.

The Burgerveen-Leiden verdict was mentioned explicitly in the first important general debate of the new Parliamentary year, the 'General Review' (Algemene Beschouwingen). During the discussion, VVD Parliamentarian Mark Rutte stated that he sided: '[...]*with people that claim that it is necessary to exclude pressure groups from the possibility of participation who do not have a direct interest but always oppose road expansions such as Milieudefensie*'.[55]

52. ABRvS 200602152/1, tracébesluit Burgerveen-Leiden R.O. 2.29.
53. ABRvS 200602152/1, tracébesluit Burgerveen-Leiden.
54. Kamerstukken II 2007-2008, 30 175 nr. 32, p. 3.
55. Handelingen II 2007-2008 nr. 2, p. 39.

He called this environmental pressure group a bunch of 'professional trouble makers',[56] a label that would stick. It was mentioned in 2008 (Cobouw 22-03 2008) and in Parliament a number of times again.[57] As late as 2012, Bernard Wientjes, chairman of the employers association VNO-NCW, mentioned the Burgerveen-Leiden verdict, and stated publicly that Milieudefensie was not a trustworthy negotiation partner (de Volkskrant 12-11 2011).

The tone changed after the verdict, even among members of the pro-health camp. At the end of 2007, socialist Parliamentarians complained of the deadlock in which the country found itself. In a November 2007 discussion by the Second Chamber's transport committee, PvdA Parliamentarian Vermeij complained that decision making was 'running like syrup' because of the obstacles and the consensus culture concerning infrastructure projects.[58] Figures of speech like syrupy decision making and putting an end to the consensus culture had belonged previously to the repertoire of the pro-infrastructure camp. This development showed that the pro-infrastructure camp had largely succeeded in transforming the problem from a health issue into one caused by over-regulation, and that even elements of the previous pro-health camp now adhered to the storyline of the pro-infrastructure camp, namely that environmental policy should not get in the way of development. In the years following the air quality clash, the subsidies paid to Milieudefensie decreased, and, according to Duyvendak, this was done to punish them for their legal obstruction (Duyvendak, interview).

In addition to this rhetorical backlash against the environmental movement, the verdict led to the installation of a commission set up to investigate the possibility of speeding up procedures and providing answers regarding the 'intense juridification'[59] of infrastructure projects. The new V&W Minister Camiel Eurlings[60] asked former head of the DSM Company Peter Elverding to chair the Commission 'Sneller Besluitvorming Infrastructuur' (Quicker Decision Making Infrastructure), (V&W 2007). I consider the institution of this Commission and its report to be one of the results of the air quality clash; it is discussed further in section 7.4.2

7.3.4 *The adoption of the Air Quality Law 2007*

The Air Quality Law 2007 was adopted by Parliament on 11 October, and in November 2007 it appeared in the 'Staatsblad' as the Law of 11 October amending the Law on Environmental Management (Air Quality standards), the air quality law for short

56. Handelingen II 2007-2008 nr. 2, p. 39.
57. Handelingen II 2007-2008 nr. 74, p. 5164; Handelingen II 2009-2010, nr. 24, p. 2158.
58. Kamerstukken II 2007-2008, 29 385 nr. 15, p. 2.
59. Kamerstukken II 2007-2008, 29 385 nr. 15, p. 2.
60. Eurlings took over as Minister of Transport and Water Management from Carla Peijs in February 2007, at the start of the new CDA PVDA and Christian Union Cabinet.

(Stb. 2007, 414). As the title of the law indicated, it was essentially an amendment of the Dutch Law on Environmental Management.

The Law on Environmental Management was expanded with a new title – 5.2 – regarding standards for air quality. The new title in the Law on Environmental Management was a framework law that made establishment of the programmatic approach possible but did not stipulate the contents of the plan itself. Article 5.12 provided the legal ground for the administration to implement the programmatic approach known as the NSL. Under article 5.16 sub 4, the Government was also granted the power to establish that some smaller projects would not be included individually in the NSL. Harm relating to these projects was considered to be of 'not to a significant extent', as they did not contribute more than 3% of the air pollution concentrations in certain areas.[61]

With the Air Quality Law, however, the problems were not resolved. Though it gave the Government the competency to establish the programmatic approach, as yet no such plan existed. Moreover, standards still were being exceeded the throughout the country, and the measures foreseen in the NSL would only take effect some years later. In the meantime, it was still possible for the administrative courts to cancel projects, as had been demonstrated by the Burgerveen verdict.

To finally put an end to the possibility of delays and annulments, the Government needed a derogation from the EU, meaning that the European Commission would approve giving the Netherlands extra time to meet the standards. The Dutch attempts to obtain this are discussed in the next chapter. After the Commission granted the derogation based on the NSL, the Dutch Government would formally adopt it. This repeal was finally granted on 8 April 2009.

The programmatic approach called NSL was the primary outcome of the PM clash in terms of institutional legal change. Therefore it is discussed elaborately in the next section, together with other outcomes of the clash, such as the Elverding Report and the General Administrative Order on sensitive destinations.

7.4 The period after the Air Quality Law 2007

The Air Quality Law paved the way for an administrative solution to the air quality clash. I call it an administrative solution, because it was not certain that the programmatic approach would in fact clean the air. It would depend on the measures foreseen in the NSL and on whether these measures would be implemented correctly.

61. The General Administrative Order, which stipulated the exact percentage a project should contribute to affect air quality 'to a significant extent', was valid from 30 October 2007 (Stb. 2007, 440).

Nonetheless, the NSL was the main ingredient as regards the Air Quality Law and the eventual solution to the air quality clash, and for this reason, it is the first item under discussion in the present section.

The second development considered here is the report of the Elverding Commission. This report indicated a way forward to reach the goal if quicker decision making in infrastructural projects. This Commission was installed as a reaction to the delays that among others the Emergency Law on Road Expansion had encountered during the clash.

A third development was the promulgation of the General Administrative Order on Sensitive Destinations (GAOSD),[62] which was an outgrowth of discussions on the Air Quality Law in 2006. It stipulated that the Minister would establish distances from the highway within which schools and other areas where children and elderly persons lived could not be built. The GAOSD entered into force in January 2009 and is discussed to highlight the way in which the highway had come to symbolise a dangerous place.

The Elverding Report indicated the possibility of repairing the Emergency Law on Road Expansion. This proposal is the subject of 7.4.4. I consider it and important symbolic development, indicating that by the time it entered into force in April 2009 infrastructure development was nearly back on track.

In 7.4.5 the end of the clash and the regulatory changes brought about by the acceptance of the new programmatic approach are recounted.

7.4.1 The NSL

The NSL was not well received when it was presented to Parliament. Wijnand Duyvendak was still not convinced that it would do anything for air quality (Duyvendak, interview), and Milieudefensie was also still against the programmatic approach. Joris Wijnhoven felt it was just an exercise to 'calculate the air clean' (Wijnhoven, interview). This very description also appeared in a more recent monograph on the resolution of the PM clash by Kees van Oosten, one of the founders of a pressure group campaigning for air quality measures (Van Oosten 2012).

The NSL was indeed a complex piece of regulation. As indicated earlier, the method used in the NSL was reminiscent of the 'balancing' approach introduced in the Air Quality Order 2005. The difference was that it allowed for balancing on a much wider scale. To that end, the country was divided into regions and agglomerations, and for each of these the balance sheet with projects and measures was drawn up that should insure that the region complied with the air quality standards at a given time. The VROM Ministry was responsible for this extended programme, and it was up to them to change or amend the NSL if needed.

62. Besluit Gevoelige Bestemmingen (Stb. 2009, 14).

The NSL also played a large part in persuading Brussels to be lenient with the Netherlands in terms of the air quality standards. The Netherlands intended to use the NSL to show that it was doing everything possible to reach the standards. The NSL therefore served three purposes: reaching the standards; facilitating infrastructure projects; and functioning as a proposal with which the Netherlands could win derogation from the standards.[63] This derogation had already been taken into account in the NSL, and thus it could not be formally adopted until after the EU granted the Netherlands extra time. After the postponement ended, air quality should be good enough to meet the standards. In practice, this meant that the NSL had to make sure the PM10 standards were met in 2011 and the NO_2 standards in 2015.

It was clear from the start that the NSL would be a daunting project; a complete balance sheet of projects and measures that had to guarantee that air quality standards would be met within 3 to 6 years. Local measures were introduced in the NSL, but generic measures were also included that would have significant impacts throughout the country.

To be able to make all the necessary calculations, a tool was developed by Goudappel Coffeng, a professional consultancy in the realm of transport and space. In this computerised application, air quality hotspots were made visible by using a map of the Netherlands, on which coloured areas indicated possible hotspots. The tool shows the situation in 2008 and the projections for 2011, 2015, and 2020.[64] The NSL was a fine exercise in calculation. To make these predictions, a number of variables had to be included, and it had to make calculations and predictions relating to air quality concentrations from a countrywide scale all the way down to street level. Minister Cramer stated that the calculation tool was unique in its combination of large-scale assessment and detailed calculation.[65] In order to perform these calculations and predictions, scenarios compiled by the Dutch Central Bureau of Statistics (CPB) concerning economic development and climate, energy, and air pollution policy were used.[66] This detailed level of analysis lent credence to De Krom's suggestion that it might create a 'paper reality'. It also made the plan impenetrable to outsiders, which fuelled the suspicion in some quarters of the pro-health camp that it was just a strategy to calculate the air clean.

The NSL contained a number of ambitious and politically salient measures, such as a raise in diesel taxes and, most importantly, road pricing. Since the programme fell under the responsibility of VROM, it granted this Ministry a potentially powerful weapon. The road- pricing scheme via a charge per kilometre was particularly controversial.

63. Kamerstukken I 2006-2007, 30 489 nr. J.
64. The tool can be found on the website saneringstool, last accessed 03-11 2009.
65. Kamerstukken I 2007-2008, 30 489 nr N, p. 19.
66. Idem, p. 17.

Perhaps the change of Cabinet from centre right to centre left made it easier to include controversial pro-environmental measures in the NSL such as road pricing. However, the inclusion of this measure had already been foreseen in September 2006 and was mentioned in the debate during that month,[67] so also in this case there was continuity between the policies of Balkenende 4 and the earlier Cabinets.

Although Parliamentarians were hardly enthusiastic when this programmatic approach was first proposed, a large majority accepted it. From the bigger parties, only GroenLinks and D66 voted against.

7.4.2 *The Elverding Report and speeding up procedures*

After the Burgerveen-Leiden verdict, the environmental movement came under fire for exploiting the law to resist politically desired infrastructure projects. The Government instated two commissions – Verheijen and Elverding – to investigate the legal problems surrounding infrastructure development and ways to speed up procedures. Both Commissions were fruits of the air quality clash.

The Verheijen Commission was asked to investigate the consequences of uncertainties in modelling and calculating air quality for court cases.[68] Air quality was tested against limit values that were treated as absolute minimum standards by the courts, while scientists pointed towards the considerable uncertainty to which air quality measurement and modelling was subject. These uncertainties had significant consequences, because it was difficult to guarantee whether a project would stay within the boundaries set by the standards.

The Verheijen Commission concluded that uncertainties in the air quality file were a 'fact of life', and the tendency to calculate to increasingly detail would not lead to certainty. The Commission pleaded for continued review and for keeping a package of measures in reserve for when the estimates proved to be overly optimistic. The measures could then be introduced to reach the required quality standard. Similarly, some measures might turn out to be not necessary after all, because air quality was better than expected. In those cases, it should be easy to scrap measures.

These conclusions were in accordance with those of the Elverding Commission, but the Elverding Report was broader in scope and contained recommendations to speed up decision-making procedures with regard to infrastructure development. The Commission was established to investigate ways in which decision making on this issue could be improved. In addition, the commission was assigned to investigate the 'link' between spatial planning and environmental standards, and how to curb the

67. Kamerstukken II 2006-2007, 30 489 nr. 25, p. 10.
68. Kamerstukken II 2009-2010, 30 175 nr. 98.

juridification of infrastructure development.[69] The Commission was also expected to look into the role played by the Council of State in procedures.[70]

The pro-infrastructure factions had high hopes that the Elverding Commission would produce a favourable report from their perspective. According to the VVD, the Commission had to combat 'rule terrorism' and 'make sure the Netherlands was not blocked anymore'.[71]

The Elverding Commission did not recommend all the pro-infrastructure camp had hoped for, but it did advise a critical rethinking of the way Dutch policy handled environmental spatial conflicts. According to the report, the core of the problem was the increasing complexity of society, the emerging juridification, the growth of economy and mobility, and the tendency of empowered citizens to fight for their own interests, such as a good living environment (Elverding Commission 2008, p. 4).

According to the Elverding Commission, we often desired to 'reconcile the irreconcilable' in the Netherlands, and this led to inertia or to the impossibility of executing a decision that had already been taken (Elverding Commission 2008, p. 4). It was considered that our consensus culture spawned this desire, which often led to delays and protracted procedures. Elverding proposed to let go of the broad margins of uncertainty and detailed calculations with which projects had to be justified and instead proposed to work with rules of thumb and additional flanking measures if later in the process projects tended to have more negative effects on environmental quality as earlier anticipated.

As regards participation, the Elverding Report proposed limiting the various access points in which participants and pressure groups could present their point of view. Access to the court would not disappear, but the number opportunities for taking a decision to court during the whole process would be diminished. The conclusions of the Elverding Report authorised the administration to limit the influence of the administrative judge by restricting the opportunities for pressure groups to use the courts as a political arena.

7.4.3 The General Administrative Order Sensitive Destinations

This GAOSD was important from a discursive perspective because it sanctioned the notion that the highway was a dangerous place. In December 2008, the Government issued this General Administrative Order based on the air quality law that stipulated that schools, kindergartens, and homes for the elderly may not be built within 300 metres of

69. Kamerstukken II 2007-2008, 30 504 nr. 4, p. 4.
70. Kamerstukken II 2007-2008, 30 175 nr. 42, p. 6.
71. Kamerstukken II 2007-2008, 29 385 nr. 15, p. 1.

the highway and not within 50 metres of provincial roads if air quality standards in the area were exceeded. The third article 5.16A of the Dutch Law on Environmental Management stipulated that the administration should determine rules to prevent the exposure to air pollution of people who were sensitive to it, by making sure new building projects close to sources of air pollution were constructed some distance from the road.

This regulation was a small victory for the pro-health camp, because Jan Pronk, responding to the problem in Overschie, had already proposed increasing the distance of schools and other buildings from roads. I include this discussion of it here, however, because it is an odd type of regulation in the context of the programmatic approach. It is reminiscent of the older type of regulation on environmental standards, viewing them as absolute limits. I consider the GAOSD important because of its symbolic significance. It portrayed highways as dangerous places, and sanctioned the ideas put forward by epidemiologists that living and working near highways was unhealthy. In contrast to the programmatic approach, it was an example of the notion that development and environmental wellbeing excluded each other.

7.4.4 *The repair of the Emergency Law on Road Expansion*

One of the recommendations from the Elverding Commission was to repair the Emergency Law on Road Expansion. To that end the Cabinet adopted a proposal from Ministers Eurlings and Cramer to change the Emergency Law on Road Expansion on the 11th of July 2008. By changing the Emergency Law of 2003 in ways that Elverding had recommended and by anticipating on the acceptance of the NSL in Parliament the Cabinet hoped to start work on the road expansions as soon as possible.

The Council of State voiced criticism against this proposal as well. It feared that the wording of the proposal could be interpreted as limiting the judge in choosing what kind of research data on air quality he or she would base her judgment in appeals against road expansions. In his cometary on the advice Minister Eurlings tried to assuage this concern by insisting that appeal remained open. In Parliament the proposal was supported by a broad majority and accepted by the second chamber of Parliament on the 13th of January 2009.

The emergence of the repaired Emergency Law on Road Expansion was a clear victory for the pro-infrastructure camp. The Emergency Law from 2003 was the most obvious victim of the air quality clash and the prestigious road expansions were halted for six years to the dismay of parties such as CDA and VVD. This time parties belonging to the pro-health camp like D66 and the SP voted in favour of the Emergency Law, only GroenLinks voted against. Eurlings had even intended to broaden the scope of the Emergency Law beyond roads covered by the law from 2003, but that attempt was thwarted by opposition from coalition partner PvdA and other parties. After the First Chamber in Parliament approved of the law with a similarly large majority on the 31st of March 2009 it entered into force in April 2009 as the Law of

April 2nd 2009, on changing the Emergency Law on Road Expansion and the Trajectory Law in relation to the simplification of the research requirements (Law on speedier decision making road projects). The acceptance of this law with a broad majority shows that during the later years of the clash the pro-health camp was on the defensive and lost ground

7.4.5 *The acceptance of the NSL and the end of the clash*

The new programmatic approach laid down in the NSL had three main goals, making sure the limit values for air quality would be met in the Netherlands, making sure spatial planning could be put back on track and to persuade Brussels to give the Netherlands a derogation in order for it to be able to implement the NSL. By and large the NSL delivered what it was intended to accomplish, in any case it achieved the last two aims. The NSL was accepted by the European Commission as a sufficiently persuasive clean air plan to grant a derogation based on the new CAFE Directive promulgated on 21st of May 2008. In April 2009 the derogation was officially granted and the champagne popped at the VROM Ministry (Interview Marjan van der Giezen).

Moreover, on the 31st of March 2010, the NSL stood the legal acid test when the Council of State accepted the NSL as sufficient evidence that the air quality standards had been observed by the administration.[72] The Council's judgment indicated that the NSL will not be extensively tested in court, but only a very marginal test is applied.[73] The acceptance of the NSL as justification in court took away the remaining fears among project developers and politicians that infrastructure development could be halted by legal procedures. With the acceptance of the NSL in court, the air quality clash was finally laid to rest. The NSL's political and legal effects were its most important outcomes in regard to regulatory change.

The programmatic approach is important for two other reasons besides ending the air quality clash: firstly, it included controversial environmental measures; secondly, the NSL represented a departure from the usual way that spatial/environmental conflicts (De Roo 2003) were dealt with in Dutch environmental spatial law. Apart from a highly controversial road-pricing scheme, measures foreseen included a premium when old cars were traded in for new ones and a tax differentiation when cars were purchased, providing a tax benefit for buying a cleaner car. Moreover, municipalities would be granted the power to establish zones within which polluting trucks and lorries were not allowed to enter. In addition, an increased tax on diesel fuel prices – generally opposed by pro-infrastructure parties and their allies – was included. These measures were substantial, and many had been on the wish lists of the pro-health camp. They can be listed as political gains for this camp, even though the measures themselves still had to be introduced.

72. ABRvS 200900883/1, 24 Oktoberplein, 31 March 2010.

73. Website STAB, last accessed 27-09 2015.

Institutionally, the NSL led to three important changes on spatial/environmental policy and law. Firstly, infrastructure development became possible when it fitted within a prospective programme. Projects were no longer judged as solitary events, but were considered within the context of other projects within an integral plan. The Dutch tradition to test projects on a project-by-project basis (WRR 1998, p. 112) had been left behind. The programmatic approach was much more integral, long-term oriented and forward looking. Such an approach was normal in environmental policy making, the NMP being the most obvious example, but now it was imported into infrastructure development, traditionally the responsibility of other Ministries. For instance, it forced the Ministries of VROM and V&W to closely coordinate their policies, since VROM was responsible for the NSL.[74]

The NSL precipitated further legal change as well. Traditionally, environmental quality standards led to picket lines being formed around certain areas. If quality standards were transgressed in an area, it became barred from further polluting activities. An area was considered to have a certain 'environmental space for use' ('Milieugebruiksruimte'). If the environmental space was being used, it could not be used by other polluting activities. With the NSL, these consequences were avoided. Projects could be continued in places where standards were exceeded, provided that in the end the standards would be met, because their harmful effects were counter-balanced by beneficial measures.

According to Niels Koeman, this new programmatic approach provided the necessary flexibility while making sure the quality of the environment remained intact, and would play a role in future environmental/spatial policy and law. The VROMraad, the advisory council of the Ministry of VROM, voiced similar opinions in a recommendation provided in 2009 when it proposed introducing the 'environmental spatial plan' as an addition to ordinary land use plans (VROMraad 2009). This forward-looking programmatic approach was reminiscent of the NSL. The lower administrative body drew up a score sheet and set environmental and spatial targets that it would like to realise in the future. The polluting activities of projects, industry, and transport were placed on one side of the balance and environmental improvements on the other. In this way, it was clear how much 'environmental space' there was left before the standards were exceeded. An environmental spatial plan could make further demands on companies to use cleaner production processes, for instance, or environmental space could be gained by constructing tangential thoroughfares and thereby unburdening roads leading through the area.

Many of these recommendations were indeed adopted in newer environmental/spatial legislation. The first of these regulations was the 'crisis and recovery law' (Crisis en Herstelwet) in March 2010 (Stb. 2010, 135). Moreover, elements of it were also incorporated into the Nature Protection Law (Niels Koeman, interview).

74. The two Ministries have been merged in a new Ministry for Infrastructure and Environment (2010), but the question of whether the NSL and further laws based on the same concept played a part in the decision to merge these policy fields lies outside the scope of this study.

While not embraced initially, the NSL opened up new ways of thinking about the relationship between spatial planning and environmental law. This new relationship was one of balancing interests and flexible decision making as well as of long-term planning and comprehensive regulation. It was itself an outgrowth of a legal battle, the struggle between the Council of State and the Government regarding the possibilities of repairing the AQO 2001.

7.5 Concluding remarks

In this chapter the emergence of a counter-claims maker in the form of the pro-infrastructure camp in April 2005 has been discussed. The clash between these camps was resolved in 2010 by the acceptance of the programmatic approach and by the

Table 9

	Pro-health camp	Pro-infrastructure camp	Government	Council of State
Actors	GL PvdA SP, D66, Milieudefensie, Scientific institutes and health professionals	CDA (partially), VVD, LPF, pro-infrastructure interest groups (Bouwend Nederland, NEPROM, VNO-NCW, etc.)	Ministry of VROM/ Ministry of V&W	Advisory Division / Administrative Jurisdiction Division
Spokespersons	W. Duyvendak (GL) D. Samsom (PvdA), J. Wijnhoven, Milieudefensie, RIVM / GGD	P. De Krom (VVD), L. Spies (CDA) J. Eerdmans (LPF). E. Brinkman (Bouwend Nederland) C. Oudshoorn (VNO-NCW)	P. van Geel (CDA)	N.A.
Interest	Maintaining public health.	Maintaining adequate infrastructure for mobility; jobs.	Finding a solution to the infrastructure blockade that would pass the Council of State.	Keeping policy in line with legal requirements.
Storyline	With its irresponsible policies, the Government endangered the health of its citizens and skirted the limits of European rules. It is violating the rights of its citizens.	The EU standards and the Dutch interpretation of them are unworkable and bad for business. They are a heavy economic burden and threaten 100,000 jobs.	The Netherlands is doing all that it reasonably can to keep the air clean. The European standards are difficult to work with and should not be interpreted overly strict.	The Govt. is not above the law. It has bound itself to strict European standards and therefore the European arena is the place to renegotiate. It should better examine whether the implementation of EU law fits within the national system.

acceptance of the NSL by the Council of State. Now it is time to make a comparison between the four main actors in the air quality clash, the pro-health camp, the pro-infrastructure camp, the Council of State and the Government. This comparison is valid for the years of the air quality clash. This caveat is made because an actor like the Government is present throughout this study and the position of the Government of the late 1980s for example markedly differed from its position in the air quality clash.

Table 9 above makes clear that not one but two conflicts are present in the air quality clash. The first conflict is the one between health interests versus those of infrastructure development. The second struggle is between policy institutions and legal ones. The Council of State was pitted against the Government over the question to what extent Council of State as Advisory Division and as Administrative Jurisdiction Division could influence the course of policy making. In this final section both these conflicts will be analysed further. In 7.5.1 and 7.5.2 the conflict of health interests versus infrastructure development takes central stage. The result of the clash is interpreted as a return to a weak form of ecological modernisation, a compromise between these two traditionally opposing interests. Moreover, perceptions regarding auto-mobility and the use of space have altered because of the air quality clash. In 7.5.3 and 7.5.4 we will discuss the conflict of law versus policy. The role of legal professionals is elaborated on in 7.5.3 and the position of the Council of State is reviewed. In 7.5.4 the position of the Council of State is analysed in relation to the legality of precaution.

7.5.1 *Weak ecological modernisation as a compromise strategy*

The air quality clash was characterised different terms by the pro-health and the pro- infrastructure camp. The pro-health camp put forward the storyline that poor air quality was causing health problems. The more extreme factions in this camp argued that the issue of air quality should put a stop to the further development of auto-mobility. Environmental NGOs belonging to the camp put forward a narrative in which good air quality was presented as a basic right, and the legal verdicts as well as the text of the Air Quality Order were presented as proof of this notion.

It is conspicuous that the parties that made up the pro-infrastructure camp did not confront these claims head on. They did not deny the health effects of mobility. In the beginning of the clash they had done so to a certain extent by arguing that the problem would be solved over time and that stringent measures now were not reasonable. This line was unsuccessful. It invited the charge that the Government did not care about the health of its people. In the spring of 2005 the pro-infrastructure camp chose a different strategy, it recast the debate from a debate about health policy to a debate about over regulation. They started using a storyline in which air quality regulation was considered as an administrative economic problem that was bad for business The pro-infrastructure camp pointed out that a stringent implementation of the Air Quality Directives was 'bad for business', a storyline that took off following the press release generated by lobby groups of the construction and service sectors.

The 'bad for business' storyline managed to wrest the terms of the debate away from the pro-health camp by presenting the air quality clash principally as an issue of bad regulation: namely, infrastructure development in the country was blocked, owing to rules drafted in Europe without regard for the national situation. It was felt that constraints could be lifted by releasing the link between standards and administrative decisions. This storyline started gradually to dominate, and it shifted the emphasis from the health effects, thus blunting the thrust of the pro-health narrative.

The storylines used by the two camps were both not amenable to ecological modernisation. The pro-health version argued for a strongly precautionary approach, and favoured limits on pollution for reasons of public health. The human rights storyline claimed that environmental quality concerns should supersede all other interests because a fundamental right was at stake. The 'bad for business' storyline meanwhile argued that environmental regulation was costly, and its impacts should be minimised. This line too rejected the idea that environmental protection and economic development could coincide. Only the Government used more reconciliatory narratives. Nonetheless, both camps had to accept a compromise that was framed in eco-modernistic terms and that bore the hallmarks of eco-modernistic policy. The programmatic approach was presented by the Government as a showcase of the notion of smart solutions. It argued that both environmental interests as well as those related to infrastructure could be accommodated by way of balancing projects and measures-based scientific calculation.

The air quality clash showed the necessity of maintaining the balancing act between environmental and economic interests. An eco-modernistic discourse was still the most credible candidate to reconcile these interests. When the Balkenende Cabinets did away with ecological modernisation all but in name, the environmental movement and progressive political parties rallied against the plans and started to use air quality as a crowbar to force more environmental measures. However, when the verdicts of the Council of State made infrastructure development highly problematic, conservative parties together with the construction and transport sectors began to pile pressure on the Government. The eventual compromise conformed to weak ecological modernisation again, because it assumed the possibility of co-development of infrastructure and environmental protection within one and the same plan. Moreover, the plan was drafted after participation and consultation took place involving stakeholders, and the plan was heavily based on science, management, and an integral approach, all of which are eco-modernistic characteristics. In fact, the direction taken by environmental/spatial law after the programmatic approach was more in conformity with ecological modernisation than the old system that inked quality standards directly to administrative decisions was.

Apparently neither the interests of environment and health nor infrastructure development and economy could lay claim to a position of dominance – both mustered powerful discursive and political resources. By framing clean air as a right, the pro-health camp managed to establish a discourse coalition between itself and the Council of State. However, the economic figures presented by the pro-infrastructure camp exerted a powerful influence on the Government. In the current Dutch political constellation it is necessary to cater to both interests at the same time. When the economy appears in jeopardy the public and politicians become anxious and wish to lift economic barriers as soon as possible, but when health claims are made, they cannot be brushed aside either and appearing not to take them seriously is a very risky political strategy. The two interests hold each other in a kind of headlock and therefore a solution along eco-modernist lines is still a necessary one.

The eco-modernist plan re-established consensus, but not a wholehearted one. The NSL resembles exactly the sort of plan that Elverding warned against; an attempt to reconcile the irreconcilable by relying on very detailed calculations and consensus. The environmental movement and Government policy makers retained their wary attitude towards each other, and this 'new' consensus did not convey the same progressive enthusiasm that had characterised ecological modernisation under Winsemius. The weak variation currently holding sway lacks the kind of optimism that characterised ecological modernisation earlier on. It depends on a very meticulous form of book keeping, constantly making sure the balance will not tip the wrong way. In that sense it fits the fearful 'moderate' forms of policy currently in vogue (Arnoldussen 2009). Risks should not be taken and we have to be constantly weary not to miss a threat to either our health or the economy.

7.5.2 *Discursive positions and changing perceptions*

The air quality clash resulted in important discursive changes. The foray into mobility was of importance from the perspective of the broader environmental conflict between environmental and health interests on one side and economic interests on the other. Mobility had hitherto been shunned as a target for environmental campaigns, NGOs were afraid to lose the sympathy of the public when they broached the subject, and the many car owners in the Netherlands had considerable lobbying power. For this reason, the attempt to modernise transport and mobility ecologically had failed. However, the issue of air quality opened up space to attack the mobility policy and to argue for restrictions on cars, which indeed appeared in the form of environmental zones where lorries were not allowed, reductions of the speed limit, and court-ordered annulments of road expansions.

More generally, the notion took root that mobility was not only a collective good but a collective bad as well. Highways became considered dangerous places alongside of which it was not responsible to build schools and other facilities. Air quality itself evolved from being an obscure subject on the mind of decision makers to an important factor to be considered in drafting plans and permitting projects. Air quality was now in the forefront of debates on infrastructure development, and such a prominent position would have been unthinkable without the air quality clash.

Moreover, the idea that the Netherlands was a dirty country with high environmental pressure became commonplace. The air quality situation in the Netherlands became emblematic for a storyline that it was an environmentally dirty country with a Government that did not do enough. Such notions can be found on the GroenLinks website, and even in 'The Economist' (GroenLinks website, The Economist Website).

Finally, sustainable mobility appeared more prominently again on the agenda, and the legal arrangements on spatial planning were overhauled. In the sphere of mobility, policy was forced to move in a more long-term and comprehensive direction. Projects needed to be researched more thoroughly with regard to their environmental effects. The administrative courts demanded more from the administration than a short-term and piecemeal investigation – it called for long-term guarantees that a project would not do damage to the environment and to public health.

These discursive changes forced policy makers to rethink the way the Netherlands would develop structurally in the long term, and with integral consideration concerning environmental as well as economic factors. In summary, the air quality clash took environmental matters to the forefront of the political discussion in much the same way acid rain had done in the 1980s. In contrast to the case of acid rain, however, health issues and mobility issues were now at the forefront of the debate. Demands were made to curtail economic development, not because of concerns for eco-systems but out of health concerns. During the air quality clash, the pro-health camp demanded guarantees that economic development would not harm the health of citizens. The crucial importance of public health arguments in this conflict indicate that in the early 2000s, people were less inclined to sacrifice public health for economic gains, while in the 1980s the eco-systems argument carried the day. Moreover, mobility was put in the spotlight, while in previous conflicts industrialization had been considered the prime environmental enemy.

From a sociological point of view this discursive development is interesting because it indicates a deepening of the environmental conflict. In the 1980s it played out on the level of the collective. The environment as a whole was pitted against industrial production, a collective good. Health – in the end an individual good – clashed with (auto)mobility, also an individualised way of transportation. The collective dimension of environmental conflict is still important, consider the problem of climate change, but the PM case shows it also acquired an individual dimension. The air quality clash aroused public opinion because their individual interests were at stake. People were concerned over their own health, instead of over the more abstract issue of the health of ecosystems. Similarly, the interests of individual car owners were on the line. They did not like the 80 kilometre speed limit because it meant they had to drive slower. This too is a much more concrete interest than industrial production of economic growth per se. I conclude that the air quality clash is important among other things because it brought the environmental conflict back to the behaviour and interests of individuals, making it tangible.

From a methodological perspective, there is another noteworthy theoretical aspect of the air quality clash regarding both discourse analysis and the actor centred perspective, the two social constructivist methodologies used throughout the study. The Government's storyline did not alter markedly when the parties that made up the Cabinet changed. The socialist PvdA had always belonged in the pro-health camp, but when PvdA Minister Cramer took over from her Christian Democrat colleague, the tone of the Government remained the same. Cramer even seemed to indicate that she was willing to discuss uncoupling the link altogether. Just as the previous Cabinet had done, she complained about interest groups using legal procedures to score political points. This was all the more surprising because Minister Cramer had been a Milieudefensie chairperson in the 1980s. Apparently discursive commitments could depend on the political position that parties held, which suggested that storylines were at least to some degree tied to positions, and that it did not matter much which actor held the position in question.

Discourse theoreticians generally hold that the opinions one can form depend on the belief systems and discursive commitments to which one is bound. More actor centred theoreticians such as Joel Best regard rhetoric as a part of the strategy that actors used to get their interests represented in the media and eventually in legislation. However, the findings in the air quality clash suggest that neither is the case. Actors do not use rhetoric at will. Minister Cramer was and is a noted environmentalist, however, when she became a Minister she had to adopt the storylines used by the Government even from before the time she took power. However, storylines also do not shape the position one takes. Discourse theoreticians are generally convinced that the way an actor is positioned depends on his or her discursive commitments and these are tied to believe systems. Instead, positions or roles shape the storylines one can choose from.

The relevance of a shift of attention towards positions and roles is that social problem construction would not primarily be about actors strategically putting forth their interests, or of storylines and discourse coalitions gaining adherents. Social problem construction would depend on the organisational set up of social practices. To give a concrete example, the problem of crime should not be considered as the result of the actions of actors demanding private property to be protected, or storylines about what behaviour is inherently malicious, but by the creation of a police force, of criminologists and of prison wardens. Likewise, environmental problems will emerge on the wings of modern measuring equipment, a class of environmental scientists and on the organisation of the Ministry of the environment. To be sure, such an insight would still fall within the bounds of discourse theory, but would downplay the importance of narrative devices and draw attention to social practices and the material conditions under which social problems emerge.

7.5.3 *The role of legal professionals and the function of legal discussion*

Even though legal professionals have played a prominent role in the search for a solution of the air quality clash, they have not been included in the table on page 175. There are numerous reasons for not including them. The legal professionals remained isolated individuals, Niels Koeman and Chris Backes are mainly discussed in this chapter. Later on they were joined by others, who added to the legal discussion, for instance Tieman et al. (2007). They did not operate as a group however or a sub group of a larger institution. They did not agree among themselves as to what the best solution would be, so it cannot be said that they shared a similar storyline or subscribed to one of the storylines above. I also feel that it would be artificial to include them in one of the two camps. It is true that they offered their advice to the Government mostly to find a solution, but they did not represent infrastructure or health interests as such.

Even though they did not belong in one of the camps, did not share a storyline and had different legal perspectives on problem, the legal professionals were instrumental in finding a solution that could be accepted by the Council of State, and appease the pro-health and pro-infrastructure camps.

The way in which these legal professionals managed to offer fertile ideas for a solution displays the fruitful function of dogmatic legal discussion. Through contradiction and discussion a legally acceptable solution was eventually reached. In the air quality clash the law shows itself as a tenacious, but also resilient social practice with a high degree of self-rejuvenating power. The Council of State, itself a legal institution, cut short the attempts of the Government to skirt the legal consequences of the implementation of the European directives in the Air Quality Order. This tenacity created much dismay among policy makers, but it did provide space for a whole new way to handle environmental spatial conflicts to develop.

The development of this new programmatic approach can by and large be ascribed to the discussions between legal professionals, Government representatives and of course the tenacity of the Council of State, both the Advisory Division and the Administrative Jurisdiction Division. The law created the blockade, but also removed it and in the process created a policy instrument that was more viable than the old link. The air quality clash shows that lawyers should not easily budge to the wishes of policy makers. Even though it may lead to delays, the fruits of legal progress are remarkable as well. For these fruits to mature though, a dialogic and patient approach is needed. Law progresses by the gradual sharpening of arguments within the limits of legal discourse. There is much to be said for letting it progress in this way, because it ensures legitimate solutions, limits the Government in pushing for a practical but controversial solution and ensures procedural justice. Juridification is seen as an important problem of contemporary decision making procedures and that is understandable. However, it may be a boon as well as a poison, because, as the air quality clash shows, the legal order has a way of reinventing itself and offering more durable solutions than the quick fixes offered by politicians.

7.5.4 The Council of State's position in the air quality clash

When residents and pressure groups took recourse to the law to force the administration to heed their demands, the highest administrative court became a key actor in the air quality clash. The legal strategy of the environmental NGOs resulted in a second conflict that played out in the shadow of the first one, and it had to do with the question as to which institutions had the final say in matters of governance. In the shadow of the clash between the pro-health and the pro-infrastructure camp, the Government and the Council of State struggled over the question of whether the link between administrative decisions and environmental quality standards should be kept. Eventually, the Council of State yielded partially to the wishes of the Government, but for years it resisted the Government's plans to weaken the standards – only after a wholly new policy instrument was drafted did the Council give up its resistance.

In terms of actor centred social constructivism, the courts are seen as social problem workers, because they determine the influence of policy on the ground. The interpretation by the courts may itself lead to further rounds of claims making. This dynamic is visible in the air quality clash as well. The strict interpretation of the AQO 2001 led to counter-claims by organisations belonging to the pro-infrastructure camp. However, more is at stake here. In the air quality clash the Council of State acted as a policy maker itself. It forced the Government not only to amend the Air Quality Order, but to come up with a wholly novel way of conducting spatial planning. The new programmatic approach is as much the result of the tenacity of the Council of State as it is from the initiatives of the Government and the acumen of legal experts.

This was remarkable in a country where primacy was considered to lie with politics, and where the Council of State was often criticised for being too Government friendly. Two questions come into play here. The first has to do with how the Council could put up such strong resistance. This question can be answered by pointing towards the characteristics of the Dutch legal system and the characteristic composition of the

Council of State, respectively. The second question has to do with why the Council of State resisted the wishes of the Government in this particular case. This question is harder to answer, but a reflection on utterances by the Council of State Advisory Division may provide an answer.

The powerbase of the Council of State

The Council of State was able to withstand political pressure to agree to a quick solution by pointing towards the European nature of the air quality standards. The origin of the Dutch regulation lies in Europe, and European law takes precedence over national law in the Dutch legal system. This precedence of European Law gives the Council of State a considerable amount of leverage, because it means that the Council of State Administrative Jurisdiction Division could decide to apply the European air quality standards directly if the Government were to amend the regulation in a direction deemed unlawful by the Council. The Council of State did not indicate that it would do so, but it did denote that it would need to pose prejudicial questions to the European Court in Strasbourg. Such a procedure would take a long time and would likely result in more delays – a scenario the Government intended to avoid. The Council of State maintained the storyline that the issue of the regulatory blockade of infrastructure development in the Netherlands should be resolved in the European arena, precisely where the problem had originated.

The second reason that the Council of State had a large powerbase was that it consisted of two divisions: the Administrative Jurisdiction Division – that acts as the highest administrative court and the Advisory Division – that acts as the Government's highest advisory council. The Government was caught in a bind, because it feared that negative advice could foreshadow that a proposed amendment would fail when examined by the Administrative Jurisdiction Division in court cases. Formally, the two divisions are strictly separated, but the extent of this separation is not clear, since they remain different parts of the same organisation and they do share expertise. When the Council of State advised against a release of the link, for instance, Van Geel became very cautious because he feared that the Administrative Jurisdiction Division would terminate his amendments before the court if the matter were brought to its attention.

The power of the Council of State had its limits though. When the political pressure mounted during the later years of the air quality clash it could not maintain its defiant stance.

The Council of State and a legality of precaution

This study provides answers to the question why the Council of State would oppose the Government in this instance, while in other cases it was considered supportive of Government policy. Firstly, the Council intended to protect access to the courts for citizens in environmental matters. Cutting appeal procedures also means that the highest administrative court would have less influence. Secondly, the Council of State is slowly moving in a precautionary direction.

The Council already opposed the curtailment of legal procedural rights when the Advisory Division advised on the Emergency Law on Road Expansion in 2002. The release of the link would force the Council to relinquish a powerful tool with which the Court could keep Government policy in check. This political explanation is compelling, but there are indications that more was at stake.

In regard to the second point mentioned above, during the air quality clash, the Council of State asked more from the administration than it had done previously – this included much more thorough research into the effects on air quality before administrative decisions were given the green light. The requirements pertaining to research subsequent to decision making were so demanding that in the early years of the air quality clash only one research institute was able meet them. I take this heightened burden of proof as an indication of a more precautionary approach to questions of air quality. There are further instances in which the Council of State used precautionary arguments outside of the field of air quality: for instance, the cockle fishing judgment of September 2004 (Pieterman 2010).[75]

However, the Administrative Jurisdiction Division did not couch its decisions in precautionary terms, nor did it take recourse to the precautionary principle, or to other typically precautionary elements. In fact, it did not have to, because it could simply provide a strict legal interpretation of the text of the Air Quality Order. In any case, the Administrative Jurisdiction Division did not cut the Government any slack, and it disregarded the Government intentions while interpreting the Air Quality Order.

Because they contain policy advice, the convictions of the Council of State Advisory Division are easier to assess. Within the domain of policy advice, it is less difficult to spot precautionary elements, and in a number of places the Advisory Division took a more precautionary line than it had done earlier.

In its advice on the 2006 budget, the Council of State remarked that both the Dutch Government and the EU had a one-sided focus on economic wellbeing. Moreover, it considered that Western European economies, including that of the Dutch, presupposed high economic growth as a given. The Council of State Advisory Division noted that structural weaknesses could make it necessary to forego these growth ambitions, and it named Particulate Matter as one of the weaknesses. I interpret such statements as a signal by the Council of State, especially because the budget of 2006 was treated in late 2005, the year in which the air quality clash was raging without a solution in sight. It felt that politics was one sided focussed on the economy.

75. ABRvS 200409107/2, kokkelvissers.

In its policy advice, the Council of State also expressed the need to keep access to the courts open, and it responded critically to proposed limits being imposed on citizens wanting to appeal against projects that had been incorporated in the programmatic approach. It is no surprise that an institution like the Council of State would uphold access to court; however, it is also an indication that it was concerned about the rights of citizens to appeal against infringements of their participatory rights, even if NGOs used the Courts to further their own goals and if this strategy clearly stagnated economic development. The increased demand for research into the effects of projects on air quality, the concern over the rights of citizens to be heard in infrastructure development, and the concern with the sustainable economic development of the Netherlands fit within a trend towards a legality of precaution in which the prevention of environmental and health damage was emphasised. The precautionary concerns of the Council of State were not very openly stated, however, and I consider it a partial explanation of its defence of the link in addition to the institutional one of securing access to the administrative courts. This link is an instrument that makes sure precautionary considerations remain an important factor in infrastructure development, however.

We will now leave the Dutch arena and return to Europe. The Dutch NSL by and large restored consensus at the national level, but in order for it to be successful the Dutch Government needed a postponement of the air quality standards, because without it, the threat of annulments would remain.

Timeline air quality clash and its resolution 2005 – 2010

2005 April	Mobilisation of pro-infrastructure camp in and outside Parliament. Van Geel promised work on new Air Quality Order
2005 May	RIVM published new figures; possibly 18000 people dying 10 years earlier because of PM.
2005 May	AQO 2005 entered into force
2005 June	Coalition of lobby groups issued press release asking for a release of the link
2005 Aug.	Van Geel suggested 'programmatic approach' for dealing with regulatory problem of air quality
2005 Sept.	Prinsjesdag package unveiled
2005 Nov.	Council of State Advisory Division critical over first plans programmatic approach
2006 March	RIVM issued new figures, air quality concentrations overestimated by 15%
2006 March	Van Geel sends proposal Air Quality Law and Programmatic Approach to Parliament
2006 Oct	Air Quality Law accepted in 2nd Chamber of Parliament, heavy criticism of EU and Council of State during Parliamentary debates.
2007 Feb.	Cabinet Balkenende 4 took over, PvdA back in Cabinet, Cramer (PvdA) Minister of VROM, Eurlings (CDA) Minister of V&W
2007 July	Council of State Administrative Jurisdiction Division annuls decision to expand. Road trajectory Burgerveen – Leiden. ABRvS 200602152/1
2007 Sept.	Heavy criticism of environmental movement during plenary Parliamentary debate. Institution of Elverding Commission on decision making procedures
2007 Oct.	Air Quality Law accepted in Parliament
2007 Nov.	Air Quality Law entered into force
2008 April	Report Elverding Commission published
2008 June	Contents of the NSL offered to Parliament by way of a Letter by Minister Cramer
2009 Jan.	General Administrative Order on Sensitive Destinations (GAOSD) entered into force
2009 April	Emergency Law on Road Expansion repaired
2009 Aug.	NSL entered into force
2010 March	NSL accepted by Council of State Administrative Jurisdiction Division as sufficient justification for administrative decisions ABRvS 200900883/1

8 The CAFE programme: a new air quality strategy for the European Union 2001 – 2009

Introduction

European air quality regulation was partly responsible for the Dutch air quality clash, since without those strict European standards from the 1990s there would have been no conflict. However, the key to resolving the issue lay with Brussels as well. The Dutch programmatic approach would only work if the Dutch administration were to be granted extra time to comply with the standards. Therefore, much depended on a new Air Quality Directive that had already been on the agenda since 2001. This new Directive would be drafted within the ambit of an overarching EU programme called Clean Air For Europe (CAFE), the aim of which was to harmonise clean air regulation. This thematic strategy regarding air pollution would consist of a new integrated approach to air quality and acidification, and within its scope the existing Directives would be reviewed.

The importance of this new harmonisation programme merits a look at the way the Dutch Government operated in the preparatory phase of the CAFE programme and in the negotiations on the Air Quality Directive itself. During the clash, the pro-infrastructure camp accused the Government of being too lenient with regard to Brussels, and not able to defend its own interests. The Government was portrayed as trying to be 'holier than thou' when it came to air quality (Smit 2006), and the Netherlands was called 'the village dunce of Europe' (Gekke Henkie van Europa).[1] However, was it really true that the Dutch always played by the book? This research shows that this picture needs to be nuanced.

In addition a look at the CAFE programme provides insights into the development of air quality policy in the first decade of the new century in general. In the first decade of the new century air quality policy reached a level of complexity and sophistication that caused it to get mired in technicality. Moreover, the complexity of this issue caused the line between scientist and policy maker to blur. The expert committees within the scope of CAFE became politicised and Member States and pressure groups vied for influence within them. By relying on science the European Commission sought to decrease politicisation, but in the case of CAFE the strategy backfired.

1. Handelingen II 2007-2008, nr. 38, p. 2997.

A third reason why the CAFE process merits attention is that in this case a precautionary approach was thwarted by a concerted effort of pressure groups and Member States that argued for flexible regulation. In CAFE environmental precaution had to give way to economic interests, despite a cost benefit assessment showing that a more ambitious policy was cost-effective.

In section 8.1, a global outline of the CAFE programme is presented. Knowledge of the institutional setup of the programme and the goals that it was intended to achieve is necessary to understand the subsequent procedure and the nature of the eventual outcomes of the CAFE programme, the CAFE strategy, and the Air Quality Directive from 2008, the CAFE Directive.

Section 8.2 is dedicated to the review of the Air Quality Directives from the 1990s, especially Directive 99/30 containing limit values for NO_2 and PM10. This review was important for the Netherlands because, as stated in the explanatory Memorandum for the original Dutch Air Quality Order from 2001, the Government expected that standards would be adjusted downwards. In the context of CAFE, the Dutch Government tried to obtain a review that would allow it to lessen the standard's impact on the Dutch economy.

Section 8.3 deals with the actual adoption of the CAFE strategy and the Air Quality Directive 2008. The Dutch position in the debates in the European Council of Ministers and the European Parliament are considered and contrasted with the Netherlands' position on environmental issues in the 1990s. This comparison is used to determine to what extent the Dutch position changed due to internal problems regarding air quality regulation.

In section 8.4, the CAFE strategy goals are discussed. CAFE started as an ambitious attempt to draft a comprehensive policy based on the input of scientific experts, policy maker, and stakeholders, and it was to become an ambitious long-term programm within which various air pollution policies were integrated. The question is whether the high level of ambition and its complexity enhanced or reduced its effectiveness in practice. First we will take a look at CAFE's intentions and its organisational structure. The chapter concludes with an analysis in section 8.5 in which the above-mentioned considerations are discussed.

8.1 Aims and organisation of the CAFE programme

After adoption of the air quality Framework Directive and the subsequent Daughter Directives, the dust did not settle around air quality policy in the EU. After the first Daughter Directive was adopted formally in April 1999, three more followed, all based on the same Framework Directive – 96/62/EC – on ambient air quality assessment and management.[2] However, the European Commission was still not satisfied with the integration and comprehensiveness of the air quality policy.

Even before the first Daughter Directive was adopted, the idea for a new and comprehensive clean air strategy was floated in informal talks with Member States. As early as October 1998, an informal discussion document was sent to various Member States and stakeholders (Wettestad & Farmer 2001, p. 5). The new programme to integrate all policies dealing with air pollution would bear the name CAFE: Clean Air For Europe. After gathering information from Member States and other stakeholders, in 2001 the Commission unveiled its new clean air strategy.

The new approach was first mentioned publicly during the 'green week' of 2001, the biggest annual conference on European environmental policy making. CAFE was to become a major programme, with input from many stakeholders and Member States, and repercussions for numerous policy fields besides the environment. An impression of the aims of this programme within the wider context of EU environmental policy is provided in the first sub-section, where CAFE is discussed in the context of the 6th Environmental Action Programme. The overall organisation of CAFE and its various sub-committees are discussed in the second sub-section.

8.1.1 The aims of the CAFE programme and its place in EU environmental policy

The CAFE programme was proposed by the European Commission in a Communication (European Commission 2001), in which the Commission justified this new initiative by considering that there were a number of priority air pollution problems. It mentioned Particulate Matter (PM), ground-level ozone, acidification, eutrophication, and the necessity of 'keeping a watchful eye' on pollutants not currently regulated (European Commission 2001, p. 3). The attention to urban air pollution by PM and Ozone was driven predominantly by health concerns, while problems relating to acidification were prioritised because of environmental impacts.

The main aim of the CAFE programme was to develop a:

> *'long term, strategic and integrated policy to protect against the effects of air pollution on human health and the environment. As required by the Treaty, the policy will aim at a high level of environmental protection based on the precautionary principle, taking into account the best available*

2. The Framework Directive 96/62/EC was followed by the previously mentioned Directive 2000/69/EC of the European Parliament and of the Council relating to limit values for benzene and carbon monoxide in ambient air, on 16 November 2000; Directive 2002/3/EC of the European Parliament and of the Council of 12 February 2002 relating to ozone in ambient air from 12th of February 2002; and Directive 2004/107/EC of the European Parliament and of the Council relating to arsenic, cadmium, mercury, nickel, and polycyclic aromatic hydrocarbons in ambient air on 5 December 2004.

> *scientific and technical data and the costs of benefits of action or lack of action'* (European Commission 2001, p. 7).

In addition to this main aim, specific objectives were mentioned. These were: 1) to collect scientific information about the effects of ambient air pollution and to develop modelling technology and refine indicators; 2) to review and to support the implementation of existing legislation, in particular the air quality Daughter Directives; 3) to make sure the necessary measures would be taken at the relevant level and to develop effective structural links with relevant areas of policy; 4) to devise an overall integral strategy to be revised at regular intervals and that determined future air quality objectives and cost-effective measures to reach them; and 5) to disseminate to the public the information gathered within the scope of the programme (European Commission 2001, p. 7). According to Commission official Peter Wicks, the CAFE programme would eventually integrate all EU legislation and related research on air quality under one umbrella (ENDS Europe, 26 April 2001). CAFE was intended to become an ongoing programme that would be updated and revised every five years.

The idea of a comprehensive overhaul of EU air pollution policies dated from 1998, but the general idea fit within the scope of the 6th Environmental Action Programme adopted in 2001 (6th EAP). This 6th EAP was important because it determined the direction of European environmental policy from 2002 to 2012, and the CAFE programme was considered to be one of the seven 'thematic strategies' mentioned in the plan. These strategies were a novelty, and formed the cornerstone of the 6th EAP. They covered seven policy fields: air, waste prevention and recycling, marine environment, soil, pesticides, natural resources, and urban environment, and were intended to make possible a more holistic and 'multidimensional' approach:

> *'Thematic strategies will consider the range of options and instruments required for dealing with a series of complex issues that require a broad and multi-dimensional approach, and will propose the necessary actions, involving where appropriate the European Parliament and the Council'* (6th EAP, sub 16).

The 6th EAP followed the principles and ideas of its predecessor, the 5th EAP,[3] but with differences in emphasis. The more 'Dutch' elements of the previous plan, such as working with target groups and market parties, were given a lower profile. The EU ventured towards a different approach, and expanded on two other tendencies embedded in the eco-modernist policy tradition: namely, comprehensive policy making and an emphasis on scientific input and presentation.

Whereas the 5th EAP provided a strategic roadmap as well as targets and timetables, the targets were largely absent from the 6th EAP. The newly introduced thematic strategies were frameworks within which committees of scientists and policy makers

3. Discussed extensively in chapter 4.

should set targets and objectives within the course of the programme. The thematic strategies would be subject to broad consultations of stakeholders and other social partners (6th EAP art. 4.3). Focusing on long-term and comprehensive policies, and relying on expert committees within the EU bureaucracies to supply the scientific underpinning, the strategies would highlight the multiple aspects of a single problem and treat them within the same framework.

8.1.2 *The organisation of CAFE*

This style of policy making was also evident in CAFE's organisational structure. The programme consisted essentially of a gathering of various working groups comprised of scientists and policy makers. Together they set out to formulate a coherent strategy on how to tackle air pollution in a broad and comprehensive way, and to hammer out legislative proposals and evaluations. CAFE's intention was to remain a science-based programme but involving stakeholders in order to give them a sense of ownership of the policy making process. This structure was intended to mitigate conflict among various participants. Together with the Communication on the CAFE programme itself, the Commission published a staff working paper that contained the organisational structure, SEC (2001)688. The programme's internal organisation was highly complex, and it is necessary to outline this complexity here, because the political manoeuvring in these various groups is the subject of this chapter.

The programme resided under the auspices of the Directorate General for the Environment, within which the CAFE secretariat was formed. The secretariat consisted of a small project management team that coordinated the programme as a whole. Two other groups consisting solely of civil servants from the Commission coordinated the communication between the DG Environment, other DGs and other services and bureaus of the Commission. Since they play no part in the discussion, they will not be mentioned further here.

The actual outcomes of the CAFE programme would be produced within groups that were considered external because they were not staffed by Commission representatives. According to the Commission Staff Working Paper, external groups fostered relations with stakeholders and experts to co-ordinate the technical work within CAFE, and to obtain technical contributions and input. Participation in those groups was mixed. Commission officials were included but so were representatives of the Member States. Stakeholder representatives such as those of the environmental movement and industrial and transport sectors were also present in some of these groups.

The most prominent external group was the Steering Group (SG), which advised the Commission on the programme's strategic direction. Chaired by the Secretariat, it was composed of representatives of Member States, the European Parliament, and stakeholder representatives, as well as of experts working within CAFE itself.

Technical discussions did not take place within this group, but members reviewed the data and gave policy advice. The group reported directly to the CAFE Secretariat.

The core business of most working groups within CAFE was to supply scientific data and make recommendations on a scientific basis. The technical work concerning modelling and other scientific work would be coordinated by the Technical Analysis Group (TAG). TAG consisted of a variety of representatives of DGs and expert bodies like the WHO, but neither industry nor the environmental movement had access to it. It would not do all the technical analysis and research work itself, but would be in charge of its coordination, and would make sure that the research agendas of the various sub-groups remained relevant for policy making. TAG's activities were to remain fully transparent and open to input from stakeholders via the SG and other groups within the CAFE structure. Stakeholders, however, were excluded explicitly from TAG, a feature of the programme that would lead to discussion, as recounted below.

The aforementioned groups comprised the organisational structure of CAFE, and their job was to coordinate all the work that was being done. Below these aforementioned groups, working groups conducted research on several of the aspects with which CAFE concerned itself. From the perspective of this study, these groups were as important as those discussed above. They did all the work 'on the ground', and ensured that goals outlined in sub-section 7.1.1 were achieved.

We may conclude from the CAFE website that five different working groups were set up and these groups were to supply report to the SG and to TAG.[4] The five groups mentioned are:

1. The Working Group on Implementation (WG Imp): This group was set up at the start of the process, and its goal was to assist Member States with the implementation of Air Quality Directives from the 1990s, and to help with implementation of the National Emission Ceilings Directive, which was mostly concerned with acidification;
2. The Working Group on Target Setting and Policy Assessment (WG TSP): This group dealt with the development of indicators, and drafted scenarios for the integrated assessment modelling[5] as well as conducted a cost-benefit analysis;
3. The Working Group on Particulate Matter (WG PM): This group was set up because PM was the pollutant of gravest concern. Its task was to examine the attainability of the PM10 limit values by Member States, and the scientific

4. Website CAFE docs accessed 13-12 2011.
5. In integrated assessment modelling. Knowledge of two or more research fields are brought together and combined. Because integrated assessments bring together and summarise information from diverse fields of study, they are often used as tools to help decision makers understand very complex environmental problems. (website integrated assessment accessed 25-09 2015).

information available on PM, in addition to writing a new position paper on the substance. The group was also tasked to assist the Commission in its review of Directive 99/30 by collecting information on the air quality situation regarding PM and on attainability of the targets;

4. The Working Group on National Emission Ceilings and Policy Instruments: This group concerned specifically the work on the new National Emission Ceilings Directive. It was still active by 2008, but not within the scope of CAFE. It is considered further in this chapter;
5. The Data Exchange Group: This group is listed on the website, but no documents are available. It does not seem to have been very active within the scope of CAFE, and presumably it was set up to facilitate the exchange of data from different Member States. It became a part of INSPIRE, a European Union project for sharing spatial and environmental data between Member States. The most important Working Groups for us are those on Implementation and on Particulate Matter, as the Dutch Government needed a review of Air Quality Directive 99/30, in which their concerns were presented.

Within CAFE, a number of environmental consultancy agencies were important as well, because CAFE was to be based on a cost-benefit analysis. This assessment was to establish the level of ambition the CAFE strategy could have, based on balancing the costs and benefits of the programme, such as life years gained from better health and the protection of ecosystems. The cost-benefit analysis was conducted by AEA Technology (AEA T), a global sustainability consultancy firm. The scenarios they devised may be found on the CAFE website.

Figure 5: EU Air Quality policy

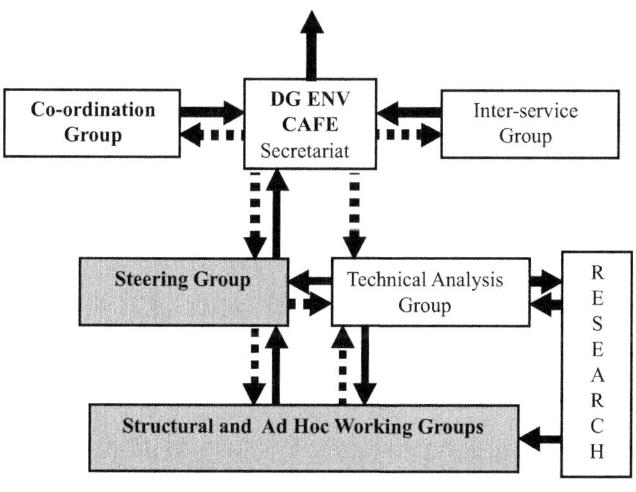

Transparency was a cornerstone of the CAFE programme, and the minutes of the Steering Group were made public as well as those of working groups on Implementation, Particulate Matter, and Target Setting and Policy Assessment. The scenarios developed in the context of the integrated assessment modelling and the cost-benefit analysis was made available as well. The internal discussions of the CAFE Secretariat and the discussions of the Technical Analysis Group were not made available though. The Communication regarding the CAFE strategy contained the following flow chart detailing the organisational structure:

It is important to note that the strategy presupposed a strict separation of policy issues and scientific discussion. In this regard, the proposal for the CAFE strategy considered:

> '*It will be important for stakeholder involvement to be managed in such a way as to ensure that it clarifies and strengthens the scientific basis of policy-making. This means that a clear distinction between technical and political issues must be made, and that the quality of scientific and technical argument remains the sole arbiter in technical discussions. It is also important that the stakeholders themselves are transparent concerning both the technical justification and the political motivation for their comments* (CEC(2001)688, p. 8)'.

This separation is of importance in light of the subsequent politicisation that occurred in the context of the CAFE programme

8.2 The review of Directive 99/30 and the politicisation of an expert-driven programme

Various tasks were undertaken within the CAFE programme, one of which was the review of the earlier Air Quality Directives. Back in 1997, The Dutch Government had lobbied to obtain a review clause in Directive 99/30, and was confident that the review promised in 2003 would lead to a relaxation of the standards. In the Explanatory Memorandum on the Air Quality Order 2001, the Dutch Government stated that it was convinced that the standards would be revised downwards when it became clear that they could not be met. The promise that the standards would be reviewed was one reason the Dutch had agreed with the Directive (Blom 1998, on file with the author).

The start of the CAFE programme in fact delayed the review, which appeared neither in 2003 or 2004. When the air quality clash started to emerge, it became even more important for the Dutch Government to obtain a review that would take Dutch concerns into account. This caused the Netherlands to become significantly involved with the working groups undertaking the review.

The struggle over the review is the subject of this section. In section 1, the concept of a review by experts in the Working Group PM is discussed. Subsequently, I recount how the Working Group Imp took over the procedure and produced its own concept review for the European Commission. In section 3, the actual review officially submitted by the European Commission is scrutinised. Section 4 proceeds with a discussion of how Dutch representatives tried to influence the CAFE Steering Group.

8.2.1 *The review of PM standards by the Working Group PM*

Within CAFE, two groups were of particular significance regarding evaluation of the Air Quality Directives. These were the working groups on implementation (WG Imp) and on Particulate Matter (WG PM).[6] The WG Imp had the task of guiding Member States through implementation of the Air Quality Directives, and helping them complete the plans and programmes that the Directives demanded. It also provided guidance on monitoring and assessment within Member States. In addition, the WG Imp's job was to make an inventory of the problems Member States encountered when implementing the Directive.

The review of the most problematic standard, the one for Particulate Matter, was part of the domain of the WG PM. Within this scope, the group had the job of assessing the air quality situation with regard to PM limit values and of collecting information on the attainability of limit values, considering contributions from long-range transport and local sources.

The Working Group on Particulate Matter was composed of experts from Departments of the Environment from the Member States and experts from both industry (BP in this case) and the environmental movement, in the form of the European Environmental Bureau (EEB). The WG PM contained representatives and expert input from the WHO, whose data had played a pivotal role in the adoption of the Daughter Directive, and whose role was also foreseen to be strong in CAFE. The group was chaired by Lynne Edwards for the UK and Bernd Seifert for Germany. The Netherlands was represented by Klaas Krijgsheld, a toxicologist at VROM. Another Dutchman, Dick van den Hout, was present as a consultant for the Commission. He worked at the Dutch research institute TNO, and was also a member of the Working Group on Implementation as well as of the Steering Group.

While the attainability of the standards was considered to be a key question, the WG PM focused more on the technical side of PM and on modelling, trends, and integrated assessment. In a meeting on 6 and 7 September 2001, the WG PM set a

6. The terms of reference of the Working Group on Implementation and links to its meetings may be found on website Working group Imp. Last accessed 13-12 2011. The terms of reference and links to a number of meetings of the Working Group on Particulate Matter are located on website Working Group PM last accessed 13-12 2011. All the documents by the CAFE working groups referred to may be found on the website of CAFÉ (Website CAFE docs, last accessed 22-06 2015).

timetable that indicated it would have a document ready by the second half of 2003. This date, however, was not met. A final draft of this document, the Second Position Paper on Particulate Matter, was presented in April 2004. The WG PM relied heavily on WHO data, but the WHO's answers to additional questions of the working group had become available only by the end of 2003.

The document was quite technical, with the bulk of it dealing with modelling, sources, and composition of PM, health effects, and the different sizes of PM, PM10, and PM2.5. In the section on attainability, the document emphasised extra measures that Member States should take. Some recommendations in the position paper were useful for the Netherlands: namely, that the Commission could consider postponing the daily concentration limit of 50 mug/m^3 not to be exceeded beyond 35 days per year (WG PM 2004, p. 168). This would be most welcome for the Netherlands, because precisely this standard was the one causing a number of problems regarding infrastructure development. However, the recommendation was not worked out, and it appears to have been little more than a side note.

Other recommendations that would be to the liking of the Dutch can be seen in chapter 11, and included a stronger insistence on source policy. The Netherlands was in favour of that, because it was in line with Dutch economic interests. The Dutch delegation insisted on source policy, because if cars and industrial processes became cleaner the quality of the air would improve. Given the geographical position of the Netherlands, the cleaning of cars and industries made sense, since the country was surrounded by industrial areas and was subjected to considerable air pollution from abroad. If these industries were forced to reduce emissions by technical means, the Dutch would benefit. The insistence on source policy was also rational from a Dutch competitive perspective, as the Netherlands had no automotive industry of its own, and Dutch industrial installations were comparatively clean. This meant it would be able to benefit from clean-up measures undertaken by other countries without having to carry out a lot of adjustments itself.

The WG PM's main recommendation, however, was to shift attention from PM10 to PM2.5, as it considered PM2.5 to be a better indicator for health damage due to air pollution. It advised the Commission to propose a limit value for PM2.5, and also proposed leaving the PM10 standard as it was. This was somewhat good news for the Netherlands, because in the Daughter Directive it was considered that the PM standard could be tightened to a yearly limit value of 20 mug/m^3. At least that possibility was averted, but other than that, the evaluation did not provide much solace. The Netherlands had already failed at meeting the 2005 limit values, let alone those foreseen for 2010. PM2.5 standards were a different matter. The Netherlands did not see much in new air quality standards. As could be expected, the Dutch Government and the pro-infrastructure camp in the Dutch Parliament were against introducing a new PM2.5 standard with binding legal consequences.[7]

7. Kamerstukken II 2005-2006, 30 300 XI nr. 10, p. 44; Kamerstukken II 2005-2006, 21 501-08 nr. 215.

8.2.2 *The review proposal by Working Group on Implementation*

Apart from the WG PM, from 2003 onwards, the WG Imp had suddenly begun to play a role in reviewing the Daughter Directive. The task envisaged for the WG Imp was to help Member States in drafting the required plans and programmes when air quality norms were exceeded. The group had a composition similar to that of the Working Group on Particulate Matter: that is, a mix of representatives from Member States, European institutions, and stakeholders. For the Netherlands, Hans Herremans was present, and the group was chaired initially by Austria.

At the beginning of 2003, Austria stepped down from the position and Herremans took over the chair, a reshuffle that changed the direction the group had been taking. During the first meeting chaired by the Netherlands, the group concluded that it could play an important role in the review of Directive 99/30, which had originally not been part of their workload. In other working groups this change of direction was noted and criticised, members pointed out that the WG Imp had no business doing policy evaluation (WG TSP 2002, November p. 2).

Within the WG Imp, however, plans were made about how to proceed. It was concluded that the best way forward was for the Commission to send out questionnaires to all Member States with detailed questions on issues that they found problematic. At a subsequent meeting, the WG Imp would offer its advice on issues for a review report, based on responses by the Member States. Such a questionnaire was to be drafted by the Commission as soon as possible, and it was agreed that the Commission would circulate the draft among the WG with a very short deadline for comments (WG Imp, Jan. 2003).

At the next meeting, Commission consultant Dick van den Hout presented the questionnaire's findings (WG Imp, March 2003), and many issues that were important for the Dutch Government were under discussion. Some Member States, among which no doubt the Netherlands, claimed that maintaining the air quality standards in residential areas was more relevant that maintaining them in areas with only office buildings or in fields beside motorways.

The opinion of the European Commission on these issues was crucial for the Netherlands, because from the time of Minister Pronk onwards, policy practice contained the distinction between populated and unpopulated areas. In places where people lived, the Air Quality Order was to be strictly observed, but elsewhere the lower administrative bodies were granted more leeway to allow for some deviation. The distinction was made in policy instructions from the Government to lower administrative bodies. As we have seen, the Council of State later rejected that distinction, but Van Geel had intended it to lessen the impact of the standards on Dutch spatial planning, just as Minister Pronk had done. Because having the blessing of the European Commission would be extremely helpful for the Secretary of State, the issue was put on the European agenda very early in the process.

A second important point for the Netherlands was that the possibility of obtaining derogation was being discussed. The findings presented to the WG Imp concluded that the 2005 standards for PM were unattainable in large parts of Europe. The drafters considered it necessary to come up with proposals to postpone the attainment date in relation to feasible EU action on reducing PM emissions. The WG Imp drafted a letter to the WG PM in order to suggest this possibility (WG Imp, March 2003, 3). This course of action was in line with Dutch interests, because from early on the Netherlands had desired the possibility of postponing the standards for PM10 and NO_2.[8]

The WG IMP drafted its recommendations for the evaluation based on answers given by the Member States, and presented these recommendations in 2004 to the Commission (WG Imp 2004). This document was different in tone and style than the second position paper of the WG PM, being shorter and focusing more on problems of the Member States.

The document did not address the question of whether the limit values should be revised, as this subject was left to the integrated assessment at the end of the programme. The document reported extensively on ambiguities within the Daughter Directive, and dealt with attainability problems. The working group clearly recommended enforcing the standards only in locations where people were expected to spend a significant amount of time. The WG Imp considered: '*The directive should not consider exceedances of the objectives at any location where relevant public exposure would not be realistic*' (WG Imp, 2004, 8). This meant that the limit values would not need to be enforced at curb side sites or at '*any other location where public exposure is expected to be short term*' (WG Imp, 2004, p. 8).

The WG Imp was of the opinion that there would be no possibility of changing the standards for Particulate Matter on such short notice, as the standards would enter into force already in 2005. However, it advised the Commission to look into the possibility of providing derogations as soon as possible. It recommended that Member States should be granted exemptions if they showed that they had taken all reasonable measures, and this possibility should be included in a new Directive based on the CAFE programme. In the case of NO_2, the working group concluded that the values would be in force as late as 2010, the possibility of derogation would be discussed, and the present Directive might be amended.

All these recommendations by the WG IMP in June 2004 were favourable for the Netherlands. If the Commission were to condone the distinction between sensitive and non-sensitive destinations, the areas where standards posed problems for the

8. Kamerstukken II 2002-2003, 27 793 nr. 8, p. 2.

Netherlands would be drastically reduced. All this took place before the air quality clash would begin in full after the Council of State verdicts from September 2004. This data demonstrates that even before the onset of the air quality conflict, Dutch policy makers had tried to gain a favourable position in Europe.

8.2.3 *The evaluation of the Daughter Directive by the European Commission*

The Commission presented its delayed review in January 2005, COM(2004)845 (European Commission 2005a), and it was one of the first policy documents to come out of CAFE, preceding the actual strategy and a new Air Quality Directive that was based on it. By that time, the Dutch problems had become acute, and the Secretary of State had already sent his letter of 30 September 2004, defending his interpretation of the Directive against the Council of State. The stalemate was complete when this letter and the proposed Air Quality Decree were rejected by the Council of State.

The review issued in January 2005 could have been a breakthrough for Van Geel, but it turned out to be disappointing. According to the Commission, however, the directive was considered a success. Under the header '*the first rather limited experience of its application is positive*', the Commission highlighted three reasons for this conclusion (European Commission 2005a, p. 4): firstly, the directive raised awareness regarding air pollution. Complaints about non-compliance of Member States found their way to the Commission and to representatives of the European Parliament, and the Commission considered this an indication that the public was taking an interest in air pollution problems; secondly, the challenging limit values forced Member States to come up with effective solutions to tackle air pollution problems; thirdly, the standards also applied to the newly acceded Eastern European Member States. Apparently the Commission considered complaints to be a sign of success, as the challenging nature of the standards succeeded in raising public awareness, and forced Member States to come up with solutions. From a Commission perspective, this conclusion was understandable, a victory for the tactic of enforcing implementation by creating 'pressure from below', the mobilisation of public pressure to force Member States to comply. This tactic was a key ingredient of the European air quality policy in the 1990s.The Commission had little to say about the practical problems Member States faced when trying to comply with the directive. Compliance was discussed, but from the perspective of helping the Member States to take clean-up measures. The Commission considered it problematic that Member States generally failed to make plans and set up programmes when they observed exceedances. No mention was made of the distinction between populated and unpopulated areas. Derogations were scarcely considered, nor were these items incorporated into the technical 'staff working paper' that accompanied the evaluation (SEC 2004, 1713).

The fact that the limit values were significantly tighter than in previous years was mentioned, but mostly in the context of their contribution to the successful raising of awareness (European Commission 2005a, p. 3). Issues such as derogation and the harmonisation of measuring requirements were considered matters to be handled within the eventual CAFE strategy itself.

The unwillingness of the Commission to be more lenient with Member States can be explained in terms of differences in interests. The Commission was concerned mainly with non-compliance with environmental standards, while some Member States, especially the Netherlands, struggled with their challenging nature.

8.2.4 *Politics within the steering group*

Through its influence within the WG Imp, the Dutch Government managed to attract attention to Member States' problems with the Directives. The position paper of the WG Imp contained a number of recommendations that pleased the Netherlands. These subjects were not incorporated in the eventual Commission evaluation, however, but they were at least now on the agenda. Subsequently, the Dutch Government tried to warm members of the Steering Group to the notion of derogations and postponements.

The CAFE Steering Group (SG) was the main forum for discussion on the direction and contents of the strategy. It was high up in the hierarchical structure of CAFE, situated just below the CAFE Secretariat, and comprised over 80 delegates, experts from the Member States, the Commission, various industrial sectors, and the environmental lobby group EEB.

Within the Steering Group, Dutch representatives tried to persuade the different Members to side with them. After the WG Imp had composed its report, Hans Herremans presented it to the Steering Group. During the discussion, the possibility of derogations was tabled by Herremans, but he was countered by Commission official Stefan Jacobi, who explained that this issue went beyond the support expected of the WG Imp. Postponements and the extent of the limit values were subjects that would be part of the thematic strategy itself. In order 'to save time', the matter was not discussed further (Steering Group May 2004, p. 7).

The Steering Group was composed of experts, but in practice they played a double role. They gave expert advice about the scientific data presented, but they could also present the views of the Member State or organisation they represented. According to Willemijn Tuinstra, many participants had difficulty describing their role as either policy maker or scientist (Tuinstra, 2007, p. 438). This finding is in line with the work of Sheila Jasanoff that indicated that the demarcation between scientific advisor and policy maker is often a thin one (Jasanoff 1990). In the case of CAFE as well, political discussions about how much flexibility should be left to the Member States or how much source policy the Commission should enact, crept into the Steering Group discussions, especially during the meetings in 2005.

In the SG, the Dutch participants supported by those from Germany and the UK reiterated calls to make the standards mandatory only in places where people were expected to spend time. The Commission explicitly opposed this doctrine, considering it open to abuse, and maintaining that it would create divergent approaches in Member States. The SG considered it 'highly undesirable' (Steering Group Feb. 2005, p. 5), which effectively put an end to the idiosyncratic Dutch interpretation and the Netherlands' efforts to have it accepted in Europe. This did not prevent the Government from trying to get it accepted in the Netherlands itself. It was part of the proposal for a Ministerial Decree to replace the AQO 2001 discussed in February and March 2005.

The Dutch representatives also addressed the necessity of flexibility, and warned that the Netherlands could not accept further limit values on PM2.5 if the Commission could not explain how to meet the standards in practice (Steering Group April 2005, p. 5). The Netherlands found allies in Poland and Italy, who also opposed new standards for PM2.5. In general, though, the proposal for new limit values for PM2.5 was not strongly opposed within the Steering Group.

8.3 The establishment of the CAFE Air Quality Directive

The evaluation of existing policy was an important part of CAFE, but its main aim was to unify existing EU approaches to pollution control and air quality policy in an overarching strategy. Together with the strategy, a new Air Quality Directive was proposed, the goal of which was also to unify the existing air quality standards into one directive as well as to add provisions to it.

The CAFE strategy itself is of interest, because in it the future of EU air pollution policy was unveiled. This subject is under discussion in section 7.4. In the present section, the emergence of the new Air Quality Directive is analysed. Contrary to the strategy, the Directive contained binding standards, and was of more interest to the Netherlands because the Dutch Government intended to counter all extra standards that the Commission might impose. This included new standards for PM2.5. The Commission was likely to impose such standards on the Member States based on the recommendations of the WG PM and on new scientific insights. However, the Netherlands also needed a new Air Quality Directive because it could offer Member States the possibility of a postponement. The mission for the Dutch representatives was clear: namely – and as much as possible – to weaken additional obligations, and to obtain flexibility and derogations.

Discussions in the various legislative institutions of the European Union displayed the position of the Dutch in the debate, but they also revealed which arguments managed to dominate the discussion among European law makers.

8.3.1 *The proposal for a new Air Quality Directive COM(2005)447*

The proposal for Directive COM(2005)447 (European Commission 2005c) on ambient air quality and cleaner air for Europe, was the first legislative proposal to come out of the CAFE programme, and it was proposed together with the final CAFE strategy, COM(2005)446, in September 2005 (European Commission 2005b). The proposal essentially contained the wish to unify existing Air Quality Directives within one and the same directive.

Apart from updating requirements regarding reporting on air pollution and integrating the exchange of information on air quality and public information, the proposal introduced a number of novelties of special interest to the Netherlands. To begin with, a new standard for PM2.5 was indeed proposed. The Commission envisioned a dual standard, and introduced a uniform yearly standard for PM2.5 of 25 mug/m^3, to be achieved by 2010. The Commission decided to keep the standards for PM10, because the WHO had noted that this fraction of PM10 could still cause health problems.

According to the Commission, the standard for PM2.5 should not become too burdensome on Member States. It was foreseen that it would constitute a burden only in the most polluted areas (European Commission 2005c, p. 3), since meeting the limit value for PM10 would almost always mean that the limit value for PM2.5 had also been met. The proposal also introduced the possibility of derogating the air quality standards for certain designated areas. Under strict circumstances, and only in the presence of an elaborate reduction plan, meeting the standards for NO_2 and PM2.5 could be postponed to 2015 (European Commission 2005c, Art. 20, sub 1). On similar conditions, the obligation to meet the limit value for PM10 could be postponed to 2009 (European Commission 2005c, Art. 20, sub 1).

In September 2005, the Netherlands was in the throes of the air quality clash. The derogations on offer were welcomed by the Dutch Government, of course, but it was still questionable whether the Netherlands could meet the PM10 standard as early as 2009. Extra standards for PM2.5 were difficult for the Dutch Government to swallow, because in Parliament the pro-infrastructure camp considered extra standards non-negotiable, and members of these parties held seats in the Dutch Government. In December 2005, Pieter van Geel was cautioned explicitly by way of a Parliamentary motion not to accept extra standards,[9] and was therefore politically obliged to oppose the Commission proposal.

9. Kamerstukken II 2005-2006, 22 112 & 21 501-08 nr. 454 p. 6; Kamerstukken II 2005-2006, 21 501-08 nr. 215, p. 3.

8.3.2 *The Council of Ministers: proposal for a new Air Quality Directive*

The Commission drafted its proposal in late 2005, and it was first discussed informally in the Council of Ministers in December. The Council of Ministers is the EU's highest legislative body, together with the European Parliament. Such an informal discussion gives Ministers holding an environmental portfolio a chance to gauge the general feelings on the proposal before having to take a formal position.

The proposal was discussed under the terms of the co-decision procedure, referred to now as the 'ordinary legislative procedure' (see text box below). The European Parliament and the Council would both have to agree on the proposal in order for it to become adopted.

From the summary available on the European Parliament website (EuroParliament website a, last accessed 25-05 2013), I conclude that both the importance of public health and the necessity of allowing flexibility for the Member States were the subjects of debate in the Council. During a formal discussion of the draft in June 2006, the Council reached an agreement on the merits of the Commission proposal, although it intended to change a few things. It preferred a binding limit value for PM2.5 from 2015 onwards instead of 2010. The Council also proposed allowing for a three-year derogation of the PM10 standard, entering into force as of the date of adoption of the Directive. This would replace the date of 31 December 2009, the one envisioned in the Commission proposal. All in all, however, the Council's common position was close to the Commission proposal.

Contrary to the situation in the 1990s, the Dutch delegation this time was vociferous in its criticism. After the Council meeting in June 2006, Environmental Secretary of State Pieter van Geel told journalists that the meeting had been an exercise in hypocrisy: *'When you set a rule, you also have to introduce the measures to comply with the rules'* (ENDS Europe, 28-06 2006). In the Dutch Parliament, he made even more undiplomatic comments. According to a document presented by the VROM Ministry (VROM, 2006, p. 33), van Geel publicly called the Ministerial Council a 'ballentent', a pejorative term for an overrated coffeehouse or restaurant.

The Dutch delegation claimed that Poland, Greece, Hungary, and two of the three Baltic States were supporting the Dutch in the upcoming vote against the deal (ENDS Europe, 28-06 2006). The Dutch delegation was especially angry, because it had failed to persuade the Council of Ministers to make the obligation to comply with the limit values dependent on the introduction of extra source policy by the Commission. As we have seen, source policy was attractive for the Dutch because of the Dutch dependency on other countries, and because Dutch industry was already comparatively clean. This meant the Dutch would not need to incur heavy costs, but could reap the environmental revenues of cleaner industry and cleaner cars abroad.

Despite Van Geel's comments, the Council stuck to its line. It incorporated a number of amendments made by Parliament in the first reading (discussed below), but these did not change the Directive in any essential manner. The introduction of source policy was not proclaimed a precondition for acceptance of the proposal. During a meeting in October 2006, the Netherlands – supported only by Poland – voted against the proposal.

The counter vote of the Netherlands during the establishment of a common position was indicative of the change of heart regarding Dutch forerunner ambitions. It had always been seen as a leader in environmental policy, and in the 1990s the Netherlands was supportive of the Air Quality Directive despite doubts regarding attainability. After the EP and the Council reached a formal agreement on the adoption of the Directive, the Netherlands issued a statement against it, which was annexed to the Council's position and dated from 12 June 2007.

The Netherlands stated:

> *'The common position provides no assurance of European measures that are essential to enable Member States to achieve compliance with the standards. Therefore, Member States are legally bound to comply with standards and thus subject to an obligation without being in a position to control the necessary prerequisites to fulfil that obligation. Therefore, the Netherlands is compelled to withhold its endorsement of the common position'* (Council of Ministers 2007a).

In the end, the Council decided on a new limit value for PM2.5 of 25 mug/m^3, to be achieved by 2015. It did not change any of the other limit values, and maintained that the derogations could be granted up to 2015 for NO_2 and 3 years from the adoption of the Directive for PM10. The result indicated that Van Geel had not managed to secure allies with the intention of blocking the proposal or gaining further concessions.

Box 7: The ordinary legislative procedure or co-decision procedure

The proposal for the CAFE Air Quality Directive is discussed under the terms of the ordinary legislative procedure, formerly known as the co-decision procedure. This legislative procedure is similar to the cooperation procedure discussed in box 4 but there are important differences. Under the terms of the ordinary legislative procedure, the Council of Ministers shares legislative power with the European Parliament.

A Member of the European Parliament, working in the relevant Parliamentary Committee, draws up a report on a proposal for a 'legislative text' presented by the European Commission. The Parliamentary committee votes on this report and – possibly – amends it. The report and the proposal are subsequently discussed by Parliament in first reading, during which the Parliament adopts its position and proposes amendments.

At the same time, the proposal is sent to the Council of Ministers. The Council makes its position known after preparatory work has taken place within working parties made up of experts from the Member States and chaired by the Member State holding the six-monthly Presidency of the Council. This preparatory work runs concurrently with the European Parliament's activity, but the Council may only adopt a position after the European Parliament has acted. The Council finalises its position on the basis of the Commission's proposal, amended where necessary, in the light of the European Parliament's first reading and resultant amendments. If the Council does not share the views expressed by Parliament, it adopts a common position, which is forwarded to the European Parliament together with a statement of reasons.

If the Council does not adopt the amendments proposed by Parliament in first reading, the common position is referred to Parliament for a second reading. If the European Parliament endorses the common position as it stands, or fails to adopt amendments as a result of not obtaining an absolute majority of its Members, or does not take a decision within the stipulated time limit, the President of Parliament will declare that the common position is approved and the act is adopted in accordance with the common position.

If the EP does propose amendments, the proposal is discussed in the Council for a second reading. If the Council agrees to accept all the amendments of the European Parliament, the act will be deemed to have been adopted in the form of the common position thus amended (wording corresponds to the EP second reading).

>If the Council and the European Parliament can still not agree on a final text, the proposal is discussed further in a so called conciliation procedure. This procedure involves talks between members of the EP, the Council and the European Commission. Such three party talks begin informally and end in the conciliation Committee, which brings together delegations from the Council, the EP and the responsible Commissioner. A compromise is negotiated here (source: archived website European Commission codecision, last accessed 25-05 2013).

8.3.3 *Discussions in the Environmental Committee of the European Parliament*

Under the co-decision procedure, the Council did not have the final say, as the European Parliament had legislative powers equal to those of the Council of Ministers. In practice, this meant that the two institutions had to agree on the proposed legislation or negotiate with each other, and therefore the Dutch representatives would have another chance at revising the proposal in their favour. As was common, before the proposal was discussed in a plenary session it was discussed in the Environmental Committee (ENVI). Members of the European Parliament (MEPs) who had a seat in the Committee dealt with environmental issues. The deliberations in the Environmental Committee in the EP were important because the Committee informed the position of the EP as a whole (Bomberg & Burns 1999, p. 174). Generally, the MEPs in Parliament followed the Committee's recommendations.

It was common for a rapporteur appointed in the ENVI Committee to draft a final report for the EP on the proposal under discussion, and such a report contained the outcomes of the discussions. Before the discussion in the ENVI Committee started, the rapporteur supplied a draft to the members. The report in this case was to be drafted by Holger Krahmer of the liberal/conservative ALDE group in Parliament as the rapporteur. He proposed his draft report to the Environmental Committee (ENVI 2006a) in April 2006.

Krahmer's draft report was critical of the proposal to begin with, arguing for more consideration with Member States. It was discussed mainly among MEPs from the Low Countries, Germany, and Scandinavia. The Dutch members Jules Maaten (ALDE), Johannes Blokland (Independent Democrats), and Ria Oomen Ruijten (from the Christian Democrat EPP) tended to operate 'en bloc', and introduced amendments together. These amendments dealt mainly with the need to curb pollution at source as well as extra time for Member States. Other MEPs were in the same camp. Anja Weisgerber, a Christian Democrat from Germany, and others stressed the scientific uncertainty of setting binding standards for PM2.5 (ENVI 2006b, am. 38, 58, 185, and 186). Weisgerber referred explicitly to the situation with PM10 in 1999, when standards were proclaimed without significantly certain data (ENVI 2006b, am 88). Dutch Socialist MP Dorette Corbey proposed amendments that would make sure the Dutch approach to balancing harmful projects with beneficial measures was accepted (ENVI 2006b am. 42, 78). During the discussions, Dutch Euro-Parliamentarians used an RIVM report produced by Jimmink and co-workers (Folkert et al. 2004) to stress that the figures provided by the Commission were unsound.

The Dutch and German MEPs were opposed by Nordic MEPs in the ENVI Committee – notably Anders Wijkman and Satu Hassi, who stressed the dangers to health and proposed sharpening the Directive.

In Holger Krahmer's final report, the majority of amendments asking for derogations and flexibility were included (ENVI 2006c). The references to the importance of source policy and insufficiency of data to warrant a PM2.5 limit value were also reiterated in numerous places (ENVI 2006c am. 1, 3, 5). It was also mentioned that the transboundary character meant that individual Member States had only limited resources in tackling pollution, and should be allowed derogations. Taken together, the derogations proposed by the ENVI committee amounted to the possibility of a ten-year postponement for the most stressing standards such as PM10 and PM2.5 (ENVI 2006c am. 30). Krahmer also proposed an amendment that implied that limit values would be assessed only in certain places. This was an important amendment, because it implied that exceedances would not be considered in places where air quality was not assessed.[10]

10. In section 8.3.5 this matter will be briefly touched upon. Germany also made the distinction between places where people lived and places where they did not spend a significant amount of time, similar to but not totally identical to the sensitive destinations doctrine that the Netherlands initially held on to. The difference was that air quality in Germany was simply not assessed or measured in such places (Backes 2006, 11).

This report was more sympathetic with non-complying Member States than the Commission was; in fact, it was more sympathetic with the non-complying Member States than the Council of Ministers, usually an institution that is keen to represent Member State interests. The report was eventually adopted by the Environmental Committee on 21 June 2006, with 39 votes in favour, 11 against, and 5 abstentions.

This result was an unusual one. In the 1990s, the Environmental Committee of the EP had been known as an 'exporter of environmentalism' (Judge 1993, p. 199; Bomberg & Burns, 1999, p. 175; Burns 2005). In this case, however, the green concerns of the Environmental Committee were side-lined by a coalition of German and Dutch MEPs who were more concerned with the complaints of Member States, with the costs of implementation, and with the need for additional source policy than with the public health arguments put forward by the Greens and Northern MEPs. The amendments also demonstrated that national concerns featured in discussions held by the Environmental Committee. The amendments of Corbey, Maaten, and Oomen Ruiten were tailor-made for the Dutch situation.

8.3.4 *The debate in the European Parliament*

After the Environmental Committee drafted its final report, a plenary debate in Parliament was held on 25 September 2006 (websites EuroParliament b, last accessed 19-12 2011), and would result in Parliament's first reading. The rapporteur for the ENVI Committee, Holger Krahmer, and Commissioner Stavros Dimas confronted each other in the plenary debate.

During the debate, pro-health and pro-flexibility arguments were traded. Proponents of a pro-health position were Environmental Commission member Stavros Dimas and the Green and Nordic MEPs Rytta Myller and Saatu Hassi. A key argument of Commissioner Dimas was that the strategy he proposed would reduce the number of forecast premature deaths from 350,000 to 160,000 by 2020, and he added that acidification and nitrogen deposition would be reduced by 50% and 30%, respectively. He also mentioned that source policy would certainly follow.

The number of excess deaths due to air pollution was put forward as well by pro-environmental MEPs Hassi and Myller, who named figures of 360,000 to 400,000 premature deaths annually, and emphasised the WHO's position on the issue. Hassi also mentioned that there were stricter limit values in the US than in the EU, and she urged standing up to the arguments of the pro-flexibility block and *'defending the rights of our citizens to breathe clean air'*. She also accused the German Christian Democrat members of having ties with the automotive industry.

Among the MEPS who defended flexibility for Member States were ENVI rapporteur Holger Krahmer, Anja Weisgerber, and the Dutch members Oomen Ruijten, Blokland, and Maaten. All the Dutch Parliamentarians argued for strict source policy, because otherwise the standards would be unattainable. Holger Krahmer summed up his position by stating: '*The problems cities have with compliance with the daily values are no secret, and we abandon municipalities to a sort of organised helplessness*'. In an attempt to support the need for flexibility, Krahmer mentioned also the uncertainty of the data.

Overall, two storylines were consistently present in the debate: a pro-health narrative on the one hand, and a flexibility account on the other. In the pro-health storyline, the WHO's work featured prominently, as did the numbers of premature deaths. MEPs who subscribed to this storyline argued that the suggestion of flexibility was a poor excuse in the face of the widespread health damage. The WHO research was mentioned in an effort to buttress claims of health damage. Saatu Hassi mentioned clean air as a 'right' of the population. As we have seen, this legal type of reasoning was used in the air quality clash by the environmental movement Milieudefensie, and, in certain instances, even the Commissioner alluded to it (Websites EuroParliament b, last accessed 19-12 2011).

Central to the flexibility storyline used in this debate was the argument that Member States had to be able to comply with the standards, and therefore these needed to be workable. This could only be achieved if administrative bodies were granted more time, and if the Commission could enact source policy. In order to illustrate the necessity of postponements, some Dutch members like Johannes Blokland and Ria Oomen Ruijten cited the Netherlands as a specific example of a struggling Member State.

In Parliament, the flexibility line won out. Most of the amendments presented in the Krahmer report were adopted, including the controversial amendments to extend the deadline for compliance with the PM10 regulation to well beyond 2010. The Green group called the vote a scandal, and Commissioner Dimas said that it '*appears to contradict*' the objective of protecting citizens' health '*by weakening the legislation we have proposed in some key respects*' (Website Euractiv a, last accessed 24-06 2015).

8.3.5 The agreement between Council and Parliament

The amended proposal was sent back to the Council, which had reached a common position prior to Parliament's first reading back in June 2006. In the summer of 2007, the Council essentially stood its ground.

According to an internal note from the Secretariat of the Council, the remaining issues were resolved by informal contacts between the Council and Parliament before they could result in a collision between the two legislative bodies (Council of Ministers 2007b). Twenty-seven compromise amendments on minor issues were accepted, and on most accounts the original Council position was accepted. At the insistence of Parliament and the Council, the Commission promised to annex a statement to the Directive outlining the proposals it was considering for the purpose of combatting pollution at source.

This turn of events was of interest. Apparently the pro-flexibility storyline managed to dominate in the European Parliament, but was not equally convincing in the Council of Ministers, where Van Geel did not have many supporters. His argument that connected flexibility for Member States with the lack of source policy was not accepted. This was understandable, because for many countries source policy was much more invasive than air quality standards. Only the Netherlands had a structural problem with the air quality standards, owing to the air quality clash. For other countries, the air quality standards simply did not result in the legal headaches they caused in the Netherlands.

Nonetheless, pro-flexibility proponents Weisgerber and rapporteur Krahmer ultimately declared themselves pleased. Weisgerber proclaimed:

> *'The new Directive abolishes useless short-term action like the closing of roads. There will be more flexibility for towns and municipalities that have done everything they can and still do not meet the limits'* (Website Euractiv b, last accessed 24-06 2015).

In the end, both camps could claim a small victory. Flexibility was granted by the Directive, but it also included a relatively weak PM2.5 standard. The Commission promised source policy, but whether it could keep its promise remained to be seen. Other countries were objecting to more measures at source.

8.3.6 Directive 2008/50/EC and the Dutch derogation

The second reading vote in Parliament became a formality due to the outcome of informal talks. The package was accepted by a qualified majority, although the most flexibility friendly amendments presented in Krahmer's report did not make it. On 11 June 2008, Directive 2008/50/EC (Henceforth CAFE Directive) was adopted.

The Directive proposed a limit value of 25 mug/m^3 of PM2.5 to be attained by 2015, but no changes in the limit values for other pollutants. The directive introduced flexibility for the member states in the form of a derogation possibility of five years for NO_2. For PM10, Member States could acquire derogation up to 11 June 2011, three years from the adoption date of the Directive. To obtain the extra time, Member States would need to demonstrate that all *'appropriate measures have been taken at national, regional and local level to meet the deadlines'* (CAFE Directive, Art. 22 sub 2).

The directive also included an odd-looking compromise that air quality must in principle conform to the standards 'in all zones and agglomerations' (CAFE Directive, Art. 14 sub 1; Art. 16 sub 2), but it would not be assessed in places where the public had no access or where there was no fixed habitation. This amendment was originally tabled by Krahmer, and it conformed to the German practice of dealing with the air quality standards. It came down to the possibility of simply not measuring air quality in places where people were not present. In principle, however, air quality still needed to conform to the standards throughout the territory.[11]

When the Directive was finalised, the Netherlands issued a statement again, but this time it was positive in tone. The Netherlands declared that it was pleased that the Directive would enter into force in 2008. This turn from discontent to being pleased at an early adoption was due to the Dutch's own solution to the air quality clash – the programmatic approach embodied in the NSL. The timely adoption of the Directive was essential for its own plan, because for it to work, derogation from the EU was needed.

The Netherlands was the first country to apply for derogation under article 22 of the CAFE Directive, which meant the Netherlands could ask to obtain a postponement until June 2011 at the latest. The country filed its application little over a month after the Directive had been adopted (ENDS Europe 3 July 2008). The programmatic approach laid down in the NSL functioned as its air quality programme, which was necessary in order to indicate that the country was doing everything in its power to improve the quality of the air. The Netherlands had its application for a derogation approved in April 2009.

The fight over the Directive illustrated the turn the Netherlands had taken from being a front-runner to becoming a foot dragger. It was not willing to support green Member States, but was actively trying to engineer a coalition to block the new standards for PM2.5 and it tabled calls for postponements and 'flexibility'. MEPs from Member States that took a pro-environmental and pro-health stance, such as the Scandinavian countries and those from ecological parties, were highly suspicious of these calls for flexibility. They saw in it an attempt to let Member States get away with pollution and perceived in it a disregard for the threat of health damage.

8.4 The CAFE strategy

The new Directive was of importance for the Netherlands to obtain its derogation, but for EU environmental policy as a whole the CAFE strategy itself was more indicative. Because of its relevance for future environmental policy, I decided to treat it later in the chapter, even though the CAFE strategy precedes adoption of the CAFE Directive. Dating from 2005, the CAFE strategy was released at the same time as the proposal for the Directive by the Commission. The CAFE strategy was the overarching policy that would outline the new air pollution policy of the European Union, and a close look at the strategy is important in order to assess the future of EU air quality policy. The CAFE strategy was underpinned by input from a host of stakeholders as well as by elaborate

11. For the Netherlands this created the ironic situation that in order to comply with air quality standards, the Netherlands needed to stop measuring air pollution on highways and other areas restricted for the public.

cost-benefit analyses and strategic impact assessments. The question is whether this emphasis on participation and science led to an effective air pollution policy.

In the first sub-section, the aims of the CAFE strategy are discussed. In the second section, I look at the influence of stakeholders in the CAFE scientific working groups, and in the third section we review the cost-benefit analysis that supported the eventual CAFE strategy. In the final sub-section, the outcomes of the CAFE programme in total are examined.

8.4.1 Aims and policy implications of the CAFE strategy

In section 8.1.1, the CAFE programme's overall objectives were discussed, one of which was the production of a policy strategy for air pollution. The first Communication regarding the strategy was forwarded by the Commission in 2001, and a final version based on the work in the CAFE programme was delivered in 2005 together with the proposal for the CAFE Air Quality Directive.

The aim of the CAFE strategy was to set a number of air quality objectives to be reached by 2020. Reductions of the concentrations of PM2.5 by 75% of what was technically possible were foreseen, as well as a 60% reduction for ozone. The acidifying pollutants were to be reduced by 55% of what was technically feasible (European Commission 2005b, p. 5).

The Commission envisaged meeting these targets by streamlining and revising its existing legislation, and by adding a limit value for the smaller version of PM: namely PM2.5, as we have seen in the previous sections.

One of its other main proposals was a revision of the National Emission Ceilings (NEC) Directive in 2006. The NEC Directives established emission ceilings for five different pollutants, stipulating the maximum amount of pollution that could be emitted within a Member State. The Netherlands was in favour of this approach because countries would have to clean up their industries and automobiles in order to meet the limits. The characteristic approach of the NEC Directive was taken over by the EU from the UN, who used this type of pollution policy in the scope of its Long Range Transboundary Air Pollution (LRTAP) convention.

The CAFE strategy merged two different EU environmental policies: the acidification strategy epitomised by the NEC Directive and the air quality standards approach characterised by the Framework and Daughter Directives extensively discussed in chapter 4. The CAFE strategy tried to kill two birds with one stone, so to speak, by unifying these policies under the same umbrella. Within the CAFE strategy, the interdependency between various acidifying and health-threatening pollutants was depicted as follows, see figure 6 (European Commission 2005b, annex 2).

Figure 6

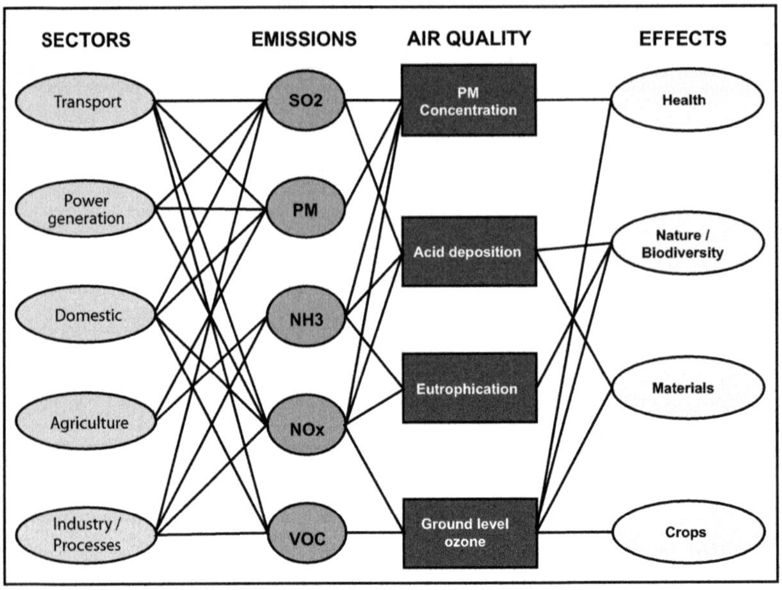

We can conclude from the picture that all five sectors and all five pollutants were considered responsible both for acidification problems and for health threats, and that each contributed both to local problems of bad air quality and to the transboundary problem of acidification. This type of framing conformed to the ecological modernistic concern with synergistic effects and a holistic perception of the environment. The twin considerations of health and environment had been rolled into one strategy and framed as a single problem for which a variety of sectors was responsible.

The CAFE strategy was partly drawn up to involve other DGs in environmental protection and the transport and agricultural sectors in particular were considered important. Regarding transport, the strategy would encourage proposals for cleaner fuel for passenger cars, vans, and heavy-duty vehicles: namely, what were termed the EURO V and EURO VI norms. The strategy would further incorporate the work done on the auto/oil agreements[12] and other policies in the sphere of product standards.

Regarding agriculture, some subsidies that had perverse environmental effects were to be abolished. The CAFE strategy focused especially on nitrate, and priority would be attached to measures and policies to reduce 'excessive' nitrogen use in agriculture (European Commission 2005b, p. 11).

12. For the auto-oil programme, see chapter 4.

The possibilities that this strategy offered for growth and jobs were highlighted, the Commission was of the opinion that:

> *'Environmental standards can be a catalyst for business growth and innovation. The EU can gain competitive advantages and exploit opportunities by focusing research and development on resource-efficient and less polluting technologies that other countries will eventually need to adopt'* (European Commission 2005b, p. 6).

The Commission also stated that countries like China and Korea looked to the EU for leadership. This endorsement was typical of the ecological modernistic storyline of win-win scenarios, because it highlighted the possibilities for innovation and the potential for business.

The policies that were proposed to reach the strategy's goals still needed to follow the normal EU legislative process; the most 'innovative' aspect of the whole strategy was the integration of two different air pollution-related philosophies. The process connected the approach to acidification and to urban air pollution, and fitted both into a framework for an integrated, comprehensive policy.

8.4.2 The UNICE lobby and the CAFE strategy

The level of ambition of the CAFE strategy was determined to a large extent by an elaborate impact assessment and cost-benefit analysis, and an examination of the way the impact assessment was used within CAFE demonstrated that these scientific instruments could be manipulated for political ends. The impact assessment promised a science-based evaluation of the various options, but it turned out that by applying political pressure, lobby groups could manipulate the eventual outcome of the impact assessment. These events illustrated the susceptibility to capture regarding such scientific committees within which environmental policy was drafted.

The goal of the impact assessment was to choose an appropriate level of ambition, and to that end environmental consultancy agencies drafted three different scenarios ranging from a very ambitious scenario C to a minimally ambitious scenario A. The assessment also included a scenario that foresaw no policy change at all.

The strategy that was finally established was forecast to deliver 42 billion Euros per annum in health benefits, and in terms of ambition it fell between scenarios A and B. The Commission implied that it had made a choice based on the calculation of different scenarios, and that it had chosen an option that *'.... is the most cost-effective level consistent with the Community's Lisbon and sustainable development strategies'* European Commission 2005b, p. 5).

The number of excess PM-related mortalities that were prevented counted for the bulk of the benefits. The benefits for eco-systems were also expected to be *'significant in terms of reduced risks and reduced areas of ecosystems that may be damaged by acidification, eutrophication and ozone'* (European Commission 2005b, pp. 5/6). The strategy was foreseen to cost 7.1 billion Euros per annum in total, divided over the whole of the EU. These costs would be incurred through the efforts Member States needed to make to achieve the air pollution standards established in the CAFE Directive and in the other policy plans outlined in the CAFE strategy.

In the annexed cost-benefit assessment, the choice of the Commission was further elaborated upon. In this document it is stated:

> *'However, the additional costs relative to benefits start to increase steeply at around the mid-range (Scenario A/B). Furthermore, the changes in ecosystem improvements between the lower (Scenario A) and mid-range scenario (Scenario B), balanced against costs, argue in favour of choosing a level between the low and mid-range that delivers the lowest levels of air pollution that can be justified in terms of benefits and costs while preventing undue health risks for the population'* (SEC(2005)1133, p. 15).

The Commission called its approach 'ambitious yet prudent' (SEC(2005])1133, p. 15). However, just before the strategy was released in September 2005, the consultancy firm AEA Technology, which had conducted the cost-benefit analysis, recommended a very different level of ambition, by stating:

> *'The cost-benefit analysis shows that the benefits of the three policy scenario levels exceed costs, significantly so for the high estimate of benefits. These conclusions are confirmed in the macroeconomic analysis: it is appropriate to choose an ambition level of at least Scenario B and possibly Scenario C'* (AEA T August 2005, VI).

An examination of why this level of ambition was not chosen revealed the influence that industrial and environmental lobby groups were able to wield. The large number of uncertainties mentioned in the impact assessment gave ample reasons to argue for a conservative policy based on the argument that the costs might rise unexpectedly.

Cost-effectiveness was an issue within CAFE from the beginning, and stakeholders from the industry intended to keep an eye on the programme's scientific input. Even before the CAFE programme started officially in 2001, discussions had arisen about who would be part of the Technical Analysis Group (TAG). Industrial groups, for instance, saw a role for themselves as experts within the TAG. The Commission doubted that the TAG was the appropriate forum, and was supported in this conclusion by a member of an environmental NGOs (Steering Group, 2001 January, p. 2). In the end, and to their dismay, industrial representatives were excluded from the TAG. Among other reasons, they felt that the CAFE programme was biased against industrial interests because of a reference to 'vested interests' in the original CAFE proposal from 2001.

The industrial group UNICE wrote letters to the DG Environment to press home the point that indicators should be robust and based on sound science (UNICE, 2002). In the Working Group on Target Setting and Policy Assessment, UNICE disagreed with the consultation procedures (WG TSP, 2003 January, p. 2). The group presented a paper on the relationship between science and policy (WG TSP, 2003 September, p. 8), complaining that it was unclear as to what was being done with its views during the meetings (UNICE 2004).

The views of the industry as represented by UNICE and various other industrial groups differed from those of the environmental groups, and a clash emerged over the impact assessment and the level of environmental ambition in the Steering Group. Both UNICE and EUROPIA, the representative organisation of the oil industry, considered that even a medium level of ambition could not be supported, owing to scientific uncertainties. In contrast, the environmental movement represented by the European Environmental Bureau (EEB) wondered why uncertainties should not lead to the highest level of ambition (Steering Group April 2005, pp. 3/4).

In spring 2005, just before the strategy was about to be unveiled, UNICE stated in a letter to Commission Chairman Manuel Barroso that they believed decisions about new targets should be based on robust scientific and economic assessment (UNICE 2005). They claimed that the model used by CAFE was highly complex, and that transparency was lacking with regard to the input assumptions and to the uncertainties of the results generated. This prompted an angry response in return from the EEB, which claimed that CAFE was *'a model of transparency'* (Tuinstra 2007, p. 440).

The letter from UNICE had the desired effect. Commission Chairman Barroso postponed the thematic strategy on air pollution, and held an 'orientation debate' about all the upcoming thematic strategies in July 2005. In the end, the DG environment was able to hold on to the ambition level it had initially chosen upon the recommendation of AEA T.

Environmental Commissioner Stavros Dimas admitted that the eventual outcome was a compromise (website Euractiv c, last accessed 25-06 2015). In particular, the Internal Market Commissioner Günther Verheugen was reported to be very sceptical of the plan (Meuwese 2008). During negotiations with other Commissioners in the presence of chairman Barroso, Dimas had to defend all seven thematic strategies against charges that they were bad for business (Pallemaerts et al. 2006, p. 57).

In the end, the intervention from the pro-business groups managed to curtail the environmental ambitions of the strategy by using the arguments of scientific uncertainty and possible additional costs. The Commission used the same cost-benefit assessment to defend the lower ambitions in the document detailing the CAFE strategy itself. This raised the question of whether such expansive cost-benefit analyses added much if they could so easily be brushed aside. In fact the clash of interests was obscured in this case by a conflict over how uncertainties should be dealt with. As regards scientific uncertainty, the pro-health storyline was pitted once more against a narrative that argued that tight environmental regulation was bad for business.

The pro health storyline implicitly featured a version of the precautionary principle, arguing that scientific uncertainty should not be an argument to refrain from measures in the event of health threats. The bad for business line maintained that scientific uncertainty should prompt cautious policies, because the costs of intervention could not be adequately assessed. To understand this struggle, we need to look at the way uncertainty was covered in the impact assessment.

8.4.3 The impact assessment and scientific uncertainty

The impact assessment conducted in the scope of the CAFE programme was hailed as a *'model of best practice'* (Meuwese 2008, p. 212). It included an examination of five scenarios, including one on an uttermost technically feasible reduction and a baseline scenario in which no extra measures were foreseen. It made use of three different models in order to predict the behaviour and flow of pollution, the impact of economic developments on pollution, the impacts of transport and traffic developments on pollution, and the impact of fluctuations in the use of energy. Nevertheless, precisely this level of detail was the reason the assessment failed to resolve the disputes between representatives of economic and environmental interests.

The detailed modelling work on such a macro scale invited scientific uncertainty. The principal model used was the RAINS model, which had been applied in the scope of the earlier UN convention LRTAP. This model had been reviewed and was considered to be adequate, but a number of uncertainties remained, especially concerning Particulate Matter (The Swedish Environmental Research Institute, 2004, pp. 22-28). The model had been developed within the scope of LRTAP and was concerned with transboundary air pollution. PM, however, was a local problem, occurring especially at street level. Uncertainties considering PM10 or PM2.5 were especially important, because the calculated health gains were due mostly to lower exposure to PM. According to the Dutch RIVM, the RAINS model was not sufficiently qualified to serve as a baseline for CAFE calculations (Jimmink et al. 2004).

Moreover, in order to increase legitimacy and involve other DGs in the process, the DG Environment used DG Transport and DG Energy models. Models such as TREMOVE for transport and PRIMES for energy were used in the context of CAFE, but these models had not been reviewed within the ambit of the programme, and contained their own assumptions and uncertainties (Mantzos 2010; Kouridis et al. 2011). The impact assessment was based on models that had ingrained uncertainties and uncertainties were simply compounded by combining them in the effort to be complete and comprehensive. This led to unpredictable outcomes.

In the CAFE impact assessment, it was stated as follows:

> *'However, costs and benefits are subject to uncertainties, some of which (on both sides of the cost-benefit equation) are significant. Knowledge of these uncertainties and the availability of information to describe them vary. Furthermore, some uncertainties are statistical and continuous in nature, others relate to discrete choices (e.g. selection of approaches for the valuation of air-pollution–related mortality) whilst yet others simply stem from a lack of knowledge. It is clear from this that it will be difficult to develop a fully consistent approach to define uncertainty across the entire CAFE analysis'* (SEC(2005)113, p. 161).

Essentially, the drafters of the impact assessment admitted here that uncertainties could not clearly be identified. In fact, there was even doubt regarding the uncertainties.

The use of impact assessment and other managerial and scientific frames for environmental policy had done the environment a great service by showing that environmental protection could ultimately be cost-effective. However, when expansive policies and impact assessments so vast like this one are undertaken, the instrument loses much of its efficacy. They begin to create uncertainty instead of reducing it. In such cases, pressure groups have the possibility of applying the 'scientific uncertainty' argument. Such arguments can be used to either demand pro-environmental measures, by, for instance, invoking the precautionary principle, or to stave off environmental measures by pointing to cases such as PM10. In the case of CAFE, the scientific uncertainty argument became an element of the flexibility storyline of unwilling Member States and of the 'bad for business' storyline used by the industry and oil lobby.

8.4.4 *Stalemate as the outcome of CAFE*

The CAFE process took seven years from the first Communication in 2001 to the CAFE Air Quality Directive in 2008, and it included more than one hundred stakeholder meetings (European Commission 2005c, p. 3). It produced one of the most comprehensive impact assessments in EU practice. It merged the acidification policy of the EU with the urban air pollution policy, and it considered the synergistic effects of all the pollutants together in the process. Originally, the Commissions intended for CAFE to evolve into an ongoing five-year cyclical programme, in which the 2005 Thematic Strategy on Air Pollution was simply a first step. In 2007, however, the CAFE stakeholder consultation process was suddenly stopped, never to be revived (Airclim website 2011, accessed 29-12 2011).

If we look at the process and its starting ambitions, the question remains as to whether it was worth the effort. We have a Directive that set new limits on PM2.5, but PM2.5

was a fraction of PM10. If the standards for PM10 were met, the standards for PM2.5 were often also met. Not much extra had to be done. This was obviously good news for the Dutch Government, but the question remains as to what the whole CAFE programme delivered.

Possibly it delivered a framework for policy integration in other sectors. Policy integration was an idea from the Fourth Environmental Action programme, but developed especially in the Fifth Environmental Action programme, and it featured strongly in Dutch eco- modernistic environmental policy as well. It is unclear, however, whether the CAFE strategy accomplished much. By the mid-2000s, the process of policy integration had been proclaimed 'effectively dead', and was replaced by faith in impact assessments (Jordan & Schout 2006 cited in Nilsson et al. 2007, p. 5). The insistence on impact assessments and on simpler, less invasive regulation was in accordance with the intentions of the 'Lisbon' agenda for growth and jobs. This emphasis on growth and jobs was viewed by environmentalists as an imminent danger, as it meant they would have to fight continuously for their case that environmental concern did not necessarily hamper competitiveness and economic growth (Wilkinson et al. 2005, p. 5).

The importance attached to the impact assessments in the CAFE programme is clear, but it did not lead to consensus between industrial interests and pro-environmental groups. On the contrary, the expert- and science-based structure of CAFE unintentionally invited quarrelling. It was a 'framework within a framework' (Hey 2005, p. 27), in which other contentious policies had to be developed. It involved multiple DGs, but that implied that those other DGs had to be granted influence over environmental policy. In this sense, policy integration backfired, because instead of environmental policy gaining in influence over other policy domains, other domains encroached on environmental policy.

The elaborate 'comitological'[13] set-up of CAFE was intended for Member States to have ownership of the process, but that did not put an end to discussions in the EU legislative institutions. It simply hid them from the public eye. The CAFE programme inadvertently displayed the limits of a scientific and holistic presentation that were both hallmarks of ecological modernisation.

The proceedings within the scope of CAFE showed that EU environmental policy may well have been 'drowning in process' as Pallemaerts argued (Pallemaerts et al. 2006). In this section, only the CAFE programme is examined, but according to Pallemaerts such

13. I use the term 'comitology' here loosely to refer to the whole procedural set-up by which the EU uses committees of experts and representatives to change and amend regulations quickly. In a strict sense, comitology only refers to those committees composed of national representatives who assist the Commission in the exercise of its implementing powers, and are regulated by the Comitology Decision (Vos 2009, p. 5) (website comitology, last accessed 22-06 2015).

criticism could be levelled at all of the seven thematic strategies from the 6th EAP. Pallemaerts and colleagues were of the following opinion:

> *'The development of these Strategies has led to a protracted policy-making process with few immediate outcomes, delaying the formulation of concrete policy proposals. In effect, the implementation of the 6th EAP has become bogged down in process at the expense of results'* (Pallemaerts et al. 2006, p. 64).

The stalemate in European environmental policy making resembled the air quality clash in the Netherlands. Groups defending environmental/health interests and those defending economic interests were similar in strength, and their argumentative storylines attracted powerful adherents to both sides. However, other than in the air quality clash, science failed to be the arbiter in the European arena. Whereas a complex policy instrument based on detailed scientific calculations, rife with uncertainty, did manage to forge a compromise in the Netherlands, the thematic strategy on air pollution could not play a similar role in Europe. Science did not manage to depoliticise the issue. The efficacy of science as a depoliticisation instrument is discussed further in the following section.

8.5 Concluding remarks

In this final section of the present chapter, the position of the Dutch representative in these European debates is further elaborated upon. However, other elements of the CAFE process merit mention as well. Of special importance from the perspective of ecological modernisation is the efficacy of scientific knowledge to abate and depoliticise environmental conflict. The CAFE programme shows there is a limit to using science to decide policy outcomes. Eventually, the inclusion of increasingly more scientific models will tend to produce complexity rather than to reduce it, thus exacerbating political conflict instead of solving it. For the Netherlands, such insights are important because its own NSL is a highly detailed science-based approach.

From a perspective of the legality of precaution, the events in Europe are of interest too. The precautionary Air Quality Directives from the 1990s lay at the root of the air quality clash in the Netherlands, and the European Commission and other institutions were considered prone to precaution. In the case of CAFE, though, we see that the precautionary concerns of the Commission were thwarted by the efforts of Member States and pressures groups demanding more flexibility. These three topics are elaborated on further.

8.5.1 *The Netherlands: from front runner to foot dragger*

During the clash, the European air quality regulation was a frequent target of criticism by the pro-infrastructure camp. The Dutch Government was criticised for implementing the regulation in too strict a manner, but also for agreeing with it in the first place. The image arose of a small European country that was trying too hard to be 'best in the class' when it came to environmental regulation, and presenting the Netherlands as Europe's 'Gekke Henkie', a pejorative name for a gullible, dim-witted person.

In light of the findings in this study, this perception of the Dutch role in the European arena needs to be adjusted. It is true that the Netherlands supported Air Quality Directives during the 1990s. The Netherlands even played a large part in the Directives' inception by involving the WHO in air quality regulation and exporting its ambitious eco-modernistic ideas. However, in the case of the CAFE Directive, the Netherlands did not operate as Europe's environmental do-gooder. It played the diplomatic game very pragmatically and with a keen eye for its own interests, and it did so even before the air quality clash erupted at the end of 2004. The Netherlands made sure it had the right people in the right committees, and saw to it that its key concerns appeared on the European agenda. It tried to water down the proposal for a new Air Quality Directive, and reverted to a positive tone only when it became clear it could obtain a postponement.

The investigation into the air quality clash revealed that when environmental standards started to hurt substantially, the Dutch Government changed tactics in the European arena, with the result that the Netherlands became a foot dragger instead of a front runner. This confirms the findings from chapter 3, namely that, for the Dutch, environmental leadership was a matter of pragmatism. These outcomes also confirm the findings of Liefferink & Birkel (2011) that the Netherlands desired to be a 'cost free' environmental leader. When it paid to be a leader, the Netherlands clamoured for tough policy measures; when these very measures became too costly, Dutch representatives changed tactics and strove to manipulate European institutions and committees as skilfully as any other. Findings in the air quality case indicated that when economic necessity dictated, the Dutch also argued against environmentally progressive policies.

In general, the investigation into Dutch air quality politics illustrated that the Dutch were most successful as front runners when principles were concerned, and the Netherlands exported its innovative policy principles from the NMP to the European Union. During that period, it also pushed for more ambitious targets. From a political perspective, this conduct was entirely rational and profitable, since in general the Netherlands is a country that benefits more than others from strong European environmental regulation. It is small and therefore dependent on the efforts of its neighbours to pursue a clean environment. The insistence on source policy during the CAFE debates should be seen in this light. The Netherlands is hemmed in between industrial areas and the density of cars in its small territory is high. It does not have an automotive industry of its own, however, and therefore cleaner cars and cleaner industries would benefit the Netherlands relatively more than it would cost. Quality standards, on the other hand, are a nuisance, because the country is highly industrialised and densely populated. In order to achieve the same environmental quality as other countries, Dutch industries and cars need to be comparatively cleaner than those of other Member States.

The air quality clash taught Dutch policy makers to be sceptical of increased European ambition in the environmental field. In the 1990s the Netherlands supported the Daughter Directive because it did not want to lose its ecological face. After the clash on the other hand, openly anti-environmental positions in Europe are no longer frowned upon. Instead, during the Parliamentary debates, policy makers expressed the wish that the Netherlands would look more pragmatically at its own interests.

8.5.2 *Science as a depoliticisation strategy*

In ecological modernisation, the use of science in the service of environmental protection plays a large part, and in the Netherlands this was no exception. The need for science is understandable from a sociological point of view, because scientific knowledge has the legitimacy to function as an arbiter and to provide closure to debates evolving around political questions. Scientific knowledge may function as an element of a depoliticisation strategy, pulling environmental conflicts away from complex value-laden ethical debates to the realm of objectivity and quantifiability.

From an environmental perspective, the increased use of science paid off, especially when it was presented in managerial and economic terms (Arnoldussen 2011). By being able to show in numerical values the cost of environmental degradation and pollution in terms of human lives and economic damage, pro-environmental policy makers and pressure groups convinced a sceptical audience of the need for preventative measures. The effect of this strategy was apparent, for instance, in the Air Quality Directives in the 1990s. A cost-benefit assessment was undertaken that showed that enacting tough standards would be extremely cost-effective, and this analysis helped the Commission to succeed in having the Daughter Directive accepted.

Perhaps science's most important role was a discursive one, as the natural and managerial sciences provided a platform upon which various interest groups were able to come to terms with one another. A kind of hybrid ecological / economic discourse was created by combining the discourses of economy and management, with the discourse of natural and environmental science. By comparing environmental degradation to the budget deficit for instance or by presenting the environment as an indispensable pool of common resources without which the economy could not function, new storylines were created that appealed to both the environmentalists and the hard-nosed representatives of economic interests. The sciences acted as the cement that bound a discourse coalition between proponents of environmental and economic interests. This discourse coalition resulted in hybrid ecological/economic policy discourse laid down in an exemplary manner in the Dutch NMP and the European 5th and 6th EAPs.

Moreover, in the Netherlands the findings of scientists and consultants were accepted by both pro-economy and pro-environmental actor as legitimate arguments for ending the debates between these representatives of various interests. The air quality clash was successfully depoliticised by the promise of sound scientific calculation. In the European Union the situation was different. A comparison between the air quality clash in the Netherlands and the EU debates over CAFE, may illustrate the promises and limitations of such a strategy.

In the Netherlands, the strongly science-oriented NSL managed to resolve the air quality clash. It was accepted by policy makers, by the judiciary, and by most members of the pro-health and pro-infrastructure camps. In this case, the reliance on science succeeded in providing closure, and those groups who did not adhere to this compromise subsequently became marginalised. In this case, a strategy of 'depoliticisation through scientisation' still proved to be successful, despite warnings by Elverding and others.

A different picture emerged from the debates about the CAFE strategy and the Air Quality Directive 2008. From the beginning, the debates within the supposedly science-oriented forums became sites of 'sub-politics', in which the line between expert and policy maker became blurred. The eventual outcomes of the CAFE process, the Air Quality Directive 2008, and the CAFE strategy of 2005 were underpinned by an elaborate impact assessment drafted on the basis of various economic and environmental scientific models, and on countless meetings with experts and stakeholders. The impact assessment was hailed as a model of best practice in the field, it but failed to provide closure to the debates. On the contrary, the impact assessment listed so many uncertainties that its findings were easily questioned. Lobby groups for the industry proceeded to do so, and their storyline that caution was needed because of uncertainties regarding costs found a willing ear among more economically oriented policy makers in the European Commission.

Apparently there is a limit to the capacity of scientific frames to depoliticise environmental issues. The success of the natural sciences in the past led to the demand to underpin policy by the most elaborate models and assessments possible, but there is a point at which these models become counter-productive: namely, when impact assessments and cost-benefit analyses start to produce complexity rather than reduce it. A beneficial impact assessment reduces the complexity of a given environmental problem by providing arguments based on scientifically backed but easily understandable numerical values. This aids policy makers in explaining their choices to a potentially sceptical audience. However, too expansive impact assessments may also produce complexity, when they inadvertently highlight the volatility of a certain problem by mentioning uncertainties on every page due to complex models and large amounts of variables. If uncertainties become that obvious, pressure groups find arguments to attack the outcomes of such an assessment, and the debates shift from being discussions regarding policy measures to the question of how to deal with uncertainties.

If uncertainties become the focus, the intended depoliticisation is frustrated, because the question of what to do in situations of scientific uncertainty is highly political. Pro-economy policy makers and groups tended to argue in such cases for conservative policies, because they felt it was unreasonable to curb economic activity in the face of uncertain threats or against uncertain costs. Representatives of environmental and health interests, however, tended to invoke a version of the precautionary principle, arguing that especially in the event of scientific uncertainty one should err on the side of caution – in this case the environment.

With regard to CAFE, the economic interests appeared to have gained the most in terms of concessions by voicing concerns over the uncertainty of the impact assessment. This concrete outcome is of interest when discussing the hypothesised legality of precaution in the following sub-section. Here, however, it is important to note that there is a limit to which science may depoliticise the environmental question and foster consensus. Its limits are reached when it produces uncertainty rather than reduces it; when politicisation starts all over again, but revolving now around the question of whether uncertainty demands or precludes precaution.

8.5.3 *The resistance against a legality of precaution in Europe*

A number of parallels may be drawn between the Dutch and European debates on air quality. Precautionary regulation was enacted in both arenas at first without much opposition, whereas subsequently, pro-economy and pro-infrastructure pressure groups managed to thwart more ambitious environmental regulation. In both arenas, therefore, the conclusion is the same: the legality of precaution had not yet established itself as the dominant type of legality.

It is argued in chapter 4 that the Air Quality Directives from the 1990s contained many elements of a precautionary approach. This approach was in accordance with the times. During the 1990s, the precautionary principle was on the rise, and in the wake of the eco-modernist turn in the early 1990s there was a strong demand for strict environmental regulation. Under the influence of scandals such as the BSE crisis in meat, public health gained a high profile as well. In law, these concerns were translated into acceptance of the precautionary principle as a cornerstone of EU policy making with the 2001 Commission Communication on the precautionary principle. In the case of PM, uncertain scientific findings quickly led to strict policy making that lay research burdens and placed result-oriented obligations on Member States in order to safeguard public health against this newly discovered threat.

The rise of the precautionary principle in the 1990s was concomitant with a large increase in environmental regulation, among which were ambitious holistic policies such as the Air Quality Directives. Conspicuous in the case of the 1990s' Air Quality Directives is the largely absent opposition from economic interests. The legitimacy of the WHO, combined with the concerns of the 5th EAP and the pro-environmental tide of the times, effectively silenced dissenting opinions. Moreover, industrial representatives and pro-economy Member States did not raise their voices to any great extent. The same pattern was revealed in the Netherlands, where implementation of the Air Quality Order from 2001 did not encounter fierce opposition from business interests, although VVD Parliamentarians tried to obstruct the process by demanding a formal law. However, these opposing voices were too weak to prevent implementation of this precautionary piece of European legislation. Even though the Dutch Government did not intend an overly precautionary line in this file, it enacted precautionary regulation without concerted opposition.

In the 2000s, the cards lay differently. Member States – such as the Netherlands – that felt the sting of the Air Quality Directives objected vociferously to new and tighter standards, and industrial lobby groups and economically inclined Directorates General disagreed as well with the precautionary concerns of the pro-health and pro-environmental actors.

The CAFE debates provided an indication that the precautionary legality that seemed to be dawning in the late 1990s was encountering opposition. In that sense, the fact that the 'bad for business' storyline by and large managed to trump the pro-health storyline in the debates was a sign on the wall. The argument that in the face of scientific uncertainty the legislator should refrain from taking measures is a classical argument against Government intervention on behalf of the environment. The precautionary principle had been proposed as an argument to undercut this line of reasoning. In CAFE, this line of reasoning was presented again, and this time it emerged on top, demonstrating that the legality of precaution had not yet risen to a position of dominance.

Nevertheless, I do not draw the conclusion that the legality of precaution was simply a temporary distortion of the old legal order, made possible by the economic prosperity and the heightened environmental concerns of the early 1990s. I consider that the advance of a legality of precaution has only been thwarted temporarily, due to stiffer resistance from economically powerful actors. These actors raised no objections at first because they were put off guard by the scientific framing of these environmental problems, and, most importantly, they did not take seriously the threat to their interests. The sleeping dogs were only awakened by the environmental regulations that really hurt their interests. Now awake, they were barking loudly at progressive environmental regulation. In that sense the opposition that legality of precaution is now encountering is a sign that traditionally strong economic interests are facing a more dangerous opponent in pro-environmental interests. The air quality clash has displayed that environmental law should not be considered solely a paper tiger, but that it can have teeth and that it can bite.

Nevertheless, any thesis claiming that the legality of precaution is already dominant should be rejected. On the contrary, the air quality clash has illustrated that a concerted effort on the part of pro-economy EU Member States and pressure groups may still bring ambitious environmental EU policy to a halt. The potency of the arguments used in the storyline of flexibility indicates clearly that this legality has not yet been firmly embraced.

Timeline CAFE

1998 Oct.	First informal talks about new approach to air quality regulation between EU Commission and Member States
1999 April	Adoption Air Quality Daughter Directive 99/30/EC
2001 May	Commission proposed CAFE strategy COM(2001)245
2002 July	Commission adopted 6th EAP
2003. Jan	NL took over chairmanship Working Group on Implementation
2004 April	Working Group on PM offered recommendations for Review of Directive 99/30/EC
2004 June	Working Group on Implementation offered recommendations for review of Directive 99/30/EC
2005 Jan.	Commission released Review of Directive 99/30/EC COM(204)845
2005 June	UNICE offered letter to Chairman Barroso complaining about forthcoming CAFE strategy
2005 Aug.	AEA T presented cost benefit analysis for CAFE and recommended a scenario between B (moderate) and C (high) levels of ambition
2005 Sept.	Commission adopted CAFE strategy, level of ambition chosen between A (low) and B. (moderate)
2005 Sept.	Commission proposed new Ambient Air Quality Directive
2006 June	Proposal for new Directive discussed in Council, NL opposed.
2006 Sept.	European Parliament critical of proposal, demanded more flexibility
2007 June	Council Reached Common position on Commission proposal for new Air Quality Directive
2007 Dec.	European Parliament adopted proposal for Ambient Air Quality Directive
2008 May	New Directive on Ambient Air Quality and Cleaner Air for Europe adopted Directive 2008/50/EC
2009 April	NL obtained derogation from the standards for air quality on the basis of Directive 2008/50/EC

9 | CONCLUSION: ANALYSING THE AIR QUALITY CLASH

In this study I have documented the history of the air quality clash. This term is used to designate a period of stagnation regarding infrastructure development in the Netherlands due to an exceedance of the European standards for air quality, notably concerning PM10 and NO_2. Between September 2004 and April 2010, numerous permitting decisions regarding infrastructure projects were annulled by the courts of law because they were found to be in breach of the air quality regulation. Concerns over the health impacts of air quality made newspaper headlines, and worrying reports were released by renowned scientific institutions.

My central research question was how and why air quality had become such a pressing social problem in the Netherlands. To find the answer, I examined Dutch and European policy documents, reports from public health institutions, and other relevant literature. In order to provide for a reality check and at times in order to fill in the gaps, interviews were conducted with specific key individuals.

Three notable developments came to the fore from an initial exploration of the literature. Firstly, it quickly became clear that in the Netherlands the conflict could only be understood in relation to developments in the European arena. The regulation on air quality that caused the annulment of infrastructure projects was European in origin. Hence, a number of chapters in this study have been devoted to developments in the European Union and the role the Dutch Government played there.

Secondly, Dutch environmental policy in the 1980s took a turn away from top-down standard setting towards a more consensus-based style of policy making. It began to focus more on creating a sense of shared responsibility with the various actors involved. Known as ecological modernisation, this consensus based policy discourse had come to dominate environmental policy discussions. In such a context, a heated debate on environmental standards was remarkable, especially since urban air pollution had almost disappeared from the public radar for almost 15 years. Therefore I examined the clash by specifically taking into account the influence of ecological modernisation on its emergence and its eventual resolution.

Thirdly, as a socio-legal scholar, I am interested in the air quality clash as an indicator for legal transitions. In the air quality clash, the legal dimension was very important because the conflict was precipitated by high-profile decisions made by Dutch administrative courts. These strict judgements may be explained by changes in the legal order that emphasise the need to prevent health and environmental threats over the economic compensation of victims. I have referred to this complex of ideas centred on the imperative of damage prevention as 'the legality of precaution'.

In order to examine these developments, three sub-questions have been formulated: firstly, how and to what extent has the interplay between Dutch and European regulation and policy contributed to the emergence of the Dutch air quality clash in 2004? Secondly, what role did the reigning Dutch policy discourse of ecological modernisation play in the emergence of the conflict; and thirdly, was the emergence of the air quality clash indicative of the advent of a legality of precaution?

I have used a social constructivist perspective focusing on both strategic social action and discursive practice. This combination of perspectives was chosen because they each expose important but different aspects of the clash. The actor-centred perspective focuses on the level of interests, and is particularly useful to examine the political manoeuvring of actors aiming to further their own interests. The air quality clash was partly due to actors making claims in order to take the opportunity to push environmental concerns on the political agenda and undermine Government policy geared to facilitate infrastructure development. The discourse-analytic perspective on the other hand focusses on the narratives used in the debates and reveals the hidden assumptions in arguments. This perspective is particularly apt for revealing the underlying social values in political and legal conflicts.

Moreover, ecological modernisation and the legality of precaution have been considered discourses in the realm of environmental policy and law respectively and ideal types of both have been constructed in chapter 1 in order to answer the last two sub questions.

In section 9.1, a bird's-eye view of the development of the air quality clash is presented as an answer to the research question how air quality managed to achieve the status of a social problem in the Netherlands. Attention is paid to the air quality clash in the context of the Europeanisation of Dutch environmental policy and Dutch influence in Europe in section 9.2. The analysis of Dutch political choices that led to the clash in the context of ecological modernisation is the subject of section 9.3. The air quality clash is discussed from the perspective of the legality of precaution in section 9.4.

9.1 History of the Dutch air quality clash: an overview

The lead-up to the air quality clash

During the 1990s, air quality was rarely discussed in this country. From 1999 onwards, the topic gradually gained public attention due to an incident near Overschie that became a public health issue. Health professionals from the Dutch Municipal Health Service in Rotterdam (GGD) claimed that breathing the air in the residential zone of Overschie was equivalent to the passive smoking of 17 cigarettes per day. Health professionals based their claim on data from investigations by Dutch and US epidemiologists, who had demonstrated a correlation between premature mortality and living in areas facing problems air pollution. This claim reached Minister Jan Pronk of the Ministry for Housing, Spatial Planning, and the Environment (VROM). Together with the Minister of Transport and Water Management (V&W), Pronk proposed lowering the speed limit in Overschie, as driving more slowly was considered beneficial for air quality. The speed limit was lowered in 2003.

Between 1999 and 2003, the topic of air quality gradually gained political salience, and various parties on the left of the political spectrum were keen to know whether there were more places where air pollution was present. These parties opposed plans for road construction in the area of Overschie, proposed by the liberal VVD. The environmental movement also responded to the situation in Overschie, and started to campaign for more speed reductions. In the present study, this alliance of political parties and the environmental movement is referred to as the pro-health camp.

This initial mobilisation of the pro-health camp around the speed limits took place at about the same time that two European Directives, 96/62 (Framework Directive) and 99/30 (Daughter Directive), were required to be implemented in Dutch law. These Directives contained strict air quality standards proposed by the European Commission under the influence of the WHO guidelines, ideas laid down in the 5th Environmental Action Programme (5th EAP), and the report '1992 The Environmental Dimension'. Moreover, the notion that high environmental quality could yield economic advantages caught on in the EU during the 1990s. This consideration is the cornerstone of ecological modernisation. The Dutch Government did much to popularise the notion in Europe that – if managed wisely – environmental protection could provide economic benefits. After the success of its own National Environmental Policy Plan (NMP), the principles of the Dutch version of ecological modernisation were exported to many countries and institutions, including the European Commission. The EU's own Fifth Environmental Action Programme (5th EAP) was influenced as well by these Dutch ideas.

The EU air quality policy was designed to integrate well with other EU pollution policies, and substantial information rights for the public were also incorporated. Raising awareness among the public was considered crucial to attract the attention of Member States to the remaining air pollution problems. Even though the Dutch Government predicted difficulties with the feasibility of these standards, it endorsed the Directives because the Netherlands was keen to retain its status as an environmentally progressive country.

These European air quality standards were promulgated in 1999, and were required to be implemented by the Dutch Government in 2001; the task fell to the VROM Ministry under socialist (PvdA) Minister Jan Pronk. Despite pressure from the VVD in Parliament to implement the standards by way of a separate law, the Government chose to apply them by General Administrative Order, a less formal regulatory instrument for which Parliamentary permission was not required. The resulting Air Quality Order 2001 (AQO 2001) was based on the Dutch Law on Environmental Management and had far-reaching consequences, because it established a link between the air quality standards and administrative decisions. This link meant that compliance with the air quality standards became a prerequisite in order for administrative bodies to take lawful decisions regarding permission for infrastructure development. Administrative bodies could not allow this development if air quality standards were exceeded in the area in question. This link was a uniquely Dutch legal phenomenon, as environmental quality standards in other countries were not that closely linked to administrative decisions. Administrations elsewhere needed to strive to achieve an air quality that remains within the standards, but decisions were not immediately susceptible to annulment, like they were in the Netherlands.

At that time, the VROM Minister realised that compliance with the air quality standards would be well-nigh impossible to achieve in large areas of the Netherlands, and would entail huge costs. Minister Pronk wrote a letter to the European Environmental Commissioner, Margo Wallström, explaining the predicament. He was convinced that through diplomacy the European Commission could be persuaded to be lenient, but he did not receive a satisfactory response. His successor in 2002, Pieter van Geel, also failed to obtain leniency from the Commission in the subsequent early years.

In July 2002, the ruling Dutch Government changed political colour. Jan Peter Balkenende became Prime Minister of a Cabinet that included the right-wing parties CDA, VVD, and LPF, and he inherited a polarised political landscape. Populist politician Pim Fortuyn (LPF) had been murdered in May 2002 by an extreme environmentalist, and values and interests considered to be left wing were viewed with suspicion by conservatives.

The first Balkenende Cabinet chose to cut environmental spending, and embarked on a clear pro-infrastructure and pro-economic course of development, marginalising environmental interests. The environment as a policy field was relegated in political importance. A Secretary of State – a more junior position – rather than the Minister of VROM himself became responsible for the environment. The environment belonged to the portfolio of Secretary of State Pieter van Geel. The Government proposed an emergency law in 2002 that made road expansion significantly easier by cutting procedural requirements of participation and environmental impact assessment. Combatting congestion on the roads was its main priority, since congestion was a nuisance to car owners, and being able to resolve the congestion problem was seen as a sign of purposiveness and of a no-nonsense attitude popular with voters at the time.

The proposal for the Emergency Law on Road Expansion was sent in 2002 to the Council of State. In October 2002, the Dutch Council of State Advisory Division advised negatively on the above-mentioned proposal on the Emergency Law. It frowned upon the scrapping of public participation procedures and the curtailing of access to the court, and considered this policy to be contrary to the aims of environmental policy. Air quality standards from the AQO 2001 were mentioned as one consideration that could prevent the speedy realisation of these road expansions, because according to the Council of State they could not be suspended. The left-wing opposition, especially the ecological party GroenLinks, demanded guarantees from VROM's Secretary of State that the air quality standards would be complied with completely, despite the provisions of the Emergency Law. Pieter Van Geel indeed made that promise during the debates on the Emergency Law on Road Expansion.

The air quality clash

Apart from an early conflict in 2001 over the building of a new residential area in Ypenburg, the air quality standards themselves did not cause much administrative and societal upheaval until the later months of 2004. In September 2004, the decision to expand a road near the town of Vught, based on the Emergency Law on Road Expansion was annulled by the administrative court. The highest court of appeal, the Council of State Administrative Jurisdiction Division, considered that the decision did not comply with the provisions of the Air Quality Order 2001. Earlier, in May 2004, a road expansion near the town or Barneveld had already been blocked by the Administrative Jurisdiction Division. The Vught expansion was the second one in a row to be impeded because of air quality concerns, indicating that road expansions were in jeopardy. A week later, the Council of State Administrative Jurisdiction Division annulled the development of a commercial zone in the town of Hendrik-Ido-Ambacht, and other infrastructure development appeared to be blocked as well. The air quality clash had started.

Despite the earlier case of Ypenburg along with the negative advice given by the Council of State Advisory Division, the termination of road expansions and other infrastructural projects was unexpected. The implications were not noticed by the Government, by conservative parties in the Government, nor by lobby groups for the transport and construction sector. Some legal scholars had predicted problems, and the environmental movement was aware of the strict standards, but on the whole the important legal role that air quality would start to play was not seriously considered.

The verdicts given by the Council of State Administrative Jurisdiction Division proved to be a game changer, as they had a number of political effects. Firstly, the pro-health camp became increasingly active, and in Parliament it argued successfully for the need for speed reductions and extra taxes in order to clean the air. Secondly, the environmental movement started campaigning vigorously, and called for public support. The environmental pressure group Milieudefensie used the legal system to strengthen its campaign by asking people to lodge appeals against administrative decisions that permitted infrastructure development. Between 2005 and 2010, air quality played a part in over 1000 cases judged by the Council of State.[1] The legal campaign delivered some important legal successes for Milieudefensie, most notably in 2007, when it was decided to terminate the decision to expand the highway between Burgerveen and Leiden.

In October 2004, the pro-health camp was aided by the appearance of satellite images of the Netherlands, in which the country was depicted as one of the dirtiest in Europe. These images made the problem visible, and constituted a point of reference for subsequent campaigns by the environmental movement. Moreover, the pro-health camp used the excess deaths mentioned in scientific reports by the RIVM to draw attention to the problem of air pollution as a public health issue. In Parliament, the pro-health camp was most successful during those early years. In 2005 and 2006, the budget for quality measures was increased substantially, and even more media attention was paid to the topic of air quality.

The involvement of the court remained the pro-health camp's most important trump card, and the environmental movement began to argue that clean air was in effect a human right because it was in the law. In Parliament, the pro-health camp accused the Government of incompetence because its policies had backfired and had proven to be legally invalid.

For infrastructure developers, economy-oriented political parties, and the Government, these court orders were highly problematic. The country was blocked in regard to infrastructure development at a time that the Government intended to give a high profile to the development of roads and other infrastructural works. The blockade of infrastructure development led to a counter reaction. During the spring of 2005, a

1. For a review of the cases between 2005 and 2009 see Janse (2008).

number of lobby groups primarily representing infrastructure developers united, and were joined by the political parties CDA and VVD in calls for a quick regulatory fix to remove the block. I have called this alliance of interest groups and pro-development parties the pro- infrastructure camp. This camp protected the interests of construction and mobility, and its prime concern was to weaken the legal impact of the standards. It demanded a regulatory solution to the problem that infrastructure development was facing. The interest groups presented worrying economic data that implied that 7.7 billion Euros could be lost because of the stagnation in infrastructure development, and that 100,000 jobs in the construction sector were on the line. Because of the pro-infrastructure camp's activities, air quality became increasingly perceived as a problem of poor regulation rather than primarily a health problem. Especially after 2006, the pro-infrastructure camps' framing of the problem began to dominate.

The Government tried to cater to the wish of the pro-infrastructure camp, but could not risk totally alienating pro-health interests. Moreover, its attempts to seriously alter the air quality legislation were thwarted by the Council of State Advisory Division, because of doubts over the legal validity of these changes from a European Law perspective. The Council of State Advisory Division sent the message that obligations entered into on the EU level were legally binding in the Netherlands. Additionally, it gave political signals that the Government should take environmental standards seriously. The posture of the Council of State had considerable impact on policy making, and it was a number of years before any effective regulatory solution was found.

The regulatory solution

Already in the spring of 2005, legal scholar Niels Koeman argued that the European Directives did not require blocking infrastructure development. The impediment had been a result of Dutch law, because it directly linked air quality standards to administrative decisions. If air pollution existed in certain areas, any administrative decision not leading to immediate improvements could be annulled by the courts. This provision, however, was Dutch, not European. In a well-known financial daily newspaper, Koeman recommended disconnecting the link between quality standards and administrative decisions. This appeal was promptly taken over by the pro-infrastructure camp, and together with other legal professionals, Koeman was invited by the Government together with other legal experts to lend his expertise in finding a regulatory solution to the air quality problem. Together with civil servants from the Ministries of VROM and V&W, they tried to resolve the impasse. A compromise was needed that would weaken the link sufficiently but would not be rejected outright by the Council of State.

The civil servants and legal experts came up with a complex solution that would be fleshed out further over the years 2006 and 2007, with a proposal consisting of a programmatic approach to the air quality standards. Projects that could harm air quality significantly would be counterbalanced by measures that would have beneficial effects. Over time, this balance of beneficial measures and harmful projects would lead to air quality that would comply with the European standards. This long-term plan was broad in scope, and necessitated detailed calculations regarding the effects on air quality regarding projects and on the counter measures. It was met with criticism from both the pro-health as well as the pro-infrastructure camp when the gist of the plan was revealed in Parliament. Both sides commented on its complexity. The pro-infrastructure camp felt that simply breaking the link would be an easier solution. During the debates, the pro-infrastructure camp criticised the EU and the Council of State for obstructing this simpler solution. The pro-health camp saw in this approach a way to 'calculate the air clean' without having to tackle the 'real' public health problem.

The Government stuck to its approach, however, and in the course of 2006 and 2007 the proposal gained in Parliamentary support, though there was still a great deal of scepticism. The political process was speeded up especially in July 2007 when the Council of State's Administrative Jurisdiction Division annulled the decision to expand the highway between Burgerveen and Leiden. This decision displayed that the air quality clash was far from over. In October 2007, the Law on Environmental Management was amended with a chapter on air quality standards: namely, the Air Quality Law, which would enable promulgation of the programmatic approach. The plan itself became known as the National Cooperation Plan Air Quality (NSL).

The plan itself had to be drafted and presented to the European Commission for acceptance, as the measures proposed in the NSL would take time to be effective. A temporary postponement – a derogation – from the air quality standards was necessary to bring spatial development back on track. In 2007, there was no possibility yet of asking for such a derogation, but it might become possible under a new Directive that was being negotiated at the time. In Europe, the Clean Air For Europe (CAFE) programme was launched in 2001, with the aim of updating existing air pollution policy. Within the many sub-committees that made up CAFE, the Dutch VROM Ministry lobbied strongly for the possibility of giving Member States extra time to meet the standards. The EU issued a new Air Quality Directive in 2008, the CAFE Directive that allowed such a derogation if certain conditions were fulfilled. On the basis of this new Directive, the Netherlands would be able to obtain extra time if it submitted a credible plan to improve air quality. The Netherlands submitted the detailed NSL in June 2008, and was granted the desired derogation in April 2009.

The NSL was a policy plan rather than a package of proposals that could be enacted immediately. It contained a number of controversial policy proposals, such as a tax on diesel fuel and environmental zoning to restrict polluting lorries. It also included the introduction of road pricing. In particular, the last proposal was and still is highly controversial. After being removed from the political agenda during Balkenende's first

Government, it returned because of the air quality clash. The hold that air quality had on Dutch spatial planning for over five years was released with the emergence of the NSL and its acceptance by the Council of State as a sufficient justification for infrastructural projects in March 2010.

This historical outline answers the first part of the research question pertaining to the history of the air quality clash, namely *how* air pollution by PM became a pressing social problem. However, in order to understand *why* air quality became a pressing matter and explain the significance of the clash for Dutch law and policy, the conflict needs to be analysed from the perspectives of Europeanisation, ecological modernisation, and the emerging legality of precaution. In section 9.2, The emergence of the air quality clash is explained in the context of the interplay between Dutch en European environmental policies, while section 9.3 discusses the influence of various Dutch policy choices analysed from the perspective of ecological modernisation. In section 9.4, the role of the emerging legality of precaution is examined.

9.2 The air quality clash and the construction of European environmental policy

One of the main causes of the clash can be found at the European level. The European Union issued new air quality regulation in the 1990s, and the air quality Daughter Directive from 1999 laid down significantly stricter standards than previous European rules. Together with elements from Dutch and UK policies, the guidelines created by the World Health Organisation formed the inspiration for this rigorous regulation. Moreover, recent epidemiological research findings on the risks of air pollution coupled with pressure by UK representatives provided the political sense of urgency to develop these new Directives quickly. The Netherlands was obliged to implement the regulation in National law, but throughout the country the standards were exceeded. As a result of this, and on the basis of the Dutch implementation of these European rules in the Air Quality Order, a large number of high-profile construction projects were terminated.

If the strict European rules had not been in place, there would have been no air quality clash, and the Dutch Government used this fact to blame Brussels for its occurrence. However, the Netherlands contributed significantly to the realisation of the European air quality policy. This interplay between Europe and the Netherlands is analysed in section 9.2.1. In section 9.2.2, I focus on policy making in Europe. European policy has an amalgamous character and is influenced by different ideas, Member States, and institutional interests. This complex character makes it a difficult arena for Member States to manage. In section 9.2.3, we will look at the consequences of Europeanisation for the Dutch legal order as well as at the balance of power between Dutch policy makers and the judiciary.

9.2.1 *Influences that shaped the 1990s Air Quality Directives*

Even before the Netherlands became known as one of the EU's environmental front runner States, it had tried to convince other countries and institutions of its policy ideas. This lobbying occurred out of understandable self-interest, since the small country needed to rely on other countries to keep its environment clean. Moreover, the Dutch administration had to make sure it did not create an overly heavy burden on its own industries, because that would harm their competitive position.

One of such instances in which the Dutch Government exported its own ideas was the involvement of the World Health Organisation in air quality policy. In the early 1980s, the Dutch Government intended to draft air quality standards for the Netherlands, and Dutch civil servants asked the WHO to draft appropriate guidelines. By involving the WHO, the Netherlands considered it would be certain that standards would be based on sound scientific knowledge, and that it could ascertain legitimacy for its own regulatory proposals. Moreover, other countries might also adopt the WHO guidelines, thereby ensuring that Dutch industries would not be at a disadvantage. By 1987, the air quality guidelines were completed, but the Dutch Government had lost interest in setting its own standards in this regard, and it followed those set down by the EU instead. Within the EU, however, the WHO air quality guidelines began to feature prominently as a benchmark for regulation. The 5th Environmental Action Programme from 19993 (5th EAP), which determined the EU's environmental policy agenda for the 1990s, contained the target that air quality standards should be based on those set by the WHO. The eventual directives from this decade, Directive 96/62/EC (Framework Directive) and Directive 99/30/EC, (Daughter Directive)[2] were based on these recommendations and on other ideas contained in the 5th EAP.

I have not been able to determine any direct Dutch influence on the air quality target contained in the 5th EAP, but on the whole it was an ambitious environmental policy plan on which Dutch policy makers had exerted a strong influence. The 5th EAP followed the innovative market-based approaches to environmental policy pioneered by the Dutch Environmental Ministers Pieter Winsemius and Ed Nijpels (both VVD). Their approach of responsibilisation of target groups, and especially the notion that strong environmental standards may be beneficial to economic growth, gained currency in EU policy-making circles.

By involving the WHO and exporting their own ambitious environmental policy, Dutch Government paved the way for the Air Quality Directives. This explains why the Dutch Government by and large supported the Directives, and at times even pushed for more ambitious regulation. Only in 2001 did Minister of the Environment

2. There were three more Air Quality Directives, but only these two are of concern for the purpose of this study.

Jan Pronk ask to renegotiate some of the Directives' more problematic points, but he did implement the Directives strictly in Dutch law.

In the policy-making process of Directive 99/30/EC – the Directive containing the problematic air quality values themselves – the UK representative wielded a strong influence. The UK assumed ownership of the air quality file by sending to the European Commission a representative who had worked on the UK's national air quality strategy. The British considered that this strategy could act as an example for that of the EU. The UK had adopted this strategy in the wake of a nation-wide scare concerning childhood asthma during the 1990s. In 1997, the Labour party had come to power in the UK, and the young Tony Blair was keen to demonstrate his environment-related credentials. In 1998 he made the topic of the environment a spearhead during the British EU presidency, and tried to achieve a common position on many pieces of environmental regulation, including the Air Quality Directives.

The EU strategy was indeed modelled after that of the UK, but there were important differences. One of the most significant was that the UK strategy allowed for a weighing of interests between environmental quality and pressing economic concerns, but the EU strategy did not. According to UK respondent Richard Mills, the main writer of the UK strategy, the EU tried to "one up" the UK strategy by making it more rigorous. He referred to the desire of the European Commission to posture itself as a strong environmental champion, daring to outperform the Member States in environmental ambition.

The fourth influence of note is the presence of an elaborate economic study that showed that the air quality standards would be highly cost-effective. However, this evaluation had painted a far too rosy picture of the abatement measures Member States needed to take in order to comply with the Directive. It had taken into account projected future benefits for air quality of other EU policies and concluded that the extra burden caused by the Directive was little. However, these future benefits failed to materialise and the extra burden that the standards became steeper very quickly.

The European Air Quality Directives were shaped by a set of diverse aspects, but Dutch influence was present in at least two of them: the involvement of the WHO and the eco modernistic policy philosophy. The Dutch Government was never simply a bystander, but was a prominent player in the European environmental policy arena. On the face of it, the strategy to involve the WHO seemed to have been a masterful move, as the Netherlands gained what it had intended – ambitious air quality standards. However, due to the changing nature of EU politics and to the long period of time in which the development took place, the Netherlands did not come out ahead, but instead finished last. To understand this development, we need to look at the dynamic nature of EU policy making.

9.2.2 *The bricolage of European environmental policy*

In order to understand how the Dutch lost control of the file; we took a closer look at the way air quality policy is made in Europe. The European policy arena is one in which many different pressure groups, Member States, and institutions play a part. All these actors have their own reasons for trying to influence the course of European environmental policy.

The institutional set up of the EU itself is complex and this allows actors to influence the policy making process through various points of access. The two main legislative bodies of the EU, Council of Ministers, and the European Parliament adopt legislation by co-decision, meaning that they need to come to terms with each other. However, these bodies do not always see eye to eye. In the Council of Ministers for instance, various Government representatives, often Ministers of the policy field in question, negotiate about the eventual polices to adopt. They have national interests close to heart. The European Parliament is organised around political parties representing dominant ideologies, such as social democracy, Christian democracy, green etc. the Members of the EP conform to the line of their respective political party and ideology. The European Commission is the main executive branch of the European Union, but it also proposes regulation and therefore it has considerable agenda-setting power. The European Commission is an institution that safeguards the interests of the European Union, its ideals and institutions as such. It tried to give the EU a more prominent profile on various social issues in order to display the Union's added value to the public and to present it as more than just an economic club. It developed ambitious long term policies on subjects as consumer protection, women's rights and also environmental protection. The transboundary character of environmental pollution makes it a subject on which the European Union can score points with concerned environmentally aware citizens.

Often actors try to influence the Commission, but also through the Council of Ministers and the European Parliament claims makers can get direct access to the EU's legislative machinery. Throughout this study, we witnessed several different Member States trying to influence both specific regulation and the policy philosophy on which EU environmental policy is based. The Netherlands exported elements of its own environmental policy to the EU level in the hope that its own innovative, market-based philosophy would also prove successful there. The UK as well as the WHO played a strong part as drivers of the Daughter Directive that contained the limit values for air quality. The UK did so in a variety of ways, but one way was by sending a UK official to aid the European Commission to draft the proposal for the Daughter Directive. When the CAFÉ programme was on the way in the mid-2000s, pro-economy pressure groups, Member States such as the Netherlands and other Directorates General used their influence to obtain a more flexible policy that allowed for more leeway in terms of the standards. The Netherlands for instance tried to wield its influence in the many sub committees comprising the CAFE programme and in the European Parliament.

In the case of air quality, this complex configuration of players and interests meant that there were many opportunities for ambitious actors to export ideas to the European level in the hope of shaping EU policy to their own needs. However, it also meant that ideas exported to the European level were subject to numerous pressures and tended to change over time.

The emergence of European air quality policy in the 1990s illustrates the complexities of this dynamic. The EU started to play an ambitious role in environmental policy, and used elements that conformed to Dutch policy, such as providing information to the public, the responsibilisation of both producers and consumers, and the integration of various policy instruments. However, it also incorporated elements that deviated from the Dutch line. Attention to air quality standards, for instance, was fashionable in the Netherlands in the early 1980s, but by the 1990s the Dutch Government had lost its enthusiasm. In the EU, however, environmental quality standards were preferred because they allowed a harmonisation of standards between countries, and so it was considered that quality standards should be mandatory and valid for the whole territory of the EU. This resulted in Directives containing provisions about public information in order to raise awareness of the issue of air quality as well as inflexible standards that were proclaimed for the whole of the EU.

This use of quality standards was alien to the way they were perceived in the Netherlands. Dutch policy makers considered that air quality standards should be used as flexible instruments, tailor-made for certain areas and based on sound scientific data.

Dutch pressure groups began to use the issue of air quality standards in court in order to force the annulment of infrastructure projects with which they disagreed. For the EU Commission on the other hand, the large number of complaints in Member States indicated that the Directive had succeeded in raising awareness about air pollution. For the Netherlands, it resulted in economic damage and stagnation. This is an example of how interests of European administrative bodies and Member States may conflict. In the air quality clash and the run up to it, these conflicting interests became manifest.

The European Commission and Member States had diverging interests, and even if they promoted the same kind of policy instruments, they may have had quite different intentions and ideas as to how to use them. The European Air Quality Directives were a bricolage of different influences, philosophies, and opportunistic moves by politicians and pressure groups. Dutch ideas provided some of the justifications for the Directives, but the UK influenced the EU to create a strategy that was well suited to the UK's own national policy. This strategy had been drafted to combat a perceived surge in childhood asthma attributed to pollution from motorized vehicles. The UK had been concerned about urban air pollution ever since it did battle with deadly smog in the 1950s, and it is easily conceivable that the UK public was pleased with the EU air quality strategy. For Dutch politicians though the early activism on air quality regulation struck like boomerang, decades later. Over the years, the original ideas had been altered, combined with other ideas, translated and interpreted in a way contrary to Dutch interests and intentions.

The same diffuse and often contradictory set of pressures and interests characterised the debates on the CAFE Directive and Strategy from the middle 2000s. The European Commission intended for an ambitious long term strategic program and a Directive that updated the 1990s' standards. Member States and pro-infrastructure pressure groups were out to obtain derogations and flexibility, resulting in disagreements between the Commission and various Member States, disagreement between the various Directorates within the European Commission and between the Council of Ministers and the European Parliament. The outcome was a compromise between these various interests that was presented as the best scientifically supported outcome, but was in fact a similar kind of bricolage of different philosophies and ideas as the 1990s' Directives had been.

This research particularly focused on the construction of Air Quality Directives, but if these findings are representative of decision making in the European arena, than the image rises of a patchy policy arena in which it is easy for actors to influence the process, but difficult to propose a policy or an idea without it being watered down and changed significantly.

9.2.3 *The Dutch backlash against Europeanisation*

From the 1980s onwards, the Dutch Government had always seen the EU as a desirable arena to push for environmental policy and to voice demands for a greater environment-related initiative. The European project was about harmonisation in the service of the common market, and it began to extend itself to the environmental field in the 1980s when it dawned on European policy makers that such a market would need the harmonisation of environmental quality. With its largely eco-efficient industry and its open economy, the Netherlands welcomed this project, and was prepared to transfer some of its sovereignty to Europe in exchange for a level playing field. The thinking was that Dutch industries would not need to worry about their competitive position if industries in other countries were subject to the same environmental laws. This process by which more and more national legislation becomes influenced by European law is known as Europeanisation.

The Netherlands had accepted Europeanisation willingly in the past because it saw more opportunities than threats in such harmonisation. The air quality clash, however, illustrated the disadvantages of European regulation for Dutch policy makers.

The level playing field suddenly became a burden, because the Netherlands was under considerable environment-related pressure, and had to do more than other countries to maintain environmental quality. When the costs started to rise, the downsides of participating in the project and transferring sovereignty became apparent. When the Dutch Government realised that the cost of compliance with these EU rules would be high, it tried to get rid of them. However, the regulations proved to be far more tenacious than the Dutch Government had originally envisioned.

The reason that the Government could not easily change the Air Quality Order had to do with the changing balance of power between policy makers and the judiciary. European law has a prominent place in the Dutch constitution: before the courts, European laws rank higher than domestic laws, and they may even be invoked directly if they are not implemented in time or in the appropriate manner. The standards laid down in the Air Quality Order were European in origin, and the highest administrative court – the Council of State Administrative Jurisdiction Division – argued that it would be forced to ask questions of Brussels if Dutch policy makers were to come up with a law containing their own interpretation of the Directives, which would have incurred significant further delays. In this fashion, the Council of State Administrative Jurisdiction Division would be able to block the introduction of new laws that reinterpreted the Air Quality Order in a manner more amenable for the Dutch Government. This threat added weight to the opinions of the Advisory Council. According to Duyvendak politicians feared that the advice of the Advisory Division functioned as a prediction of what the Administrative Jurisdiction Division would do.

The pro-infrastructure camp used this state of affairs to criticise Brussels and to portray the European Union as an inflexible bureaucracy that ran roughshod over Dutch interests. It also provided the ammunition to criticise the Council of State over its stance that the law could not be amended easily because European law had precedence over national law. Conservative parties played on these sentiments to urge the Government to resist European demands for more air quality regulation.

The pro-infrastructure camp began to present the air quality regulation itself as the real problem, and in the end the camp's perception of the issue started to dominate in policy-making circles. The regulation from Brussels was gradually presented as undesirable meddling in Dutch affairs, and fuelled the resentment some politicians already felt towards Europe. The episode caused the Dutch Government to cease lobbying for strong European air quality laws and instead to push for relaxation of the rules. During the subsequent CAFE process in the 2000s, the Netherlands was one of the prominent foot draggers: namely, those Member States that put the brakes on environmental policy development in the EU. Apparently, it was easily forgotten that the Netherlands bound itself to these European regulations and as a small, highly populated and highly industrialised delta country benefitted rather more than most from the environmental level playing field prescribed in Brussels.

9.2.4 *Conclusion: Europeanisation – from opportunity to threat*

The Europeanisation of environmental policy has been of utmost importance for the Netherlands. The Dutch themselves were major players in this process by uploading their own policy philosophy to the EU level, as did other environmentally forward States of the time like Germany and the Scandinavian countries. As a result, the EU became an enthusiastic player on the environmental front. The EU overtook the Netherlands in environmental ambition and this ambition is one of the reasons why air quality became a sudden social problem in the Netherlands, even though the Dutch Government fanned the flames of ambition in the first place.

EU policy making was steered by extremely diffuse sets of interests and actors. Member States wielded significant influence, the various institutions of the EU itself made their voices heard, the scientific community influenced EU decision making, and a whole set of pro-business and pro-environmental pressure groups lobbied the various institutions and even within the expert committees in which policy was developed. This combination of factors made the EU a potentially unpredictable source of policy and law. Its ambitions were high, but the philosophy on which it bases its policies and the actors that drive certain policies was diffuse.

This constellation had never deterred Dutch policy makers from trying to use the European political arena as a place to upload their own nationally preferred policies, which they did successfully in the 1980s. However, the emergence of European policy and law meant a relinquishment of national sovereignty. When no conflict existed between European and Dutch intentions and regulation, this relinquishment caused little concern. However, now that European environmental ambition exceeded national environmental policy plans, the situation was different. Once in the forefront, the Netherlands became a foot dragger, and the air quality clash played an important role in this transformation. The Netherlands experienced first-hand the conflicts that could arise from a mismatch between European and national policy, and the safest option therefore appeared to lie in limiting Europe's ambition, which is what the Dutch did during the CAFE programme. However, a retreat from the European arena in the environmental field now would mean that the Netherlands will have even less influence of the European policy outcomes than it already has.

9.3 Developments in Dutch policy and ecological modernisation

In the following sections I investigate a number of Dutch political choices that contributed to the emergence of the air quality clash are under scrutiny. Already from the late 1970s, onward Dutch spatial planning had become legally intertwined with environmental quality standards. This link meant that no new polluting activities could be carried out in an area where standards were exceeded, and this ruling was strict because environmental quality standards had precedence over other administrative plans. The eventual implementation of the European air quality rules in Dutch national law in 2001 disregarded the inherent problems this intertwining could pose when combined with strict environmental standards.

Moreover, Dutch policies embraced the formation of a strong and active subsidised environmental movement during the 1980s and 1990s, with the aim of creating a countervailing power with regard to economic interests. The movement was intended to act as a representative of environmental interests and to convince other segments of society, notably industrial interests, of the importance of environmental protection. During the 1980s, the Dutch Government gradually created an ambitious environmental policy based around the idea of shared responsibility and consensus. These consensus policies, however, met with both success and failure. The environmental movement chose negotiation instead of action in the 1990s, but the activist Milieudefensie reverted to more conflictual tactics in the 2000s. Moreover, policy failed to achieve such a broad consensus in the field of transportation and mobility, as mobility interests had no love for the new environmental direction. When Prime Minister Balkenende openly adopted a pro-mobility policy in 2002, the environmental movement and left-wing political parties reacted to this change. Air quality was already in the picture due to the small upheaval regarding air pollution in Overschie, but it now became a serious issue over which to criticise Government policy, especially after the Council of State Administrative Jurisdiction Division began to terminate infrastructure projects.

9.3.1 *The turn to ecological modernisation*

In order to understand the choices made by Dutch politicians, it is necessary to examine them in the context of the turn to ecological modernisation in the 1980s. Ecological modernisation is an environmental policy discourse that gained ground in the Netherlands and established itself most firmly with the National Environmental Policy Plan of 1989. The use of eco-modernistic discourse represented a move away from earlier environmental discourse that was prevalent in the 1970s, and was characterised as 'limits to growth'. In this 'limits' discourse, environmental pollution was considered to be the result of economic development and industrialisation, and therefore it implied setting limits to economic growth as well. Economic growth should be sacrificed for ecological wellbeing, and permits were considered the primary way of regulating the environmental behaviour of polluters.

The policy discourse changed after modern managers and modern management theories were introduced to the Ministry of the Environment (VROM) after a reorganisation of the Ministry in 1982. Minister Pieter Winsemius tried to combine effective environmental protection with the ideals of deregulation that the ruling Cabinet at the time espoused. 'Aided' by the emergence of new transboundary environmental problems that required a more comprehensive approach, he embarked on a policy to foster responsibility for environmental protection in the important industrial sector, in other Ministerial departments, and among the public at large. He chose a more horizontal approach rather than the 'command and control' regulatory methods applied in the 1970s. Environmental policy became organised according to the problems that needed to be tackled instead of to the different environmental compartments. Rather than regulating emissions to the air, water, and soil separately, for instance, it was considered that regulation should target the problem of acidification, an issue that involved both air and water pollution. Likewise, targets were set to combat the spread of dangerous substances, waste control, and so on, in consultation with industry and other powerful economic sectors. These targets were often promulgated in covenants – agreements between the Government and industrial sectors to meet certain targets.

In order to make it appealing to a sceptical audience, ideas were developed to make environmental protection economically attractive. If industries could be convinced that environmental protection did not necessarily entail economic sacrifice, they would then be more willing to engage in it. Part of this idea was that by preventing environmental problems to arise in the first place, it would be possible to save clean-up costs later. Because it was cast in these economic terms, a preventative environmental policy became an attractive option to the representatives of economically important sectors. Moreover, the possibility of technological innovation to help the environment and to create a profit was highlighted in order to create a market for environment-friendly technology. By applying these strategies, the Ministry tried to achieve consensus and shared responsibility.

The notions of consensus-based policy, market-based approaches, prevention, and the possibility of win-win trade-offs between the environment and the economy form the bedrock of ecological modernisation. The eco-modernistic turn in the Netherlands reached its summit with the National Environmental Policy Plan (NMP) in 1989. In the NMP, a policy philosophy was combined with actions to improve the environment and with a programme to change the environmental behaviour of various target groups, such as the public, industry, transportation, and even foreign countries and institutions like the EU. It established the Netherlands reputation as an environmental front-runner in Europe until at least the year 2000.

Ecological modernist policy making conforms to both the idealistic leanings of Dutch policy as well as to its pragmatic profit-oriented side. This dual nature is captured in the Dutch expression 'the reverend and the merchant', which denotes both the Dutch penchant for preaching to other countries as well as its tendency to put profit first when push comes to shove. Ecological modernisation is appealing, as it promises that both the reverend and the merchant may coexist peacefully and even cooperate.

Not surprising in this light, ecological modernisation is a discourse with two faces. On the one hand, it may be used to argue for an ambitious far-reaching environmental policy with the intention of restructuring the economy along sustainable lines. In accordance with authors like Christoff (2000), I have called that branch strong ecological modernisation. On the other hand, however, it may also be used to legitimise existing economic structures on the premise that economic growth is good for the environment because it fosters technological innovation. In this weak form of ecological modernisation, behavioural change is not necessary, but economy and ecology should be managed in ways that make their coexistence possible.

The first political choice to be discussed in relation to this discourse is the implementation of the air quality standards and the incorporation of the link between quality standards and administrative decisions. Subsequently, we discuss the failed ecological modernisation of transport, the relationship between pressure groups and Government in ecological modernisation, and the return to ecological modernisation as a result of the clash.

9.3.2 *The implementation of the Air Quality Directives and the link*

During the air quality clash, the connection between spatial planning and environmental standards became a controversial issue. After the Council of State started to annul decisions to permit infrastructure projects, legal scholars pointed out that the reasons for the annulment should be sought in the Dutch legal system itself. Under Dutch environmental/spatial law, administrative decisions should 'observe' the limit values laid down in quality standards, and that meant that permits should be refused if they were to lead to lower environmental quality. The pro-infrastructure camp argued that this connection must be broken in order to pull spatial development back on track.

The link itself preceded the Dutch turn to ecological modernisation. In fact, it bears the hallmark of the limits to growth discourse from the 1970s, because it imposed clear limits on development. It made further polluting activity impossible in areas where the limits were being exceeded. However, the connection was kept during the turn to ecological modernisation. One of the aims of the VROM Ministry was to integrate environmental considerations in other areas, and the connection ensured that the environment was taken into account where spatial planning in the Netherlands was concerned. To that extent, the link was compatible with ecological modernistic policy making. However, the role played by quality standards in Dutch environmental policy making was gradually transformed in the 1980s. Quality standards were considered useful especially in 'area specific policy making', and could be used, for instance, to safeguard natural areas against development. As mentioned above, quality standards needed to be used flexibly and not create too much friction between developmental interests and environmental protection. They had to be based on strong scientific data, and economic interests needed to be taken into account when standards were to be imposed.

The European air quality standards, however, did not exhibit these qualities. They were binding countrywide, without feasibility being taken into proper account, and because of the Dutch link, they had the effect that large tracts of the Netherlands were suddenly closed off from further development. The Minister in charge of the implementation in 2001 Minister Jan Pronk, however, had not intended such strict consequences, and he himself and later his successor Pieter van Geel chose a pragmatic solution. They would not construct new infrastructure in 'sensitive destinations', those areas where people spent a considerable amount of time. In non-sensitive destinations, however, areas where people did not live and did not spend time, the reasoning was that construction could take place. It was felt that the standards protected human health, which was not at stake in such areas. Moreover, in the explanatory Memorandum, the responsibility for the policy that related to the most problematic pollutant, Particulate Matter (PM10), was taken away from municipalities and lower administrative bodies, and laid down at state level. This would release the lower administrative bodies from having to comply with the strict consequences of the link, and only require them to apply the As Low As can Reasonably be Achieved (ALARA) principle. Such a provision came down to a permission to weigh and balance economic and environmental interests.

Nevertheless, these pragmatic solutions were rejected by the Council of State Administrative Jurisdiction Division in its three seminal verdicts, and by the Advisory Division in answer to a letter by Pieter van Geel dated 30 September 2004. The Court's interpretation restored the link as a strict limit regarding the ability to pollute in contaminated areas. It pointed out that the European Directives mentioned the whole of the EU territory, and that neither the Directives nor the Air Quality Order made the distinction between inhabited or uninhabited areas.

The foregoing raises the question as to why the Dutch Government chose such a strict implementation in 2001 and incorporated the link. The argument to exempt lower administrative bodies from responsibility for PM10 was explicated in the Explanatory Memorandum (though not in the law itself) but the distinction between sensitive and non-sensitive destinations was not. The reasons for this implementation lay in political decisions made by Minister Jan Pronk as well as in the pragmatic de-politicising nature of ecological modernisation.

The Air Quality Order, drafted in 2001 by Jan Pronk as Minister of VROM, was a lower regulatory instrument that needed to be based on a particular law: the Dutch Law on Environmental Management. Because it was a lower regulatory instrument, however, the Order did not need to be approved by the Dutch Parliament.

Pronk explained that the strategy was to implement the European air quality rules dutifully and to renegotiate them later. He felt that his hand would be much stronger if he could show the European Commission that the Dutch had good intentions but were not able to comply with the standards. Moreover, as Environmental Minister, he was also trying to organise a deal on climate change measures in Europe that he considered important for the Netherlands. By showing that the Netherlands implemented European environmental law faithfully, he hoped to strengthen his position on this file as well. He might also have been influenced by what he saw in Overschie. He considered air quality an important subject in terms of public health, and a strong air quality order might bring about measures to alleviate the situation in areas suffering bad air quality. These political motives explain the strict implementation to some extent, but do not account for everything. For instance, a more careful implementation without the explicit link could have prevented a number of problems later, since the link as such was not required by European law.

A more structural explanation for the implementation is provided by referring to the nature of Dutch environmental policy. From Winsemius onwards, ecological modernisation became entrenched in it. This discourse emphasised the need to resolve environmental problems through policy integration and through consensus by shared responsibility. Firstly, the link was retained because it ensured that air quality considerations were integrated with spatial planning. In the Dutch system, integration was considered crucial and even Van Geel defended the link by stating that it ensured quality considerations were taken into account early on in the process.

Secondly, the link was retained because no one thought that such a strict rule would be applied so rigorously. This counter-intuitive notion has its roots in ecological modernisation and in the way it was employed in Dutch policy. Characteristic of Dutch environmental policy is that none of the parties involved should feel that unreasonable demands are being made upon them. The silent agreement that parties should 'do all that is reasonably possible' is considered a prerequisite for the consensus and for the sharing of responsibility in tackling environmental questions. In principle, such an understanding is possible, because ecological modernisation considers that environmental problems may be solved with adequately clever management and a combination of economic growth and environmental innovation.

During the 1990s, many parties subscribed to this policy of gradual reform and its promise of win- win solutions. Dutch policies met with success abroad, and domestically they ushered in a kind of environmental polder-politics in which former adversaries started to discuss their conflicts at the negotiation table rather than through aggressive public campaigns. The consensus policies led to a depoliticisation of environmental conflict in many areas, and the consensus approach seemed successful.

Moreover, Dutch ecological modernisation was internationalist in outlook. It considered large-scale transboundary problems to be more urgent than the local problem of urban air pollution caused by transport and traffic. Solutions to environmental problems could only be achieved through international cooperation – for instance, within the EU.

Both developments caused policy makers to respond phlegmatically to new research that indicated PM10 was possibly a far greater threat than had previously been considered. When results of this research became available in the 1990s, PM10 became a policy priority, but concrete policy never got off the ground in the Netherlands, compared to, for instance, in the UK. Policy was promised in the form of a memorandum on air quality, but this memorandum never materialised. A cost-effective policy could not be devised due to the uncertainties surrounding PM10, and as a consequence, successive Dutch Cabinets chose to leave policy making to the EU. Even when the European rules were agreed upon, no sense of urgency was evident, and policy making was left to the implementation of the Directives in the Air Quality Order.

The same dispassionate approach characterised the Air Quality Order 2001. A strict implementation was not intended, but at the same time it was not prevented, the reason being that no one could imagine air quality would lead to such a heated debate. Van Geel for instance defended his policies by stating that it would be unreasonable to demand that the Netherlands comply in full with the air quality rules. Among Dutch policy makers, the notion prevailed that full compliance with the standards could not be required, and therefore the standards could not fully apply. The Explanatory Memorandum asserted confidently that the standards would be revised downwards upon evaluation. In hindsight, however, such an assertion was naïve, since the EU would not revise standards if only the Dutch were seriously bothered by them. This illustrates the reigning Dutch mentality of the time, which held that such standards could not plausibly be considered mandatory.

9.3.3 *The failed ecological modernisation of transport and marginalisation of the environment*

In the 1990s and early 2000s, the prevailing consensus on environmental policy lulled policy makers to sleep in the case of air quality policy. However, the environmental consensus was also being gradually undermined on a terrain that would prove to be crucial for the air quality clash: transport and traffic. In the early 1990s the Government tried to introduce ecological modernisation in the areas of transport and traffic. Infrastructure development to accommodate traffic growth would be restricted, while public transport and rail transport would be accommodated.

At the same that that the NMP was released in 1989, the Dutch Government made public a first version of a plan in which future developments for transport and traffic were outlined. This plan was called the Second Structural Scheme for Traffic and Transportation (SVV2 1989). The final version of the SVV2 saw the light of day in 1991. This plan included numerous concepts and ideas that could be traced to the NMP. I consider it an attempt to modernise traffic and transport in an ecological manner. The Government did not intend to curb the expansion of mobility per se, but considered that this expansion could be achieved by extending public transportation and by other environmentally friendly means. The aim of the SVV2 was to solve congestion problems without expanding roads but by diverting mobility towards public transportation. One of the core ideas was to have motorists pay for the use of roads by way of road pricing.

Road pricing was an eco-modernist idea because it used the pricing mechanism to encourage behavioural change in an environmental direction, without limiting freedom of choice. However, this idea met with fierce opposition by the ANWB and in the 1990s by the VVD. The motorist's association vehemently opposed it, and though the idea was mentioned in the final version of the plan, its concrete implementation was postponed.

However, other measures to restrict auto-mobility were introduced. Road construction could only take place after a thorough process of stakeholder consultation, after parking space became restricted, and after the use of public transportation had been encouraged. By and large, however, the SVV2 was not a success, as motorists felt that their interests were being neglected, and they remained wary of road- pricing schemes. The SVV2 had an environmental dimension but – as behoves eco-modernistic policy – it also had an economic aspect. While driving ought to be discouraged, the Netherlands should nevertheless become a main hub for European transportation. Transport should preferably be undertaken by river or rail, but it most certainly should still grow.

During the 1990s, the environmental ideals of this new transport policy were gradually relinquished. To achieve a sustainable transport policy, the SVV2 aimed at a behavioural change on the part of motorists, but this ambition was gradually dropped. Successive Cabinets paid more attention to the rival goal of transforming the Netherlands into a strong transportation country and expanding Rotterdam's main ports and Schiphol Airport. In 2001, the last attempt at introducing some form of road pricing in the successor to the SVV2 was torpedoed in Parliament by the VVD. This last act marked the failure of ecological modernisation in the transport sector. Where other economic sectors adapted, transport and traffic remained remarkably resilient.

When Jan Peter Balkenende took over the helm of a new conservative Cabinet in 2002, he made an unambiguous choice in favour of auto-mobility and against road pricing: namely, subsidies for cleaner cars were abolished, and roads were to be expanded again. Eco-modernistic rhetoric about a sustainable transport policy remained, but was used primarily for an audience of foreign institutions such as the EU. The Netherlands has no automotive industry, and EU regulation that made cars cleaner could still count on the Netherlands unequivocal support. Nationally, however, having more asphalt was the preferred solution to the social problem of congestion, an issue that resonated strongly with the political parties in Balkenende's Cabinet, the liberal conservative VVD, and the populist LPF. Large scale infrastructure development plans were announced, mostly to stimulate the Dutch transport sector and to appease motorists who resented road congestion.

These policies upset the reigning environmental consensus based on the concern to combine environmental and economic values. Conflict between health and environmental interests and transport and traffic interests had previously never led to such a clash, because the environmental movement never openly selected this area as a high-profile target. The flagship of this Cabinet's mobility policy, the 2003 Emergency Law on Road Expansion, elicited angry responses from environmental parties that demanded that the provisions of the Air Quality Order remain in full. The Council of State Advisory Division observed as well that this policy went against the grain of traditional Dutch policy and Dutch international obligations regarding climate change. The Council of State reiterated that the provisions of the Air Quality Order would remain valid in full.

The clear-cut choices for mobility and budget cuts for the environment strengthened the opposition against the Government. The situation in 'Overschie' had already prompted campaigns for speed reductions, but the environmental movement now started to campaign seriously against Balkenende's policies.

The conflict was fuelled by the failed ecological modernisation of the transport sector along with the subsequent choice of the first Balkenende Cabinet to relinquish eco-modernisation in this policy domain. Auto-mobility had in the past been excluded from heated debate, mostly because so many Dutch people owned cars. Environmentalists feared that if mobility were touched, they risked losing the support of the population. However, when the search for win-win solutions was dropped, the environmental movement and the pro-health parties no longer had anything to gain from the environmental consensus, besides marginalisation. They rallied against Balkenende's policies, portraying them as grossly negligent, given there were so many victims of air pollution. After the court cases emerged in late 2004, the tide turned in their favour, and the parties were able to accuse the Government of legal incompetence in addition to negligence.

Seen in this light, the air quality clash was a reaction against the marginalisation of environmental interests and the unbridled expansion of mobility and infrastructure development during the first decade of the new century. This study illustrates that the air quality clash is a significant event in the history of environmental conflict, because it drew transport and mobility – which had managed earlier to escape unscathed – into the ambit of this conflict. In the air quality clash, auto-mobility did not emerge totally victorious. The public became aware that living along a busy road could be dangerous, and this conviction was translated into policy in 2008. It became illegal to construct schools and homes for the elderly close to highways if the air quality standards were exceeded. Moreover, as will be discussed further, the bar for research into the harmful effects of road construction and expansion was raised significantly.

9.3.4 *The relationship between pressure groups and the VROM Ministry*

The eventual mobilisation against the Government's policies was actually in line with the philosophy governing eco-modernist policy. Within ecological modernisation, consensus is sought to resolve environmental conflicts. Such a consensus that would deal equitably with the interests of economic development and the environment could only be achieved if representatives of those two interests reached a certain balance of power. Therefore, Dutch eco-modernist policies included the active sponsoring of a countervailing power in the form of a strong environmental movement. Through subsidies and granting environmentalist movements a seat at the table, this power was both strengthened and co-opted within the decision-making structures of the Dutch corporate consensus democracy. Dutch law provided ample opportunity for NGOs to stand as representatives of environmental interests.

When the Balkenende Cabinet broke the environmental consensus by choosing squarely for mobility and infrastructure and the Council of State Administrative Jurisdiction Division started to annul infrastructure projects, this active and professionalised environmental movement decided to mobilise the public and other pressure groups to start court proceedings against infrastructural projects. Without this strategy, the pro-health camp could never have garnered enough political support for its clean-air measures. This strategy was highly successful, both in terms of having projects annulled and in securing extensive media coverage of the clean-air problem and the environmental campaigns against it.

Policy makers, however, were not amused. After a significant legal defeat regarding a road expansion project near Leiderdorp in 2007, politicians complained strongly about what they saw as an unfair strategy. The pro-health interests could not win in Parliament, and so they sought a second opportunity in the courtroom. The environmental movement earned the reputation of being 'professional trouble makers', and was considered the culprit as regards the sluggish decision making around infrastructure.

When the clash drew to a close in 2008, political parties on the right (CDA and VVD) demanded a cut in subsidies for environmental movements, one of the main reasons being their protests against road construction. In particular Milieudefensie – the environmental organisation that had pushed the legal strategy hardest – was targeted.

These events demonstrate problems with the eco-modernist line of cultivating a counter movement. On the one hand, this line was facetious because the environmental movement was supported as long as it remained on the VROM Ministry's leash. The countervailing power was created precisely to give environmental interests a voice, but when it did speak out it was silenced. This called into question the Ministry's intentions of financing these pressure groups. On the other hand, it took away the independence of pressure groups because they became 'addicted' to subsidies, which might have influenced their choice of campaign topics. The idea of raising a strong environmental group actually reduced their independent capacity for action, because they became dependent on the interests that it should have been combating. The strategy for funding environmental pressure groups turned out to be that of divide and conquer rather than allowing for the emergence of a true representative of environmental interests. Environmental responsibilisation was cultivated within the context of ecological modernisation, but only to the extent that it did not bite the hand that fed it.

Open access to the courts for environmental movements was similarly criticised. The pro-infra parties hoped that the Elverding Commission, set up during the clash to investigate decision making, would lead to a reduction in points of access at which environmental movements could initiate legal proceedings. The outcry following the actions of the environmental movement in this case shows the catch-22 character of such policies. A critical pressure movement is funded as long as it is not critical and does not begin to apply actual pressure. This is reminiscent of a classic scene from the book Catch-22, where privileges to question policy were granted only to the people who never asked any questions. Apparently, open access to the courts and the creation of a countervailing force was seen as beneficial, but only until it was actually used to resist Government policy.

9.3.5 *The resolution of the clash and the return of ecological modernisation*

Even though the environmental consensus was shattered, the eventual resolution consisted of a return to eco-modernistic consensus policy. From 2005 onwards, the Government sought a regulatory solution to the problem, but it would only be able to celebrate this in 2009. After four long years of stalemate, the National Cooperation Plan Air Quality (NSL) was submitted to the EU in June 2008, and the Commission accepted it a year later as a sufficiently sound policy plan. This acceptance was necessary to be eligible for a Commission sanctioned postponement of the obligation to comply with the standards. Postponement was granted the following April.

The NSL quite literally balanced the interests of infrastructure development with improvements in air quality. In this long-term plan, projects considered harmful would be offset by measures that had beneficial effects. This move would result in meeting the standards for NO_2 in 2015 and for Particulate Matter in 2011. The NSL approach was not popular when it was originally proposed and discussed in 2006, but gradually it became seen as the only viable solution.

In terms of ecological modernisation, the plan contained many features that conformed to this philosophy. It delivered flexibility to administrative bodies because they would no longer be confronted with areas in which construction remained impossible due to air quality exceedances. It also conformed to the assumption that good environmental quality could be achieved without making development impossible, and – if correctly managed – beneficial measures could compensate for the harm done by development. The desire for a pre-emptive approach to environmental policy was met by the NSL as well, because before development could take place, it needed to be determined by which measure the project would be compensated.

From the perspective of environmental protection, the NSL represented a retreat from the earlier, less flexible legal consequences of air quality standards. Before the programmatic approach was introduced, construction projects in areas where air quality was poor were simply annulled by the Council of State Administrative Jurisdiction Division. Environmental quality standards were considered to represent absolute limits below which no further deteriorating was allowed under any circumstances. However, in the face of the annulments caused by the air quality standards, these legal consequences became untenable. The NSL may therefore be interpreted as a plan aimed primarily at making infrastructure development possible again, and that is precisely how the environmental movement saw it.

However, an interpretation that considers the outcome of the clash a complete victory for construction and mobility interests, does not take into account the strong shift in air quality policy during the clash. Before the conflict, no funds were available, because the Ministry of VROM was facing budget cuts. Environmental protection was treated with animosity by the first Balkenende Cabinet of 2003, and its interests were marginalised. Only after 2005 did concerns regarding clean air and sustainability issues return to the top of the political agenda. Large amounts of funds for air quality protection were made available and the sustainability chapter of important policy documents such as the mobility memorandum was strengthened. Even the controversial issue of road pricing was debated again as a possible solution and this policy proposal was even included in the NSL.[3]

3. Though as we now know, it was not implemented with the NSL, again due to severe opposition from motorists and pro-mobility parties and the VVD.

The NSL represented an important policy change. It was essentially a more comprehensive way of dealing with clashes between economic development and environmental wellbeing, because the harmful effects of projects had to be considered not in a piecemeal fashion but in light of the comprehensive pattern of development. Environmental considerations needed to be taken into account when the whole developmental agenda of the Netherlands was devised, instead of only on a project-by-project basis.

Ultimately, the emergence of the NSL was a victory for the pro-infrastructure camp, but in the years preceding it environmental considerations had trumped economic considerations. This balance of power was unprecedented in the Netherlands, and could not last in the end. Socially and politically, there was no support for such a sweeping environmental turnaround. Moreover, it was a victory with strings attached, because when the Government nowadays decides to develop infrastructure it must ensure that it has a place in the overarching plan.

9.3.6 *Conclusion: ecological modernisation as cause and resolution of the clash*

When viewed over a long period, the trend in Dutch environmental policy is aptly characterised by the pendulum metaphor. During its inception period in the 1970s, Dutch discourse on the topic was starkly antagonistic and eco-centric, and it contained apocalyptic ideas of environmental ruin if economic growth was not limited. Ecological modernisation weakened the rhetorical apocalyptic edges, and by allowing room for economic considerations, managed to provide a more amenable and arguably more effective policy discourse. Ecological modernisation was a successful environmental discourse, and it gradually picked up in ambition again to reach a peak lasting from 1989 until 1992. Ecological modernistic arguments were now employed to challenge seriously the existing economic structures. Not only could economic progress and environmental protection coexist but economic growth also became conceptualised as being dependent on environmental wellbeing. This resulted in large-scale policy plans on numerous terrains including industrial development as well as transport, mobility, and spatial planning. In the field of mobility, for instance, attempts were made to collectivise transport to a large degree, and to persuade motorists to leave their cars behind and commute by train.

During the 1990s, the ecological modernisation of the transport sector was put to the test and gradually the intention to change the behaviour of motorists was relinquished. Economic considerations became more important and congestion became perceived as a nuisance but also as a threat to economic prosperity. Under Balkenende's first Cabinet, environmental intentions were by and large forfeited, and only lip service was paid to ecological modernistic concerns. In practice, this policy had been abandoned, and the Government introduced an ambitious programme of road expansion and infrastructure development.

The air quality clash was a backlash against the forfeit of environmental ambitions in the Netherlands. Pressure groups, pro-environmental parties, and to some extent the Council of State as well rallied against their relinquishment. This backlash could only be solved by a return to ecological modernistic storylines that emphasised consensus and the possibility to reconcile economic and ecological concerns.

The return of a weak form of ecological modernisation was a step forwards again from an environmental point of view. In 2010 it apparently still had the potential to reconcile competing interests in environmental conflicts. However, debates during the air quality clash illustrated that weak ecological modernisation was embraced owing to the lack of anything better. Many actors complained of the triteness of ecological modernistic win-win storylines, such as those regarding environmental protection as an opportunity for growth, or improving competitiveness by setting high environmental standards. The programmatic approach was a legal novelty, but was not popular initially with either the pro-infrastructure camp or the pro-health camp. Politicians from the pro-infra camp complained about the detailed calculations that became necessary, and the pro-health camp considered the NSL an 'ultimate smuggling tactic' in order to 'calculate the country clean'. Though necessary, the NSL's compromise was not wholeheartedly embraced. It was a different kind of consensus than the one ecological modernisation had managed to forge in the 1980s between environmental and economic interests and it lacked its earlier beguiling vision.

9.4 The air quality clash and transitions in the legal order

A final crucial development in the emergence of the clash was the legal interpretation that the Council of State gave to the Dutch Air Quality Order. This interpretation was strict, and focused on the letter of the law, ignoring the intentions of the Government that had drafted the AQO. The termination of high-profile infrastructure projects, notably road expansion and commercial and residential zones resulted in air quality becoming an important topic in policy making circles and among economically important sectors of Dutch society.

In addition, the Advisory Division of the Council of State thwarted attempts by the Balkenende Cabinets to make quick amendments to the AQO. It blocked the easy way out, and necessitated an altogether new approach to air quality standards. This programmatic approach served to restore the consensus, and functioned as the official application to the European Commission to grant the Netherlands the required postponement so that it could get its infrastructure development back on track.

The judgements of the Council of State fit within a pattern of increasingly strict drafting, implementation, and interpretation of air quality rules. The European Union issued strict air quality regulation, the regulation was implemented by the Dutch national Government even more rigorously, and, finally, the regulation was interpreted strictly by the Council of State. This pattern raised the question of whether the air quality clash could be viewed as signs of an emerging trend in the legal order to prioritise precaution when dealing with environmental and health hazards. I have called this putative new legal way of handling pollution and other hazards the legality of precaution. This term signifies a discourse within the legal order in which damage prevention is deemed so important that preventative measures should go even beyond cost-effectiveness. In this type of legality, considerable attention is paid to the potential victims of environmental and health threats, as they are seen as victims of mismanagement by the authorities, and are considered having a right not to be exposed to damage. A typology has been made of the legality of precaution, and to make comparison possible it is contrasted with an earlier legal discourse – that of risk and compensation. These typologies may be found in table 2, in chapter 1.

In order to answer the question of whether the demand for precaution is increasing within the legal order, we need to assess where we find precautionary elements in the acts, verdicts, and policies of the institutions that make up the legal order: in other words, where did this legality become visible before, during, and after the air quality clash? Below, I will discuss the emergence of a legality of precaution in the EU, the Dutch Government, the pro-health camp, and finally the Council of State.

9.4.1 *Precaution in the EU air quality policy*

In a discussion regarding precaution in the EU air quality policy, two policies merit consideration: the Framework and Daughter Directives of the 1990s, and the Directive proposed in the context of the CAFÉ strategy. In regard to the first Air Quality Directives, the findings indicate that many elements of a precautionary legality were present in these policies and the way they were drafted. The precautionary principle was never mentioned as such, but the standards were based on science that was uncertain at the time. The knowledge concerning particulate pollution came from epidemiological studies that showed a correlation between air pollution and premature mortality. This means that premature mortality and high level of pollution were often found to occur together in the same population. However, this connection did not in itself display a causal relationship, but the correlation became quickly regarded as causative. Science at the time was – and actually in many regards still is – in the dark about the exact mechanism through which PM10 harms the body and which of its components are most harmful. The UK pollution watchdog COMEAP used precautionary considerations to argue for a rigorous standard, and its judgement that it would be imprudent not to treat the correlation as a causative relation was swiftly embraced in the EU as well.

The standard for PM10 was actually a compromise, as there were indications that PM2.5, a smaller fraction of PM, was more harmful than PM10, but there was even less information regarding it. This quick regulation of a pollutant about which so little was known, but which was feared for the damage it might do, is consistent with a legality of precaution: namely, threats must be thwarted before there is scientific certainty of their exact nature.

The Directives were defended by pointing to the long-term effects of exposure to air pollution and to the insidious effects it might have in the long run. Short-term damage was determined to be negligible, but the prevention of long-term damage was considered. In order to argue for the Directive, the Euro-Parliamentarians drafting the report for the European Parliament invoked the image of the European citizen left to breath poor-quality air. The responsibility of European institutions to put an end to this situation was highlighted, and the situation was considered shameful.[4]

The Directives relied on the insights of the 5th Environmental Action Programme, which called for a more holistic approach to policy making as well as for the raising of public awareness. The policy constituted by the 1996 Framework Directive, and fleshed out in subsequent Daughter Directives, took into account more pollutants than had previously been considered. Instead of only setting minimum standards, it also stipulated strategic requirements, among which was the provision of information. The raising of public awareness was a key element of this policy, as citizens should be knowledgeable regarding the remaining air pollution problems.

As far as I know, citizen's groups were not represented at the negotiation table or in the consultation phase of the Directive, but environmental groups were present in force. Together with representatives of environmental Ministries, they formed a block of actors speaking on behalf of environmental interests. Accordingly, even the environmental lobby group EEB conceded that environmental interests were strongly represented, in excess of economic ones.

The air quality policy was not the only one in which precautionary concerns were taken seriously. In the late 1990s, new and ambitious holistic approaches were established as well as policies intended to raise awareness among various actors. Moreover, in 2000 the Commission released its milestone communication regarding the precautionary principle (European Commission 2000b). I consider that the ecomodernist ideas that gained a firm foothold within the 5th EAP made it possible for such policies to appear.

4. Incidentally, the Parliamentarian drafting the report was from London, the city most discussed during the asthma scare that hit the UK at the beginning of the 1990s. For a complete description of this process, see chapter 3.

Even though many precautionary elements may be pointed out, a number of elements in the Directives fit better in the legality of risk and solidarity. Public participation was encouraged, but the Directives themselves relied on expert knowledge. Moreover, Directive 99/30/EC was defended by a cost-benefit assessment. Conducting this type of analysis to determine policy was paradigmatic of a risk and solidarity approach, in which cost-effectiveness was a major consideration. In this instance, a number of things went wrong with the analysis, and it provided much too rosy a picture of the possibility for reduction; nevertheless, it was still an element better suited to risk and solidarity than precaution.

On the whole, a legality of precaution was clearly gaining ground in the EU's air quality policy from the 1990s; however, an opposite trend is detected when we review the CAFÉ strategy of 2005 and the concomitant policies. The proposal for CAFE from 2001 still spoke of the precautionary principle, but it could not be found in the final text from 2005. Instead, the cost-benefit element was much more prominent and important than in the establishment of the Directives in the 1990s. Outcomes of the cost-benefit analysis were used here by groups representing economic interests as arguments for postponing the process and adjusting the level of ambition downwards. The role of scientific expertise was reinforced in CAFE by an intricate and complex set-up of committees dominated by experts and policy makers. Political concerns began to dominate eventually, but expert science was considered to be decisive. Public participation was encouraged, but only to provide the programme with extended legitimacy.

All in all, in the case of CAFE, sceptical Member States such as the Netherlands as well as industrial and business groups managed to prevent further precautionary policies. One of the arguments brought forward by actors arguing for less environmental ambition was that the uncertainty of the data should dissuade the Commission to take far-reaching measures. This squarely contradicted the precautionary argument that uncertainty should not be a reason not to take measures. The storyline that flexibility was needed in order to avert high costs managed to carry the day eventually, a clear sign that the precautionary legality is struggling nowadays in Europe.

9.4.2 *Precaution in the Dutch air quality policy*

On the basis of findings in this study, a number of interesting conclusions on the Dutch Government's political and discursive commitments regarding precaution may be drawn. In the field of air quality policy, successive Dutch Cabinets have not demonstrated a clear propensity towards precaution – on the contrary. On the national level, the Dutch Government dealt with the issue in pragmatic ways. For instance, ambitious policies were proposed in the early 1980s, but when the EU became involved the Dutch Government trailed the EU developments. More ambitious standards were sometimes proposed, but always to an extent that they could be easily met. It is notable that when the new scientific findings regarding PM reached the Netherlands, the Government did not revise its policies immediately. It proclaimed PM10 a priority, but when no cost-effective policy could be found it decided to wait for Europe. The Dutch Government tended to save precautionary rhetoric for the international arena. The Dutch strategy had been one of enlightened self-interest from the start, and by and large it remained that way. The Netherlands displayed a 'merchant and reverend' kind of approach, preaching for strong environmental protection, but always making sure it did not harm economic interests. A legality of precaution demands more, because it argues for prevention against all – or almost all – costs.

The Netherlands did support the 1990s Directives, because it did not want to lose face in terms of its ecological interests. However, when problems began to loom on the horizon, the Dutch Government turned quickly to favour renegotiation and postponement of the standards. Even Jan Pronk, a noted proponent of the precautionary principle, advocated postponing implementation of the standards due to his concern regarding the cost-effectiveness of the policies.

In general, the Dutch Government showed few signs of a precautionary legality in its air quality policy. The closest to precaution the Government came was with the promulgation of the Air Quality Order 2001, and Jan Pronk chose an implementation of the EU Directives that was more strict than necessary. This line was chosen partly because the Minister himself became involved in the Overschie situation, and believed that lowering the speed limit and halting the expansion of roads was a good idea, in addition to his other strategic considerations.

These considerations notwithstanding, it is clear that the Air Quality Order was de facto a precautionary piece of regulation, because it was based on the precautionary European standards and added the forceful link between air quality standards and administrative decisions. The link was precautionary because it prevented further activities in polluted regions, regardless of economic costs. The Air Quality Order mentioned this link explicitly, and placed an onerous burden on lower administrative bodies. These precautionary aspects, however, were not intended as such. They were unintended effects of stringent EU regulation, hastily implemented without taking into account the particularly Dutch legal regime in this area, and with too much confidence that dire legal consequences would not emerge.

When we look at the level of discourse, strong ecological modernisation has overt precautionary tendencies. Weak ecological modernisation has precautionary elements as well, but to a lesser degree, and it is compatible with pragmatic, less ambitious environmental policy. The National Environmental Policy Plan and the Second Structural Scheme Traffic and Transport (SVV2) were examples of strong eco-modernist discourse, and they included precautionary elements. The strong eco-modernist elements of behaviour changes recommended by the NMP were not taken very far, however, and the SVV2 failed to achieve the ecological modernisation of transport and mobility.

The Dutch Government returned gradually from the more participatory, communal, and holistic policies unveiled in those early documents to a more pragmatic and calculative way of policy making in the 1990s and 2000s. The resolution of the air quality clash was a case in point. The balancing exercise that was the NSL fitted much better in the legality of risk and compensation than in that of precaution. It had elements of both, but the calculative framework stood out, and that was the most characteristic feature of the legality of risk and solidarity.

Furthermore, it is conspicuous that on the level of rhetoric, but also on the level of legal and regulatory practice, the limits to growth discourse was more precautionary than the ecological modernisation approach. The limits to growth discourse did not lead to ambitious environmental results, because the actual standards proposed were not overly stringent, due to sustained opposition from economic interests. The rhetoric, however, and even the specific character of Dutch environmental spatial policy, was more precautionary than the more flexible practices currently favoured. On the whole, the Dutch Government did not embrace a precautionary legality, either on the level of air quality policy or on the level of policy discourse. It preferred solutions that fit in the legality of risk and solidarity.

9.4.3 Precautionary legality and the pro-health camp: the case for a right to clean air

It is no surprise that precautionary arguments are most forcefully used by actors belonging to the pro-health camp, since they serve its partisan interests; their impact, however, should not be underestimated.

In 1999 municipal health services had already framed the issue of air quality in Overschie in a precautionary manner. The metaphor it used was that living in Overschie was equal to the passive smoking of 17 cigarettes a day. A small risk was made immediately tangible and threatening, by associating air quality with smoking. This metaphor infused the situation in Overschie with the sense of danger and harm that the presentation of numerical values could not have done.

During the air quality conflict itself, the pro-health camp used a variety of arguments against the Government, accusing them of harming the health of the citizens, having no eye for the plight of residents living near highways, pursuing 'sickening' policies of road expansion, and of not doing everything possible to minimise damage. This message contained precautionary elements such as the idea that the occurrence of damage was a disgrace, that citizens were being treated with indifference, and that the Government should prove it was doing everything possible to protect Dutch citizens.

These elements were all to be expected in a political struggle over an environmental issue, but the pro-health camp also used specific legal argumentation to drive its point home. According to the environmental movement, citizens had a right to clean air, and the Air Quality Order as well as the verdicts of the Council of State proved as much. Using this storyline, the environmental movement mobilised residents and other pressure groups to start proceedings against administrative decisions that allowed infrastructural development. The Council of State never framed its annulments in such a fashion, but the law and the verdicts could be seen to lend credence to such an argument. It was a precautionary argument, because framing clean air as a legal right promoted health concerns over others. From legal scholar Ronald Dworkin we know that rights act as trump cards, and this argument suggests that clean air would eclipse other concerns. This storyline is not new, as it had been proposed by human rights activists and environmental activists, but here we find it being used outside of academia. In the context of the Euro-Parliament debates on CAFE, the notion was also reiterated by left-wing MEPS and even by environmental Commissioner Stavros Dimas.

The mobilisation of people behind the storyline that a clean environment is a right represented a step in the formation of a precautionary legality, even if it was not yet recognised as such in court. If this argument were to enter the mainstream and become embedded in policy documents or in law, it would come down to the acceptance of the precautionary assumption that ecological harmony should trump economic development.

9.4.4 *Precautionary legality and the Council of State*

The air quality clash occurred in part because of the Council of State's termination of infrastructure projects. The Council justified its annulments of the administrative decisions by pointing out that too little research had been conducted into the effects on air quality. In situations in which the standards were exceeded, administrative bodies should show that its solution of the problem would not further deteriorate air quality. It followed from the 2004 Vught Ekkerswijer decision that administrative bodies were required to show that their decisions conformed to the need to redress the air quality situation as quickly as possible.

The main question is whether we should view these decisions of the Administrative Jurisdiction Division and the opposition of the Council of State Advisory Division against Balkenende's mobility policies as heralds of a precautionary legality. The question carries extra weight because the different Divisions of the Council of State act as the highest political advisory council as well as the highest administrative judge. Both the acts of the Council of State Advisory Council and the Council of State Administrative Jurisdiction Division have a profound influence on shaping the legal order.

In regard to the verdicts of the Council of State Administrative Jurisdiction Division, it is difficult to select overtly precautionary arguments: the Court did not state that preventing damage was the paramount principle of the legal order; it did not indicate that the citizen was victimised by the administration on account of its decisions; and it never conceded to the 'right to clean air' storyline presented by the environmental movement. In its verdicts, and in the yearly reports in which it commented on its decisions, the Council of State pointed to the way the Netherlands had chosen to implement the Directives. It also remarked that the practices by which the Dutch Government had tried to soften the impact of the Air Quality Order did not conform to the wording and intent of the Directives or the wording of the Air Quality Order. The Air Quality Order entailed a strict implementation of the Directives, but did constitute a faithful interpretation of them. If the Government needed more leeway it should negotiate with the European Commission. This storyline line amounts to the acknowledgement of the primacy of European law and that the administrative court 'merely' applied the law as it should, and that it did not consider its task to include taking into account the social consequences of the regulation. If the Council's verdicts could be determined as precautionary, that would then be a consequence of the precautionary nature of the regulation.

There are, however, also indications that the Council did take a more precautionary line in excess of the Directives' demands. The Administrative Jurisdiction Division asked for much more thorough research into the effects on air quality than it had done previously with regard to administrative decisions. The requirements on research subsequent to decision making were so difficult to fulfil that in the early years of the air quality clash only one research institute was able to meet them.

I take this heighted burden of proof as an indication of a more precautionary approach to questions of air quality, but it is hard to judge. The Administrative Court could couch its verdicts in legalistic argumentation and did not need to expose its convictions. I conclude that the Court intended to draw a line in the sand and to indicate that the administration could not run roughshod over the laws that protect people, and especially not over European obligations.

The convictions of the Council of State Advisory Division are easier to assess, because they are worded in publically available policy advice. Without the veil of legalistic argumentation, precautionary elements stand out more clearly. In a number of places we may find that the Advisory Division takes a more precautionary line. It acted as a champion of the Air Quality Order when the Emergency Law on Road Expansion was forwarded to the Council of State, by stating that these provisions would remain valid in full. Moreover, in that same advice it stated that the Emergency Law on Road Expansion ran counter to the grain of Dutch sustainability policies and international obligations.

In its advice on the 2006 budget, the Advisory Division commented that both the Dutch Government and the EU had a one-sided focus on economic wellbeing.

Moreover, it considered that Western European economies, including the Dutch one, presupposed high economic growth as a given. The Council of State noted that structural weaknesses could make it necessary to forego these growth ambitions, and it named particulate emissions as one of the environmental weaknesses. I interpret such statements as signals from the Council of State, especially because the budget of 2006 was treated in late 2005, the year in which the air quality conflict was raging with no solution yet in sight.

In its policy advice, the Council of State also expressed the need to keep access to the courts open. It responded critically to proposed limits on starting appeal proceedings laid down in the Emergency Law on Road Expansion, and also maintained the rights of citizens to appeal against projects that had been incorporated into the NSL. It is no surprise that an institution like the Council of State would uphold access to court, but it is also an indication that it was concerned about the rights of citizens to appeal against any infringement of their participatory rights, even if use of the Courts resulted in the stagnation of economic development. The Council of State, for instance, withstood considerable political pressure levelled against it to be more restrictive towards environmental organisations.

A further reason I consider that a precautionary approach is visible in the considerations of the Advisory Division involves the tenacity with which it held on to the interdependence between spatial planning and quality standards. The Council of State did not intend to release this link, and argued that it was not at all clear whether it was even allowed by European law to release it, even though other countries did not have such a strict connection. Despite thorough opposition from politicians belonging to the pro-infrastructure camp, the Council defended the strict applicability of environmental standards in spatial decision making. I consider this relationship between standards and decisions to be precautionary because it gave absolute priority to environmental standards over economic feasibility. No weighing of interests was allowed. Combined with strict standards, as was the case for air quality, this link set limits to growth, and fits with the most precautionary of discourses considered in this study.

Nevertheless, the later year of the clash from the year 2008 and onwards illustrated that there was a limit to how strongly this precautionary legality could take root. After the air quality law, and especially when the NSL was promulgated in 2009, it became very difficult to win an appeal using air quality arguments. The Council of State Administrative Jurisdiction Division accepted the NSL, which itself is no example of precautionary regulation, but remained firmly rooted in the legality of risk and solidarity.

9.4.5 *The Council of State as policy maker*

The air quality clash was a confrontation between environmental and economic interests, but also a battle between institutions claiming over the demarcation if their domains. The Council of State is comprised of two sections that have different and strictly delineated competencies. Because of the constitutional demand of the separation of powers, its political function is separated from its judiciary function, even more since the changes made to the Act on The Council of State in 2010. However, the air quality clash showed that the Council of State wielded significant power because of its dual role as policy advisor and administrative judge. Advice given by the Advisory Division was taken very seriously indeed, probably largely because if the Government did not heed the advice, it felt that it would lose its case in the Administrative Jurisdiction Division. This fearful attitude on the part of the Government, and especially Secretary of State Pieter van Geel, was reproached by Parliamentarians from the pro-infrastructure camp, while the Council of State was praised by the pro-health camp.

The Council of State received vehement criticism from Parliamentarians from the pro-infrastructure camp during the debates on the account of the role it played. This criticism is understandable because in this case the Council of State assumed the role of fellow policy maker, - as understood in terms of Joel Best - rather than resigning to its roles as advisory council and highest administrative court. This is problematic from the point of view of democratic legitimacy. However, there is much to be said for the Council of State assuming such a role in the air quality clash.

One of the tasks of the Council of State Advisory Division should guard the continuity of Dutch policy. In a polarised political constellation such as the Netherlands experienced at the beginning of the 2000s, the Advisory Division should make sure the Dutch political ship stays on course. The remarks made by the Council concerning the pro-infrastructure policies of the conservative Balkenende 1 Cabinet and the Emergency Law on Road Expansion should be read in this light. The tenacity of the Council of State forced the Government to come up with a more sustainable solution to such environmental spatial conflicts.

As regards the Council of State, I conclude that in order for it to fulfil the role of 'legal conscience', it should truly be independent of the administration. During the air quality clash it refused to be a mouthpiece for the Government. Often the Council of State is criticised for exactly the opposite reason, for being too environmental friendly. I find this kind of criticism much more scathing than the reproach that it hinders speedy decision making. The air quality clash is a good example as to why the primacy of politics need to be kept in check by the rule of law. Speedier decision making all too often translates to limiting the access of civilians to courts and increasing the power of the administration over the judiciary. I consider it valuable that citizens who lost out in the political arena are granted their second chance in front of the courts of law, even though that may take time. An institution like the Council of State manages to guard long term interest, exactly because it is removed from the hustle of day to day policy making and vying for the favour of the voters.

However, the double function of the Council of State, being both the leading policy advisor as well as the highest court, is non-transparent and causes resentment. In the air quality clash, the Government found itself caught in a vise. The Administrative Jurisdiction Division anulled projects while the Advisory Division advised negatively on Government proposals to patch up the regulation. The recommendations of the Advisory Division were non-binding, but the Government could not know whether the Administrative Jurisdiction Division would accept such a patched up regulation in court. Early on in the clash Van Geel decided to ignore the Council's advice and was promptly put in his place by the Administrative Jurisdiction Division.

I consider this institutional double role not problematic per se, because an advisory council that considers questions of the legality of policy proposals should be in touch with the mood prevalent in the highest administrative court. This role should be made as transparent as possible though, because now politicians cannot know whether disagreement with the Advisory Division of the Council of State might have legal consequences before the Administrative Court or not. In short I support having an advisory Council with a considerable degree of power. It may guard over the long term interests of the country and environmental interests are long term interests. However its connection to the administrative courts should be made abundantly transparent. It should be as clear as possible whether a certain policy runs the risk of being scuttled by the administrative court when it goes against the advice of the Council.

9.4.6 Conclusion: the struggling emergence of a legality of precaution

The question of whether the air quality clash was indicative of the emergence of a legality of precaution can be answered only in a qualified form: only if we ask where in the air quality clash actors used arguments, promulgated regulation, or passed judgements that indicate the rise of such legality, can an answer be given. The strongest indications of a precautionary legality were found in European institutions before and during the drafting of Air Quality Directives in the 1990s. The policy philosophy outlined in the 5th EAP created space for holistic, ambitious, and preventative regulation, in which the raising of public awareness was a core component. Through strict standards and the provision of information, the European Commission urged a change in behaviour on the part of Member States and their inhabitants. Science played a strong role in decision making as well, but especially in order to signal potentially disastrous health threats.

However, the precautionary legality that supported the Directives waned in the 2000s when cost-benefit assessment became more important, and decision making was left to scientific experts and policy makers without a strong say from environmental organisations. These changes led to lower ambitions regarding air quality and to more flexibility for Member States that were having problems meeting the standards.

The forceful precautionary approach gave way to a cautious calculative approach of risk assessment and flexibility.

In Dutch air quality policy, precaution was never truly dominant. Rhetorically, the early Memorandums of the 1970s discussing environmental policy in general were quite precautionary. Certain legal aspects of environmental policy, notably the link between environmental standards and administrative decisions, would fit well within precautionary legality. The interconnection did not create many legal problems, however, because the actual standards were usually quite lenient. This discrepancy can be explained by pointing to the weak position of the Environmental Ministry at the time. Although it could propose precautionary principles, the actual standards were lenient because of feasibility concerns.

The move to ecological modernisation amounted to a pragmatic turnaround for environmental policy. The cautious variant explored by Winsemius in the early 1980s was marked by cool-headed calculation and prioritisation, elements that fitted better with the legality of risk and solidarity than with precaution. The strong eco-modernistic regime that evolved from it in the latter 1980s and early 1990s, however, was amenable to a precautionary legality, though actual policy apart from principles remained pragmatic. It is noteworthy that within this less precautionary discourse, environmental protection became more mainstream and arguably more effective as well. Too much ecological ambition apparently does not work for the pennywise Dutch. When environmental protection is framed as a profitable business it attracts much more attention.

Nevertheless, elements of a precautionary legality were present in Dutch air pollution policy. Through the implementation of the precautionary European air quality policies in 2001, they entered Dutch air quality law, and – despite its pragmatism – the Dutch Government found itself stuck with precautionary regulation, even reinforced by uniquely Dutch arrangements in environmental spatial law.

The most vociferous proponent of a legality of precaution was the pro-health camp, which succeeded in directing the political trend in favour of environmental protection. Through its pressure, sustainability concerns were placed higher on the agenda, and the chapters on sustainable development or air quality in relevant policy documents were either added or reinforced. That meant that these concerns would find their way into policy and into the legal order. The most conspicuous legal development was that the use of the courts in this environmental conflict was so successful. This success spawned the storyline that people had a right to clean air – a storyline that was heard often in academic discourse but was now applied 'in action'.

Elements belonging to the precautionary legality have certainly played a part in the emergence of the Dutch air quality clash. Civilians were prepared to take a stand for environmental and health matters. The courts demanded more research into deleterious effects and the large amounts of deaths attributed to PM10 made headlines in newspapers and were used in heated political discussion. In the Netherlands the hatchet appears to have been buried on the topic of air quality. It is still a concern for people living in the cities and along highways, but it does not stir emotions in the same way it did around 2005. However, new precautionary concerns might quickly arise in the shape of infectious diseases, danger from nuclear power plants, or some other potential threat. Precautionary concerns are now visible in other files such as the exploitation of shale gas reserves or drilling for gas in the Dutch province of Groningen.

These issues may well prove to be future legal battle grounds. The Council of State Administrative Jurisdiction Division and Advisory Division were more open towards precautionary arguments than the Dutch Government. This indicates that if a legality of precaution is to establish itself, it is through European law together with a judiciary that is open towards it. Dutch Government policies remained firmly in the realm of pragmatism, but European policies together with a judiciary that is prepared to interpret and enforce them strictly could bring about further changes in the legal order. The air quality clash provided an indication that such developments do indeed take place. The backlash against these policies in the EU and in the Netherlands, however, suggests that the legality of precaution is still struggling.

The above does not indicate that the advance of the legality of precaution has halted, as elements of precautionary legality have gained significantly in strength in different domains. Moreover, the clash has illustrated the potency of taking a political fight over air quality to court. The conflict put environmental concerns squarely back on the agenda from which they had almost disappeared: for instance, issues like speed limit, environmental zoning, and even road pricing. During the clash, however, the fledgling legality was forced to give way again to the more traditional legality of risk and precaution. The history of the clash reminds of Echternach's procession, with developments taking three steps forward and two steps back. Was the legality of precaution on the rise at the time of the clash? Yes it was, but its time had not yet come. Countervailing powers still held sway.

9.5 Summarising conclusion: a short characterisation of the air quality clash

How and why did air quality emerge suddenly as a problematic issue in the Netherlands? This study shows that firstly EU air quality regulation strongly contributed to the outbreak and continuation of the Dutch air quality clash. It also shows that the interplay between the EU regulation and Dutch attempts to influence it are highly important. From a long term historical perspective we may even say that the Dutch air quality clash is the result of Dutch influence at the EU level at an early stage. The strict EU norms would not have come about if the Dutch had not involved the WHO in order to promote scientifically underpinned norms. That the norms are as strict as they are cannot be solely attributed to Dutch influence. It is also the result of the influence exerted by another member state, the United Kingdom.

Another characteristic of Dutch environmental law played a crucial role as well. This is 'the link' between environmental law and permits for infrastructure development. Because of this link infrastructure could not be developed in situations where environmental norms are already exceeded. Because the EU standards were strict, they were transgressed in large parts of the Netherlands. Hence, the development of infrastructure came to a halt in major parts of the country. The fact that the Council of State Administrative Jurisdiction Division operated in quite a formalistic way meant that there was no room to consider the social impact of cancelling permits. However the Directives and the link did not give it much discretionary room to consider other options than to cancel permits because of noncompliance with the air quality regulation.

Seondly, the clash can be understood as an episode in Dutch politics where the consensus of weak ecological modernisation was dissolved. This happened first because the Balkenende cabinets from 2001 broke away from the consensus by prioritising infrastructure development. This political change infuriated the environmental movement, which then made the strategic move to turn to the judiciary and requested a stop to the one-sided developmental policies. As stated above, the legal conditions favoured their attempts and during the clash they found an ally in the Council of State. This even led to the storyline of clean air as a human right.

The fact that the pro-health camp found an ally in the Council of State prevented the Dutch Government from quickly finding an administrative solution to the situation where infrastructure permits were cancelled on an unprecedented scale. It was forced to propose a new air quality law to Parliament which contained a novel regulatory instrument, the programmatic approach.

In Parliament the pro-health camp initially had everything going for it. The Council of State Administrative Jurisdiction Division kept annulling projects, the Government was forced to make ever more funds available for air quality and the Council of State Advisory Division issued strongly negative opinions on the Governments' attempts to quickly patch up the Dutch air quality regulation. However, the pro health camp became confronted with an increasingly strong pro-infrastructure camp that made the counter-claim that the real problem with air quality was not its effects on public health, but that the regulation of it had blocked the country to further development. The storyline that it put forward was that the regulation itself was to blame and that billions of dollars and as much as 100.000 jobs were at stake.

The clash could only come to an end in the political arena by rekindling the eco-modernist consensus in the Air quality law of 2007. The consensual, managerial, technical and bureaucratic instrument that was developed out of this law, the NSL, first led the EU Commission to extend the Dutch some postponement of the standards. Second, it was accepted by the Council of State as an instrument that facilitated both air quality improvement and infrastructure development.

This research shows – to my knowledge for the first time – a case in which the Council of State as a two pronged institute was able to keep policy formation deadlocked for several years. One prong – the Administrative Jurisdiction Division – kept annulling permits with reference to EU air quality standards while the other prong – the Advisory Division – kept reminding the government that policy adaptation would have to be in line with EU regulation. In the air quality clash the Council of State acted as a fellow policy maker.

A third way to understand the clash is to look at the level to which the legality of risk and compensation and the legality of precaution can be found in the handling of the air quality clash in the legal and the political arena. During the clash strict standards based on uncertain scientific knowledge blocked important developments in the Netherlands. Only by balancing the health and environment interests with the interests of development and mobility could the clash be ended.

The swift promulgation of standards in the European Union, the quality of the clean-air-is-a-right storyline as a mobilisation device and the attachment of the Council of State Administrative Jurisdiction Division to the link are indications that a precautionary legality was cautiously on the rise. However, the strong backlash against this temporary legal preference of environmental interests over economic ones, in the Netherlands as well as in Europe, pointed out that the legality of precaution had a long way to go still before it could be considered to reach a dominant position within the Dutch legal order.

Main events of the emergence and resolution air quality clash in the Netherlands

1970	Establishment Ministry of VOMIL
1972	Report for the Club of Rome, Limits to growth issued, VOMIL issued Urgent Memorandum on Environmental Hygiene
1982	Establishment of Ministry of VROM Pieter Winsemius first Minister of VROM
1983-1992	Embrace of ecological modernisation in the Netherlands
1983	Dutch civil servants approach WHO requesting air quality guidelines
1988-1990	Take Care for Tomorrow, SVV2 and NMP strong ecological modernisation in the Netherlands
1990-1998	Ecological modernisation of transport failing in the Netherlands
1993, 1995	Harvard Six Cities and ACS studies published
1994-1994	Asthma scare in the UK
1996, 1999	Promulgation of Directive 96/62/EC and Directive 99/30/EC establishing new air quality standards
1997	UK established national air quality strategy
1999	GGD report claimed breathing the air in residential area of Overschie equalled the passive smoking of 17 cigarettes
2001 June	Air Quality Order 2001 adopted in the Netherlands, containing European air quality standards
2002 May	Speed limit near Overschie lowered because of air quality concerns
2002 July	Cabinet Balkenende 1 installed, established a strong pro-mobility and pro infrastructure agenda
2004, May, Sept.	Council of State Administrative Jurisdiction Division annuls road expansions and a commercial zone, established strict interpretation of AQO 2001
2004 Sept., 2005 June	Government Interpretation of AQO in letter of 30 September rejected by Council of State, proposal for new Ministerial Decree rejected as well. Pro-infrastructure camp mobilised in April 2005 and June 2005 to demand release of the link
2005 Jan.	EU Commission review of Directive 99/30/EC unfavourable for the Netherlands
2005 Sept.	Government promised 900 million for air quality in 'Prinsjesdag package
2006 March	Van Geel sent proposal for new Air Quality Law containing new programmatic approach to Parliament
2006 Oct.	Proposal for Air Quality Law accepted by Second Chamber
2007 July	Council of State Administrative Jurisdiction Division annuls trajectory decision Burgerveen Leiden
2007 Nov.	Air Quality Law entered into force
2008 May	New Air Quality Directive adopted
2008 June	Contents of the NSL offered to Parliament by way of a Letter by Minister Cramer
2009 April	NL received derogation of air quality standards
2009 Aug.	NSL entered into force
2010 March	NSL accepted by Council of State Administrative Jurisdiction Division as sufficient justification for administrative decisions ABRvS 200900883/1

Samenvatting: de sociale constructie van de fijnstofstrijd in Nederland

In 2007 keerde ik terug naar Nederland, na een verblijf van zes jaar in Istanbul, Turkije. Ik leerde toen al snel een nieuw woord: 'fijnstof'. Wat dat precies was, kon niemand mij vertellen, maar wel dat het een vorm van luchtvervuiling was, dat het gevaarlijk was en dat het vooral werd uitgestoten door auto's. Het hoopte zich op in drukke straten en vormde daar een gevaar voor de volksgezondheid. Deze nieuwe vorm van luchtvervuiling intrigeerde mij omdat ik had gewoond in een metropool met meer dan tien miljoen inwoners, dichtslibbende verkeersaders en rokende stoomboten. In tegenstelling tot Amsterdam echter was de luchtkwaliteit daar echter geen onderwerp van gesprek. In Nederland was luchtvervuiling toch ook al een tijd van de publieke radar verdwenen?

In dit proefschrift is slechte luchtkwaliteit als sociaal en juridisch probleem nader onderzocht en is de vraag beantwoord waarom de luchtkwaliteit in Nederland plotseling een groot maatschappelijk probleem is geworden dat speelde tussen september 2004 en april 2010. In 2004 zorgde uitspraken van de Nederlandse bestuursrechter ervoor dat de uitbreiding van infrastructuur geen doorgang kon vinden in gebieden waar de luchtkwaliteit niet voldeed aan de geldende normen. In april 2009 gaf de Europese Commissie Nederland uitstel van de normen om de zogenaamde programmatische aanpak te kunnen implementeren. Deze aanpak maakte een bestuurlijke oplossing mogelijk voor de stagnerende ontwikkeling van infrastructuur. De aanpak is eind maart 2010 uiteindelijk door de hoogste bestuursrechter, de Afdeling Bestuursrechtspraak van de Raad van State, gesanctioneerd. Gedurende deze jaren ontstonden twee zienswijzen op het probleem van slechte luchtkwaliteit. Aan de ene kant werd vanuit het perspectief van de volksgezondheid gewezen op de risico's op vroegtijdige sterfte en verhoogde ziektelast onder de bevolking. Aan de andere kant werd gewezen op de economische ontwrichting die de regulering van fijnstof tot gevolg had voor belangrijke economische sectoren zoals de bouw en transport.

Deze twee zienswijzen streden in die jaren om de voorrang in het de publieke en politieke arena. Vandaar dat ik spreek van de 'fijnstofstrijd'. In essentie ging het hier om een belangenstrijd: moeten milieu- en gezondheidsbelangen zwaarder wegen, of moet prioriteit gegeven worden aan economische ontwikkeling? Gaandeweg heeft het debat zich verbreed en rezen er ook vragen over de toenemende Europeanisering van de Nederlandse rechtsorde en de invloed die de Raad van State als adviesorgaan en bestuursrechter mag uitoefenen op de politieke besluitvorming.

Een aantal aspecten van de fijnstofstrijd is in dit proefschrift verder uitgediept. In de eerste plaats ga ik in op de rol die de Europeanisering van het milieubeleid heeft gespeeld bij het ontstaan van de problemen. Strenge Europese regulering was in Nederlandse wetgeving omgezet waardoor de luchtkwaliteitsnormen significant strenger waren geworden dan tevoren. Op basis van deze normen werden prestigieuze infrastructuurprojecten tegengehouden door de Raad van State.

Ten tweede analiseer ik het Nederlandse milieubeleid. De strengere normen alléén kunnen het ontstaan van de clash niet verklaren, aangezien die ook in andere Europese landen golden en daar speelde een dergelijke problematiek niet. Zowel de achterliggende beleidsfilosofie als de concrete beleidskeuzes hebben het ontstaan van de strijd in de hand gewerkt.

Ten derde wordt het ontstaan van de fijnstofstrijd geanalyseerd in relatie tot de opkomst van de zo genaamde 'voorzorglegaliteit', een rechtssociologische notie die duidt op een verschuiving in de rechtsorde waarbij niet langer compensatie van slachtoffers van schade het leidend beginsel is, maar het voorkomen van schade, ook als de kosten zeer hoog zijn.

Methodologie

In dit proefschrift is gekozen voor een sociaal constructivisme methodologie. In dat perspectief worden sociale problemen niet gezien als objectieve gegevens, maar als resultaten van een intersubjectief proces van betekenisgeving. Bepaalde maatschappelijk situaties 'zijn' geen probleem in objectieve zin, maar worden door mensen als probleem ervaren. Als vervolgens meerdere andere groepen mensen ervan overtuigd raken dat de situatie inderdaad onwenselijk is, spreken we in de (rechts)sociologie van een 'sociaal probleem'. Sociale problemen zijn het product van sociaal handelen door verschillende actoren.

Het begrippenapparaat uit dit proefschrift is geënt op twee typen van sociaal constructivisme. Het eerste type noem ik actor-gecentreerd sociaal constructivisme, en het tweede type noem ik discours-analytisch sociaal constructivisme.

Het eerste type gaat ervan uit dat bepaalde groepen van mensen er belang bij hebben om een situatie als probleem te definiëren en hun best ervoor zullen doen om anderen daarvan te overtuigen. Deze actoren worden 'claims makers' genoemd. Zij zullen via verschillende kanalen de publieke opinie en beleidsmakers proberen te beïnvloeden om ervoor te zorgen dat hun probleemdefinitie – hun claim – algemeen geaccepteerd wordt. De claims zelf kunnen ook weer oppositie van andere actoren oproepen. Claims makers en counter-claims makers wisselen hun argumenten uit in geschrift en voor de camera's van nieuwsprogramma's, op manifestaties en bijeenkomsten, in het parlementair debat en soms ook voor de rechter. Al die plaatsen waar argumenten worden uitgewisseld worden 'arena's' genoemd. Mocht de claim inderdaad weerklank vinden bij invloedrijke groepen uit de samenleving, dan kan de probleemdefinitie leiden tot regelgeving. Beleidsmakers zullen voorstellen indienen om iets aan het sociale probleem te doen en dan ligt regulering voor de hand.

Beleidsmakers kunnen met regelgeving reageren op claims, maar die regelgeving kan bij andere groepen weer weerstand oproepen. Is dat het geval, dan zullen er nieuwe claims makers komen die betogen dat de regelgeving zelf tot problemen leidt en begint het spel van voor en af aan.

De laatste twintig jaar doet een tweede sociaal constructivistische methode opgeld die de argumentatie die groepen hanteren om hun claims te ondersteunen als uitgangspunt neemt, de zg. . 'discursive turn'. Discours-analytici wijzen er namelijk op dat claims alleen zullen worden omarmd als zij passen binnen het geheel van heersende opvattingen en overtuigingen. Argumenten voor een bepaalde claim moeten plausibel zijn in het licht van de heersende mening. De manier waarop de heersende zienswijze op een bepaald gebied naar voren gebracht wordt en de manier waarop die zienswijze geïnstitutionaliseerd is in maatschappelijke praktijken wordt 'discours' genoemd.

Een claim wordt nooit neutraal naar voren gebracht maar past altijd binnen een bepaald verhaal. Een problematische situatie wordt gepresenteerd als een toestand met een bepaalde oorzaak, met bepaalde gevolgen die op een specifieke wijze verholpen moeten worden. Binnen het kader van dit proefschrift wordt een dergelijk verhaal een verhaallijn (storyline) genoemd. Verhaallijnen kunnen ook door andere actoren overgenomen worden. In het geval verschillende actoren eenzelfde verhaallijn omarmen wordt gesproken van een discourscoalitie. Verhaallijnen moeten aansluiten op het heersende discours, maar het discours verandert ook onder invloed van verhaallijnen. Dat maakt deze methode geschikt om het stelsel van overtuigingen te onderzoeken dat leeft onder verschillende groepen binnen een samenleving.

Ideaaltypen

De fijnstofstrijd wordt in deze studie in verband gebracht met veranderingen in het Nederlands milieupolitieke discours en het juridische discours. Daartoe wordt hij gerelateerd aan twee ideaaltypen, te weten een ideaaltype van ecologische modernisering en een ideaaltype van de voorzorglegaliteit. Het eerste ideaaltype dient ertoe om veranderingen in het milieupolitiek discours in kaart te brengen. Het tweede type plaatst de fijnstofstrijd in het kader van veranderingen in de rechtsorde.

Het begrip ecologische modernisering wordt in deze studie opgevat als een discours op het gebied van het milieubeleid dat in de jaren 80 ingang heeft gevonden in het

Nederlandse en later Europese beleid en vanaf de jaren 90 dominant is geworden tot en met de eerste jaren van het nieuwe millennium. In tabel 10 hieronder wordt een ideaaltypische weergave van ecologische modernisering gepresenteerd zoals het wordt gehanteerd in de studie. Het discours wordt gecontrasteerd met het dominante vertoog uit de jaren 70, dat ik 'Grenzen aan de groei' heb genoemd. Dit discours ontleent zijn kernelementen aan het rapport voor de club van Rome met dezelfde titel. Daarnaast is een onderscheid gemaakt tussen sterke en zwakke ecologische modernisering. Sterke ecologische modernisering vergt vergaande structurele economische veranderingen in de productiestructuur en gedragsverandering bij de consument. Zwakke ecologische modernisering daarentegen is goed verenigbaar met de bestaande kapitalistische economische verhoudingen.

Met de term voorzorglegaliteit wordt een verandering in het juridische vertoog aangeduid. Het begrip 'Legaliteit' wordt door de Nederlandse rechtssociologen André Hoekema en Niels van Manen gedefinieerd als 'het geheel van in de maatschappelijke praktijk van de rechtsorde uitgedrukte en in sociale instituties werkzame overtuigingen omtrent de juiste opbouw van de samenleving en onderdelen ervan' (Hoekema & Van Manen 2000). Voor de rechtssocioloog behelst de rechtsorde geen vast stelsel aan overtuigingen, maar wisselen dominante opvattingen over de taak van het recht en zijn maatschappelijke rol elkaar af, zonder overigens eerdere overtuigingen volstrekt te overvleugelen.

In deze studie wordt het door Weber geïnspireerde begrip legaliteit gebruikt om een dominant discours binnen de rechtsorde mee aan te duiden. Twee sets overtuigingen worden met elkaar vergeleken, de legaliteit van risico en compensatie en de legaliteit van voorzorg. Onderzocht wordt of de rechtsorde heden ten dage wordt gedomineerd door een legaliteit van risico en compensatie, maar ook of dit type van legaliteit langzamerhand zijn dominante positie zou kunnen verliezen aan een type van legaliteit gericht op het voorkómen van schade überhaupt, ook al gaat dit gepaard met zeer hoge kosten. De opkomst van deze 'voorzorglegaliteit' is ieder geval duidelijk zichtbaar op het terrein van het milieurecht. Kernelement van dit type van legaliteit is de toepassing van het voorzorgbeginsel dat stelt onzekerheid over de aard en omvang van dreigende schade geen excuus mag zijn voor het niet nemen van beschermende maatregelen.

Een vraag die in deze studie wordt gesteld is of het optreden van de fijnstofstrijd een indicatie is van de verdere voortschrijding van dit type legaliteit. In tabel 11 is een ideaaltypische weergave gegeven van de legaliteit van voorzorg gecontrasteerd met de legaliteit van risico en compensatie.

Onderzoeksmateriaal en onderzoeksopzet

Deze geschiedenis van de fijnstofstrijd is uitgevoerd door middel van een analyse van de Nederlandse parlementaire stukken waarin de term luchtkwaliteit naar voren is

Tabel 10

Beleidsdiscours	Grenzen aan de Groei	Zwakke ecologische modernisering	Sterke ecologische modernisering
Relatie tussen economie en ecologie	Gespannen: Economische groei wordt gezien als bedreiging van het milieu	Vreedzame co-existentie: Zowel ecologische als economische doelstellingen kunnen worden gecombineerd en gehaald	Positief met elkaar verweven: Een betere kwaliteit van het milieu is noodzakelijk voor het behoud van economische groei en zal die groei versterken.
Rol van de wetenschap	Natuur en milieuwetenschappen zouden het milieubeleid moeten dicteren.	Verschillende wetenschappen leveren expertise aan om milieubeleid op te baseren, natuur en milieuwetenschap, maar ook economie en management	De wetenschap wordt ingezet om mogelijke milieubedreigingen in kaart te brengen wijdverbreide inzet van milieu-indicatoren op het gebied van het verbruik van grondstoffen en vervuiling. Dergelijke indicatoren worden een essentieel onderdeel van economische politiek.
Consensus-streven	Conflictmodel: Marktpartijen zouden zich gedwongen moeten aanpassen aan het milieubeleid door middel van regulering	Consensusmodel: Het milieubeleid komt tot stand op basis van Onderhandelingen tussen de overheid en marktpartijen	Brede en geïnstitutionaliseerde-participatie in totstandkoming van beleid met de milieubeweging als volwaardig tegenwicht tegen economische belangen
Preventieve aanpak	De last voor het milieu moet tot een minimum worden beperk. Dit impliceert beperking van de economische activiteit en wantrouwen tegenover wetenschappelijke vooruitgang.	Vervuiling moet worden voorkomen door schonere en efficiëntere productie en zo nodig het toepassen van het voorzorgbeginsel bij dreigende schade.	Het milieubeleid richt zich direct tot vervuilers, zowel de industrie als de consument. Een brede toepassing van het voorzorgbeginsel en verinnerlijking van het milieubeleid door doelgroepen.
Verinnerlijking van de doelstellingen van milieubeleid	Zowel onderwijs als wetgeving moeten worden ingezet om de juiste milieumentaliteit aan te kweken.	Het bevorderen van bewustzijn via mediacampagnes, en het verschaffen van informatie, alsmede subsidiering van de milieubeweging	Actieve betrokkenheid van burgers door middel van educatie, voorlichting en het recht op inzage van vervuilingsregisters van bedrijven. Brede juridische mogelijkheden voor burgers om in het geweer te komen tegen milieubederf.
Integraal beleid	De aarde wordt gezien als één complex ecosysteem. Idealiter is het beleid integraal geformuleerd. Als dit echter niet mogelijk blijkt wordt het beleid sectoraal vastgesteld.	Verschillende vormen van milieuvervuiling moeten via hetzelfde beleid aangepakt worden, dus geen sectorale benadering.	Verschillende vormen van milieuvervuiling moeten via hetzelfde beleid aangepakt worden, gecombineerd met economische prikkels tot gedragsverandering.

Tabel 11

	Legaliteit van risico en compensatie	Legaliteit van voorzorg
Kernaspect van de rechtsorde	Compensatie van schade door spreiding van risico's wordt van het grootste belang geacht. Daarom zijn verzekeringen en risicoaansprakelijkheid essentieel.	Voorkomen van schade wordt van het grootste belang geacht. Daarom is toepassing van het voorzorgbeginsel essentieel.
Type schade dat van het grootste belang wordt geacht	Schade moet worden voorkomen als het kosteneffectief is om dat te doen. Schade optredend op de korte en middellange termijn, met een relatief zekere omvang.	Lange termijnschade die potentieel catastrofaal is, maar waarvan de omvang en het daadwerkelijk optreden nog onzeker zijn.
Morele kijk op het optreden van schade	Schade kan niet altijd voorkomen worden en wanneer die optreedt is compensatie aan de orde.	Schade is schande en een indicatie dat men nalatig is geweest in het voorkomen ervan.
Perceptie van het slachtoffer	Drager van rechten uit hoofde van zijn lidmaatschap van een risicocollectief.	Burger die schade heeft ondervonden als gevolg van falend optreden van de overheid of andere toezichthouders
Veiligheidsgaranties en bewijslast die noodzakelijk worden geacht ter onderbouwing van besluitvorming	Besluitvorming moet gebaseerd zijn op wetenschappelijke kennis en kosten-batenanalyses om optimale oplossingen te kunnen inventariseren.	Besluitvorming moet gebaseerd zijn op wetenschappelijk bewijs dat het besluit geen schade tot gevolg zal hebben. De bewijslast rust op de schouders van diegene die het potentieel riskante besluit neemt.
Wijze waarop recht en beleid functioneren als instrument van maatschappelijke sturing	De samenleving kan worden verbeterd door op maat gemaakt beleid en een instrumenteel gebruik van wetgeving. .	De samenleving kan worden verbeterd door integraal en planmatig beleid, het geven van informatie aan het publiek en verhogen van maatschappelijk bewustzijn.
Rol van publieke participatie	Klein, want burgers worden gezien als leken zonder veel kennis van risico's of complexe systemen. Beleid dient door experts te worden gemaakt.	Groot, want de burger wordt gezien als ervaringsdeskundige. Het beleid dient burgers de kans te geven hun zienswijze naar voren te brengen en daar ook naar te handelen.
Kernwaarde die het meest beschermd moet worden	Economische ontwikkeling: Groei en uitbreiding van de productie van goederen, mobiliteit, kennis en innovatie.	Ecologische harmonie: De balans tussen ecosystemen, volksgezondheid en externe veiligheid.

gekomen. Vooral de stukken vanaf september 2004 tot december 2007 – de hoogtijdagen van de clash – zijn bekeken, maar ook documenten van voorgaande en van latere jaren zijn geanalyseerd. Ik heb gekozen voor die periode omdat de fijnstofclash zich in die jaren heeft afgespeeld. De voorgeschiedenis van de fijnstofclash begint echter in 1972. Om die reden is ook het milieubeleid in de jaren 70, 80 en 90 onderzocht door middel van literatuurstudie en bestudering van de 'grote' nota's zoals bijvoorbeeld het Nationaal Milieu Beleidsplan.

De regulering van de luchtkwaliteitsproblematiek heeft uiteraard een grote Europese component. Dit maakt het noodzakelijk om ook EU documenten te onderzoeken. Verder geeft het onderzoek van alleen parlementaire stukken een onvolledig beeld. Zij zijn weliswaar cruciaal omdat elk sociaal probleem uiteindelijk door de politiek zal worden besproken, maar zij bestrijken slechts een klein deel van de maatschappelijke werkelijkheid. Naast parlementaire stukken is daarom een groot aantal wetenschappelijke rapporten, krantenberichten en zogenaamde 'grijze' literatuur bestudeerd. Daarnaast zijn interviews afgenomen met een aantal experts, politici en actievoerders. Deze interviews hebben geholpen om het beeld dat oprees uit parlementaire stukken in een breder perspectief te plaatsen.

Zoals gezegd besloeg het onderzoek een lange periode. Voor wat betreft de jaren van de fijnstofstrijd zelf, 2004-2010, zijn vooral de Nederlandse parlementaire stukken belangrijk geweest. Voor wat betreft de jaren 90 zijn de Europese stukken cruciaal. De jaren daarvoor zijn vooral onderzocht via secundaire literatuur, alsmede de belangrijkste parlementaire Nederlandse stukken op het gebied van milieubeleid.

Korte geschiedenis van de fijnstofstrijd

De fijnstofstrijd begon op het moment dat de hoogste bestuursrechter in Nederland, de Afdeling Bestuursrechtspraak van de Raad van State, vergunningen voor weguitbreidingen en andere bouwprojecten consequent ging afkeuren op grond van strijd met de standaarden uit het Besluit Luchtkwaliteit. Vooral het programma om op grote schaal wegen uit te breiden werd hierdoor zwaar getroffen. Woordvoerders van het Ministerie van Verkeer en Waterstaat klaagden dat ze geen spade meer de grond in kregen. Aan de andere kant zagen actiegroepen deze rechtspraak als een overwinning.

De risico's van slechte luchtkwaliteit vanwege fijnstof werden in Nederland in eerste instantie onder de aandacht gebracht door epidemiologen en de Rotterdamse GGGD. Amerikaans epidemiologisch onderzoek had begin jaren 90 uitgewezen dat vervroegde sterfte meer voor kwam in gebieden met slechte luchtkwaliteit. Uit Nederlands onderzoek bleek dat kinderen die naast de snelweg woonden, last hadden van verschillende kwalen, in het bijzonder aan de luchtwegen.

Hierover ontstond commotie in de Rotterdamse wijk Overschie in 1999. De wijk is dichtbevolkt en wordt doorkruist door de drukke A10. In een alarmerend rapport werd de luchtkwaliteit vergeleken met het passief roken van 17 sigaretten per dag. De risico's van passief roken zijn erg klein maar deze claim bleek sterk genoeg om de aandacht van de politiek te trekken. De toenmalige minister Pronk besloot een proef te doen met snelheidsverlaging op de A10 bij Overschie. Langzaam rijdende auto's stoten namelijk minder fijnstof uit. Deze claim werd overgenomen door politieke partijen aan de linkerkant van het politieke spectrum en er ontstond een verhaallijn rond de nadelige effecten van automobiliteit en snelwegen op de volksgezondheid.

In 2001 verscheen het Besluit Luchtkwaliteit waarin de standaarden voor schone lucht sterk werden aangescherpt. De (Europese) standaarden voor fijnstof zouden vanaf 2005 moeten gaan gelden. Drie jaar later, in 2004, vestigde de Raad van State in drie uitspraken een consistente lijn in de jurisprudentie voor de standaarden voor fijnstof en Stikstofdioxide. Onvoldoende onderzoek naar de luchtkwaliteit leidde voortaan tot het vernietigen van besluiten van bestuursorganen, of er nu mensen daadwerkelijk aan de vervuiling werden blootgesteld of niet.

Toen de RvS beslissingen bleef vernietigen ontstonden er twee kampen van actoren rond luchtkwaliteit. Met 'kamp' wordt hier bedoeld een verband van actoren dat dezelfde claims naar voren brengt en die met gelijkluidende argumenten staaft. Derhalve kan men ook van discourscoalities spreken. Het eerste kamp bestond uit wetenschappelijke instituten, actiegroepen en politieke partijen die de slechte luchtkwaliteit zagen als een gevaar voor de volksgezondheid en die structurele maatregelen ter verbetering eisten. Ik noem dit kamp het volksgezondheidskamp. Het eerste jaar na de uitspraken van de RvS had dit kamp de wind in de zeilen. Zij eisten meer geld voor luchtkwaliteit, minder wegen en Milieudefensie eiste een algehele verlaging van de maximum snelheid.

In reactie op de uitspraken kwam er eveneens een belangrijke counter-claim op gang. De luchtkwaliteitswetgeving werd gepercipieerd als het probleem volgens een alliantie van verschillende belangengroepen uit de bouw en de transportsector en politieke partijen, te weten CDA, VVD en de LPF. Dit kamp stelde dat de standaarden onhaalbaar waren voor Nederland en zouden leiden tot forse economische schade. De luchtkwaliteitswetgeving gecombineerd met de veel te strenge interpretatie ervan door de RvS had Nederland 'op slot' gegooid.

Volgens dit 'infrastructuurkamp' was het loslaten van de typisch Nederlandse koppeling van milieustandaarden aan besluiten van bestuursorganen die gevolgen hebben voor de ruimtelijke ordening geboden. Voor de fijnstofclash gold, dat als een besluit van een bestuursorgaan ertoe zou kunnen leiden dat milieustandaarden in een bepaald gebied niet werden gehaald, de bestuursrechter het besluit moest vernietigen. In andere landen werkten die standaarden over het algemeen niet zo direct door.

Een serie aan debatten en nieuwe wetgeving volgde in de jaren na 2005, en het zou een aantal jaar duren voor er een structurele oplossing voor de clash werd gevonden. Die oplossing kwam er eind 2007 door de systematiek van de koppeling gedeeltelijk los te laten en over te gaan op een zogeheten programmatische aanpak. Alle ruimtelijke projecten die significante gevolgen hadden voor de luchtkwaliteit zouden in een plan moeten worden opgenomen dat naast deze projecten ook positieve maatregelen voor de luchtkwaliteit bevatte. Uiteindelijk zou de som van positieve maatregelen de negatieve gevolgen moeten overtreffen waardoor de luchtkwaliteit structureel zou verbeteren en op termijn de standaarden gehaald konden worden. Een wet die een dergelijk programmatische aanpak mogelijk maakte werd eind 2007 aangekondigd, maar het duurde tot 2008 voordat het uiteindelijke plan er lag en groen licht kreeg van Brussel.

Bovendien bood het plan niet onmiddellijk soelaas, maar voorzag het in een structurele verbetering op de lange termijn. Het was de vraag of de Europese Commissie en de Nederlandse bestuursrechter deze oplossing zouden aanvaarden. De luchtkwaliteitsregulering was het resultaat geweest van Europese wetgeving en Nederland had door de EU gesanctioneerde uitstel van de inwerkingtreding van de standaarden nodig. Uitstel was mogelijk geworden dankzij een nieuwe Europese richtlijn op het gebied van de luchtkwaliteit, afgekondigd in 2008. De Europese Commissie accepteerde het Nederlandse plan in 2009. In 2010 accepteerde ook de Nederlandse bestuursrechter het als voldoende onderbouwing van projecten met potentieel negatieve gevolgen voor de luchtkwaliteit.

Niet iedereen bleek tevreden met de uiteindelijke uitkomst, de milieubeweging en sommige parlementariërs zagen in de programmatische aanpak nog een truc om bouwprojecten ondanks de standaarden mogelijk te maken. In het Nederlandse omgevingsrecht bleek deze nieuwe aanpak echter wel degelijk toekomst te hebben, ook buiten het beperkte gebied van de luchtkwaliteit. Verschillende omgevingswetten werden op de leest van de programmatische aanpak geschoeid.

De fijnstofstrijd in het licht van Europeanisering, ecologische modernisering en voorzorg

Het onderzoek rond de gebeurtenissen rond luchtkwaliteit in Nederland en Europa bevat voor de rechtssociologie interessante uitkomsten op drie terreinen te weten europeanisering, ecologische modernisering en de legaliteit van voorzorg. Hieronder zullen per terrein de belangrijkste conclusies worden weergegeven.

Europeanisering en gebricoleerde wetgeving

Over het fenomeen Europeanisering is in kringen van Nederlandse rechtssociologen nog niet zoveel geschreven. Europeanisering wil zeggen dat steeds grotere delen van de Nederlandse rechtsorde beheerst worden door regulering die van Europese oorsprong is. Inzicht in de wijze waarop die regulering tot stand komt en de gevolgen ervan, zijn rechtssociologisch van belang.

In het geval van luchtkwaliteit waren het inderdaad Europese standaarden die in Nederland voor bestuurlijke problemen hebben gezorgd omdat ze hier niet haalbaar bleken te zijn. In 1999 werd er een Europese richtlijn afgekondigd waarin standaarden voor luchtkwaliteit werden vastgesteld. Die standaarden zijn vervolgens één op één overgenomen in Nederlandse wetgeving door middel van het bovengenoemde Besluit Luchtkwaliteit. Omdat het hier normen betrof van Europese origine hanteerde de Nederlandse regering graag het excuus dat het was overvallen door Europa. Uit dit onderzoek blijkt echter dat Nederland al in een zeer vroeg stadium een medespeler was in de Europese arena die niet afkerig was van strenge luchtkwaliteitsnormen.

In 1994 werden plannen onthuld waarover al werd gesproken in het Vijfde Europese Milieu Actieprogamma (5^e MAP) van 1993, namelijk het opzetten van een integraal stelsel van luchtkwaliteitsnormen voor de gehele EU. Het strategisch kader werd in een eerste richtlijn neergelegd (de zogenoemde moederrichtlijn luchtkwaliteit) en de standaarden zelf werden afgekondigd in een vervolgrichtlijn, een dochterrichtlijn. De standaarden waren gebaseerd op de normen die de WHO hanteerde. De WHO werd een sleutelrol toebedacht omdat dit een onafhankelijke organisatie was die de wetenschappelijke legitimiteit bezat om de lidstaten op één lijn te krijgen. De normstelling werd gecombineerd met een tactiek om het milieubewustzijn onder de burgers te verhogen door het publiceren van informatie over luchtkwaliteit. De normen moesten streng genoeg zijn om lidstaten te dwingen tot innovatieve strategieën.

Nederland droeg zelf echter ook bij aan de totstandkoming van de luchtkwaliteitsregulering. Op instigatie van Nederlandse ambtenaren in de vroege jaren 80 is de WHO betrokken geraakt bij luchtkwaliteitsregulering. Nederland wilde zelf een stelsel van kwaliteitsnormen opzetten, maar had daar een wetenschappelijke basis voor nodig. Net als de EU zocht het een instantie die de juiste wetenschappelijk legitimiteit bezat om zowel de oppositie in Nederland te kunnen overtuigen alsook het buitenland aan te sporen eenzelfde stelsel op te zetten. Dit laatste punt had te maken met concurrentieoverwegingen. Als Nederland zijn industrie aan strenge wetgeving bloot zou stellen, dan zou de industrie uit andere landen een voordeel hebben. Datzelfde idee lag ten grondslag aan het EU voorstel om te komen tot een uniform stelsel van luchtkwaliteitseisen. De bedoeling was om een gelijk speelveld te creëren op het gebied van milieubelasting.

De Nederlandse inspanningen vonden een gewillig oor bij de WHO en Nederland bekostigde het onderzoek naar de WHO luchtkwaliteitsrichtlijnen. Toen die in 1987 het licht zagen, was het enthousiasme in Nederland inmiddels gedoofd. In de EU echter in het geheel niet en de WHO normen gingen als benchmark fungeren. De doelstelling dat de lucht aan de WHO normen moest voldoen, werd opgenomen in het 5^e MAP en dat plan fungeerde als basis voor de uiteindelijke regulering. Op het moment dat die regulering in de jaren 90 werd beklonken, bleef Nederland een voorstander van strenge normen.

In de jaren 80 bekeerde het Nederlandse beleid zich tot het discours van de ecologische modernisering. Informatieverstrekking aan het publiek, het stimuleren van milieu-innovaties en het mobiliseren van het publiek voor milieudoeleinden waren belangrijke speerpunten. Nederlandse politici in Europa drukten een stempel op de Europese milieupolitiek van die jaren. Het 5^{e} MAP was sterk beïnvloed door het Nationaal Milieubeleidsplan en name de sterke eco-modernistische toonzetting ervan.

Nederland slaagde er al in een vroeg stadium in om het Europese milieubeleid te beïnvloeden en lobbyde voor strenge regels op het gebied van luchtkwaliteit. Dat was ook in het Nederlandse belang. Nederland had een sterke milieubeweging en als klein land was het erg afhankelijk van beperkingen van uitstoot in andere landen.

De fijnstofstrijd leert echter dat die beïnvloeding heel anders uit kan pakken dan een lidstaat op het eerste gezicht op het oog heeft. Dat kwam doordat dit dossier een lange doorlooptijd had op EU niveau en de nationale politiek intussen een andere richting was ingeslagen. De betrokkenheid van de WHO was een kernaspect van de Europese regulering, net zoals de eco-modernistische beleidsfilosofie. De bedoeling was dat dit voor Nederland positief zou uitpakken, maar de normen bleken te streng. Nederland heeft dan ook in een later stadium gewaarschuwd voor de haalbaarheid, maar toen was het zijn greep op het dossier al kwijt.

Andere actoren hadden andere belangen. De Europese instellingen wilden een strenge norm die gold voor de hele EU om het gelijke speelveld te garanderen. Daarnaast heeft een Engelse ambtenaar de regulering vormgegeven op een wijze die in Engeland gebruikelijk was. Engeland zelf had net een paar jaar eerder een luchtkwaliteitsstrategie opgezet in antwoord op het nationale probleem van astma bij kinderen. Het autoverkeer werd daarvoor verantwoordelijk gehouden en nog voordat de EU haar strategie ontvouwde in 1999 had Engeland in 1997 de eerste integrale luchtkwaliteitsstrategie van Europa al ingevoerd. Daarnaast bleek uit een kostenbatenanalyse dat strenge normen bij uitstek kosteneffectief waren. Uit dit onderzoek blijkt dat foutieve aannames over de effectiviteit van andere wetgeving deze analyse hebben ondergraven, waardoor de standaarden vele malen duurder uitvielen dan de bedoeling was.

Al met al kwam Europese regulering tot stand in een arena die gevoelig was voor druk vanuit veel verschillende actoren. Het bespelen van de Europese arena bleek verleidelijk voor lidstaten. Als het Europese beleid aans zou sluiten bij het nationale beleid, dan zouden er geen kosten gemaakt hoevem te worden om het nationale beleid aan te passen. Bovendien leek het rationeel om ambitieuze milieuwetgeving op bepaalde terreinen ook in Europa gerealiseerd te krijgen, anders zit een lidstaat zelf met strengere wetgeving dan de buurlanden, wat uit concurrentieoverwegingen onvoordelig is. Dit leidde tot een situatie waarbij lidstaten elk afzonderlijk probeerden hun favoriete elementen in de EU wetgeving opgenomen te krijgen. Daarnaast hadden de Europese Commissie en het Europees parlement ook hun redenen om zich actief met milieuwetgeving te bemoeien. De Commissie bijvoorbeeld wilde dat Europa zich profileerde als hoeder van de belangen van EU burgers en daarom moest een hoog beschermingsniveau worden nagestreefd.

De veelheid aan stakeholders, de verscheidenheid van belangen die zij nastreefden en de verschillende beleidsfilosofieën waarmee zij probeerden het luchtbeleid te beïnvloeden leidden ertoe dat de uiteindelijke wetgeving niet meer leek op wat er oorspronkelijk was voorgesteld. De luchtkwaliteitsregulering bleek een lappendeken geworden te zijn, een 'bricolage' van elementen uit verschillende stelsels, met verschillende doelen en allerhande onbedoelde effecten.

Deze bevindingen hebben implicaties voor ons begrip van Europeanisering, want er blijkt van een wisselwerking sprake te zijn. Europese regulering is inderdaad de Nederlandse rechtsorde gaan beheersen, maar omgekeerd probeert Nederland ook een voor het land zo gunstig mogelijke Europese regulering af te dwingen. Daartoe worden Nederlandse beleidsconcepten en denkbeelden "geëxporteerd" naar Europa. Lidstaten, waaronder Nederland, zorgen ervoor dat zij mensen hebben op posities in Europese comités en andere overlegorganen die nationale concepten proberen in te steken bij de relevante Europese beleidsbepalers. Europeanisering is geen eenrichtingsverkeer, maar een wederkerig proces van beïnvloeding. Wél blijkt er sprake te zijn van de nodige ruis op de lijn waardoor lidstaten de eigen beleidsvoorstellen vervat in Europese wetgeving terug krijgen die in niets meer lijken op wat zij er oorspronkelijk mee hadden beoogd.

Nederlandse beleidskeuzes en ecologische modernisering

In de jaren 70 werd het Nederlandse milieudiscours nog sterk bepaald door de pessimistische vergezichten uit het rapport voor de Club van Rome, Grenzen aan de Groei. Het devies was in de eerste jaren van het milieubeleid om de noodzakelijke grenzen te stellen door middel van wetgeving. Het milieuministerie was echter in die jaren bij lange na niet sterk genoeg om die wetgeving ook daadwerkelijk door te voeren en milieu als beleidsterrein verzandde in een complex stelsel van vergunningen. Op het gebied van ruimtelijke ordening realiseerde het departement een instrument dat grenzen stelde aan de hoeveelheid vervuiling. De koppeling van RO besluiten aan kwaliteitseisen betekende dat als ergens de milieukwaliteit onvoldoende was, het niet was toegestaan om verdere vervuilende activiteiten te vergunnen. Hierdoor gingen kwaliteitseisen rechtstreeks doorwerken in de individuele besluiten van bestuursorganen.

Met de komst van het ministerie van VROM in 1983, waarbij milieu werd gevoegd bij ruimtelijke ordening en volkshuisvesting, deden moderne managementideeën hun intrede. De eerste minister van VROM was Pieter Winsemius van de VVD. Hij moderniseerde het ministerie en ruilde het pessimistische vertoog van grenzen aan de groei in voor het optimistische discours van ecologische modernisering. Volgens ecologische modernisering kon aan milieuproblemen het hoofd worden geboden door juist verder te moderniseren, in een milieuvriendelijke richting. Winsemius begon een campagne om zogenaamde 'doelgroepen' mee te krijgen in de bescherming van het milieu. Hij werd daarbij geholpen door de toenmalige grootschalige aandacht voor zure regen. Een schoon milieu werd voorgesteld als noodzakelijke grondstof voor verdere productie, milieuvriendelijke alternatieven werden voorgesteld als winstgevend, vervuiling werd een voorbeeld van inefficiëntie en milieubederf werd vergeleken met een financieringstekort die door volgende generaties terugbetaald zou moeten worden. Het milieudiscours werd kortom sterk 'geëconomiseerd'.

Winsemius legde taken op aan het bedrijfsleven maar betrok het ook bij het milieubeleid en in ruil voor inspraak werden convenanten gesloten. Deze op consensus, preventie en management gerichte milieupolitiek kende zijn hoogtepunt in het integrale Nationaal Milieu Beleidsplan uit 1989. In de jaren 90 werd het succesvol geëxporteerd naar andere landen en naar de EU.

Dit beleid kende echter een keerzijde. Het beloofde een positieve wisselwerking tussen economie en milieubeleid en dat betekende dat al te grote offers van de economische sectoren niet gevraagd konden worden. Belangrijke sectoren als de petrochemische industrie en het transport werden dan ook uit de wind gehouden. Bovendien streefde het een internationale aanpak van milieuproblematiek na, waardoor nationale maatregelen impliciet van minder betekenis werden geacht. Het betrekken van de milieubeweging bij de geijkte overlegstructuren verliep niet altijd voorspoedig en de beoogde consensus begon scheuren te vertonen in de jaren 90.

Op het gebied van luchtkwaliteit leidde dit discours na de jaren 80 tot weinig nieuwe initiatieven, ook niet toen er berichten kwamen dat fijnstof schadelijker kon zijn dan gedacht. Er werd gewacht totdat 'Europa' zou handelen. Op het moment dat Nederland wel gedwongen was tot handelen, realiseerde de regering zich niet wat voor consequenties de normen zouden kunnen hebben. Achteraf is dat naïef gebleken. Het werd echter ingegeven door de in Nederland breed gedeelde overtuiging dat normen die geen rekening hielden met economische haalbaarheid geen juridische consequenties konden hebben. De Nederlandse regering was er van overtuigd dat de normen in Europa herzien zouden worden en dat ze alleen toegepast hoefden te worden op plekken waar mensen woonden. Als het maximaal haalbare gedaan werd, kon men eenvoudigweg niet meer verlangen, zo was de aanname.

De Nederlandse regering heeft zich in de jaren 90 wel sterk gemaakt op het punt van ecologische beginselen en een strategische lange termijnaanpak, maar bleek huiverig voor het nemen van maatregelen die daadwerkelijk veel geld zouden kosten. Economische overwegingen hebben hun leidende positie in het milieubeleid versterkt.

Waar ecologische modernisering in zijn sterke variant rond 1990 een activistische discours bleek, is de zwakke variant vooral behoudend en gericht op het voeren van milieubeleid door vooral in te zetten op technologische vooruitgang. Gedurende de jaren 90 verzwakte ecologische modernisering steeds verder. Bovendien bleken niet alle sectoren even vatbaar voor ecologische modernisering. De pogingen om dit discours ingang te laten vinden op het punt van mobiliteit en transport strandden in de jaren 90 op het verzet van de automobilist die niets voelde voor het beprijzen van het gebruik van de weg via een kilometerheffing. Het parkeerbeleid werd eveneens gehekeld en de campagnes om de automobilist te laten overstappen op schonere vervoermiddelen bleken niet succesvol. De mislukte poging tot ecologische modernisering van verkeer en vervoer betekende een substantiële verzwakking van dit discours.

Op het gebied van automobiliteit voltrok zich in de jaren 2002 en 2003 een fundamentele omwenteling van het beleid. De politiek van de kabinetten Balkenende, met name Balkenende 1, was erop gericht de automobilist ruim baan te geven. Subsidies voor schonere auto's werden teruggedraaid en de automobiliteit werd gefaciliteerd, onder andere door de invoering van een spoedwet wegverbreding. Milieubelangen op het gebied van verkeer en vervoer werden gemarginaliseerd.

De fijnstofstrijd moet begrepen worden als reactie op de marginalisering van milieubelangen die op haar hoogtepunt was in de politiek van het eerste kabinet Balkenende. In de Europese arena beleed Nederland nog het ideaal van ecologische modernisering, maar binnenlands werd vol ingezet op het stimuleren van groei.

De oppositie begon sterker te hameren op het gevaar van slechte luchtkwaliteit en dat bleek een goed middel om terug te vechten, vooral nadat de Raad van State in 2004 was begonnen met het vernietigen van bouwprojecten. De brede consensus over een aanpak die zowel economische vooruitgang als ecologische bescherming zou bieden, was verdwenen. Het volksgezondheidskamp hield de regering voor dat ze blind was voor de gezondheidseffecten van meer asfalt en het infrastructuurkamp betoogde dat dergelijke normen schadelijk waren voor de economische vooruitgang. De regering meende ondertussen dat zij het 'maximaal haalbare' deed. Dit is een impliciet appèl op de eco-modernistische consensus waarin men moet kiezen voor milieubescherming als dat economisch haalbaar en redelijk is. Dit appèl vond echter geen gehoor meer.

Uiteindelijk verschafte ecologische modernisering toch de sleutel om tot een oplossing te komen, maar pas jaren later. De programmatische aanpak uit de Wet luchtkwaliteit van 2007 had veel kenmerken van een zwakke variant van ecologische modernisering. Infrastructuur en milieu worden in dit discours geacht naast elkaar te kunnen bestaan als het geheel goed gemanaged werd en werd, ondersteund door berekeningen en via een integraal lange termijnplan. De programmatische aanpak voorzag expliciet in betrokkenheid van lagere bestuursorganen, wetenschappelijke instituten en stakeholders. Het betrof een preventieve aanpak maar niet zodanig dat infrastructuurontwikkeling een halt wordt toegeroepen.

Met de programmatische aanpak keerde Nederland terug naar een zwakke variant van ecologische modernisering. Een zwakke variant van ecologische modernisering bleek het meest aantrekkelijk beleidsdiscours want blijkbaar kon het Nederlandse publiek niet zonder ecologische principes, maar was het ook niet bereid daar grote offers voor te brengen, in ieder geval niet op het punt van mobiliteit. Dit vertoog bleek in de fijnstofstrijd een geschikt lapmiddel, maar het had niet meer de ambitie of de vernieuwende kracht die het had in de jaren 80. Destijds introduceerde het een nieuwe kijk op het milieuconflict en waarmee oude tegenstellingen doorbroken konden worden. De hedendaagse variant met haar precieze balans tussen vervuiling en sanering leek eerder behoudend dan visionair.

Het gehele domein van luchtvervuilingspolitiek en ecologische modernisering overziend valt op hoe conservatief het Nederlandse vertoog in wezen is. Principes worden weliswaar sterk aangemoedigd maar in de praktijk worden er stelselmatig compromissen gesloten en geldt het adagium dat niemand overvraagd mag worden. Deze kloof leidt tot problemen wanneer actiegroepen zich op die principes gaan beroepen en zich verzetten tegen de behoudende milieupolitiek. In de fijnstofstrijd is de politiek door de rechter op de vingers getikt, omdat de rechter de regering eraan herinnerde dat zij in Europa bindende afspraken had gemaakt. De overheid reageerde door de subsidies voor groepen als Milieudefensie te verminderen. Deze wending maakte eens te meer duidelijk dat ecologische modernisering in Nederland voornamelijk een inkapselingsvertoog was, in plaats van een ambitieus milieudiscours. De opkomende rol van de Europese Unie en de bestuursrechter maakte dat de Nederlandse regering uit onverwachte hoek geconfronteerd werd met de principes die zij eerder had beleden. Maar politiek gezien bleek ecologische modernisering geen duurzaam discours.

De fijnstofstrijd en voorzorg

Aan het begin van de studie werd de hypothese geformuleerd dat het optreden van de fijnstofstrijd als indicatie kon worden gezien dat de voorzorglegaliteit in opkomst is. Uit deze studie blijkt enige steun voor de hypothese, maar er kan zeker niet zonder meer worden gesproken van een opmars van de voorzorglegaliteit binnen de rechtsorde. De casus heeft namelijk laten zien dat als een voorzorgbenadering een groot stempel op recht en beleid gaat drukken, tegenkrachten opkomen die het juist weer ondergraven.

Uit deze studie is gebleken dat Nederlandse regeringen bij dit beleidsprobleem niet erg geneigd waren om de voorzorglegaliteit te omarmen. Ondanks of misschien beter dankzij het beleidsdiscours van ecologische modernisering bleef de legaliteit van risico en compensatie dominant. De reden is dat de Nederlandse regering altijd oog bleef houden voor de kosteneffectiviteit van milieumaatregelen, geheel in lijn met de zwakke variant van ecologische modernisering. Het voorzorgbeginsel werd op dit terrein niet expliciet beleden en gebleken is dat als de kosten te hoog waren, de Nederlandse regering niet van plan was om strikte voorzorg te implementeren. Op het moment dat de luchtkwaliteitsrichtlijnen omgezet moesten worden naar Nederlands recht bij voorbeeld, schreef minister Jan Pronk een brief aan de Europese commissaris Margot Wallström dat het uitvoeren van de richtlijn voor Nederland niet kosteneffectief zou zijn. Pronk stond en staat bekend als een voorstander van het voorzorgbeginsel, maar zelfs hij ging ervan uit dat een scheve kostenbatenverhouding een goede reden was om de richtlijn niet integraal uit te voeren.

Het Europese beleid was tot in de jaren 2000 meer gericht op voorzorg. De Europese richtlijnen luchtkwaliteit uit de jaren 90 vertoonden tekenen van een voorzorglegaliteit. Zo waren zij zonder veel wetenschappelijke kennis verordonneerd, kenden zij een breed bereik en werd de burger expliciet bij de handhaving van de richtlijn betrokken. De informatievoorziening diende het doel om 'druk van onderop' op lidstaten te creëren om de richtlijn goed uit te voeren. Dat laatste bleek ondubbelzinnig uit de evaluatie van de richtlijn uit 2005.

De Raad van State zelf bleek ook geporteerd voor een voorzorggerichte benadering. Hoewel omfloerst, zijn er aanwijzingen dat een voorzorglegaliteit een rol speelde in de beslissingen van de hoogste bestuursrechter op het terrein van luchtkwaliteit. Zo legde de Raad van State de lat voor het benodigde onderzoek een stuk hoger dan voorheen. Daarnaast stelde de Raad zich op als hoeder van de strenge 'koppeling' tussen bestuurlijke besluiten en kwaliteitsnormen. De Raad is niet alleen de hoogste bestuursrechter, maar ook de belangrijkste adviseur op het terrein van de wetgeving. De Raad liet zich in beide hoedanigheden kritisch uit over beleid dat milieurechtelijke waarborgen op de helling zette, zoals de Spoedwet wegverbreding. Daarnaast bekritiseerde de Raad van State verwijzend naar luchtkwaliteit de regering: zij zou beleid voeren dat teveel was gericht op economische groei en te weinig op duurzaamheid.

De RvS bleek op dit dossier een geduchte tegenstander van de regering te zijn. Het feit dat de normen van Europeesrechtelijke aard waren en direct konden doorwerken in de Nederlandse rechtsorde, maakte de positie van RvS sterk. Zij kon het argument hanteren dat 'Den Haag' de wetgeving wel kon wijzigen, maar dat de normen uit de Europese richtlijnen direct door burgers ingeroepen konden worden. De Raad hanteerde een verhaallijn waarin het stelde dat de Nederlandse regering zichzelf had gebonden aan Europese wetgeving en dat het beleid daarom in Den Haag niet gewijzigd zou kunnen worden, maar alleen in Europa. Hiermee legitimeerde het zijn blokkade van een snelle wetgevingsoperatie om het effect van de normen te versoepelen. Deze nieuwe lijn was belangrijk want daarmee maakte de Raad van State duidelijk dat hij het Nederlandse beleid daadwerkelijk zou gaan toetsen aan Europees beleid. Via deze rechterlijke weg vond een voorzorgbenadering alsnog ingang, tegen de wens van de regering in.

Zoals te verwachten viel, gebruikte het gezondheidskamp zeer vaak op voorzorg gerichte argumenten. Het meest belangwekkend is de genese van een expliciet juridische verhaallijn aangaande luchtkwaliteit. Met name Milieudefensie speelde sterk in op het uitspraken van de Raad van State en ging op campagne met het verhaal dat de rechten van de Nederlandse burger waren geschonden omdat de regering niets aan de luchtkwaliteit in vieze straten wilde doen. Zij eisten voor de rechter dat de regering maatregelen zou nemen zoals een snelheidsverlaging op de ringen en het sluiten van de meest vieze straten voor verkeer. Die procedures verloor milieudefensie, maar het punt was gemaakt: een schone lucht is een (burger)recht. Met dit motto werden ook burgers gemobiliseerd om zelf naar de rechter te stappen. In veel gevallen speelde milieudefensie een rol op de achtergrond terwijl burgers betoogden dat de luchtkwaliteit door een bepaald project gevaar zou lopen. Daartoe had milieudefensie een 'doe het zelf pakket' ontwikkeld waarin stap voor stap werd uitgelegd hoe een procedure bij de RvS gevoerd moest worden.

Het succes van deze expliciet juridische campagnestrategie is een van de meest in het oog springende rechtssociologische conclusies. Het idee dat een schoon milieu een mensenrecht zou zijn werd wel verkondigd in academia, maar tijdens de fijnstofstrijd werd het ook ingezet in een daadwerkelijk milieuconflict. Rechters bleken er gevoelig voor en de regering werd geconfronteerd met een groot aantal juridische procedures.

Het kan niet verbazen dat het infrastructuurkamp een consequente tegenstander was van een voorzorgbenadering. Zij zette het standaardargument in dat de wetgeving negatieve gevolgen voor de economie zou hebben en dat Nederland niet 'op slot' gezet kon worden vanwege luchtkwaliteit. De harde interpretatie van de RvS dreigde het milieubelang inderdaad te laten prevaleren. De verhaallijn van het infrastructuurkamp sloeg meer aan naarmate de clash zich voortsleepte. Na verloop van tijd kwam de regering ook op die lijn en het luchtkwaliteitsprobleem werd steeds meer gezien als een probleem van inflexibele regulering. De programmatische aanpak was inderdaad flexibeler dan de oude koppeling en in zoverre kreeg het infrastructuurkamp zijn zin. Het plan behelsde echter ook maatregelen voor schone lucht die eerdere kabinetten nooit van plan waren geweest te nemen en in zoverre was het een compromis. In termen van voorzorglegaliteit kan de programmatische aanpak eerder geschaard worden onder de legaliteit van risico en compensatie, met zijn gedetailleerde balans tussen maatregelen en schadelijke projecten.

Opvallend is dat ook in Europa de voorzorglegaliteit voor wat betreft milieubeleid een halt lijkt te zijn toegeroepen. Ook in deze arena hebben lobbygroepen van projectontwikkelaars en andere economisch belangrijke actoren gelobbyd voor flexibiliteit.

Uiteindelijk is daar inderdaad aan toegegeven. De standaarden werden in 2008 weliswaar iets verscherpt, maar er kwamen ook uitstelmogelijkheden, waar Nederland van gebruik heeft gemaakt. Hier leidde druk van lidstaten, waaronder niet in de laatste plaats van Nederland, en lobbygroepen tot een relatief milde nieuwe luchtkwaliteitspolitiek. De conclusie moet dus zijn dat hoewel de luchtkwaliteitsregulering inderdaad een product is geweest van een opkomende voorzorglegaliteit, deze legaliteit echter nog niet de overhand heeft gekregen. De opkomst is er een van horten en stoten en de nationale en Europese tegenkrachten die het oproept zijn aanzienlijk.

Het Nederlandse milieubeleid is in ieder geval op het gebied van luchtkwaliteit in wezen behoudend van aard en geworteld in de legaliteit van risico en compensatie. De Nederlandse regering echter heeft echter ook de neiging te sterk te hameren op milieurechtelijke beginselen en principes, zeker in zijn betrekkingen met de Nederlandse burger, het buitenland en de EU. Dat leidt ertoe dat zij soms door rechter en Europese Commissie pijnlijk aan de principes gehouden wordt.

De fijnstofstrijd samengevat

Hoe en waarom is luchtkwaliteit plotseling een probleem geworden in Nederland? Deze studie laat zien dat Europese luchtkwaliteitsregulering sterk heeft bijgedragen aan het uitbreken van de fijnstofstrijd. Het laat ook het belang zien van de wisselwerking tussen Europese regulering en Nederlandse pogingen om deze te beïnvloeden. Gezien vanuit de lange termijn kunnen we zelfs stellen dat de fijnstofstrijd het resultaat is van vroege Nederlandse bemoeienis op het Europese niveau. De strenge luchtkwaliteitsrichtlijnen zouden er niet geweest zijn als Nederland de WHO niet bij het luchtbeleid had betrokken om zo wetenschappelijk goed onderbouwde luchtkwaliteitsnormen te promoten. De strikte normering kan echter niet op het conto van Nederlandse bemoeienis of de WHO alleen worden geschreven. Ook het Verenigd Koninkrijk heeft zich sterk gemaakt voor een ambitieus Europees luchtbeleid.

Een ander element karakteristiek voor het Nederlandse milieubeleid en milieurecht heeft eveneens een rol gespeeld. De koppeling tussen milieukwaliteitseisen en besluiten van bestuursorganen zorgde ervoor dat infrastructuur niet kon worden ontwikkeld in situaties waar de normen werden overschreden. Vanwege de strikte EU standaarden was dat op veel plaatsen in het land het geval. De ontwikkeling van met name wegaanpassingen kwam daardoor in grote delen van het land in de knel. Het legalistische optreden van de Afdeling Bestuursrecht van de Raad van State verkleinde de ruimte om de sociale gevolgen van de uitspraken mee te laten wegen. De bewoordingen van het Besluit Luchtkwaliteit liet daar ook weinig ruimte voor.

De fijnstofstrijd kan verder worden begrepen als een episode in de Nederlandse politiek waarin de consensus omtrent een milde variant van ecologische modernisering werd gebroken. De kabinetten Balkenende, in het zadel vanaf 2002, braken met de consensus door absolute prioriteit te geven aan de ontwikkeling van infrastructuur. Deze beleidswending verontrustte de milieubeweging die in antwoord daarop haar heil ging zoeken bij de rechter om deze eenzijdige wending naar economische ontwikkeling te keren. Zij werd daarbij geholpen door de hierboven beschreven juridische mogelijkheden en vond een bondgenoot in de Raad van State. De milieubeweging betoogde onder andere dat schone lucht een recht was, omdat die standaarden nu eenmaal bij wet waren vastgelegd.

Het hardnekkige verzet van de Raad van State betekende dat de Nederlandse regering er niet in slaagde snel een bestuurlijke oplossing te vinden voor de vernietiging van vergunningen. De regering zag zich genoodzaakt een wet aan te nemen waarin de mogelijkheid werd geboden om een nieuw reguleringsinstrument in het leven te roepen, de zogenaamde 'programmatische aanpak'.

In het parlement had het volksgezondheidskamp het eerste jaar de wind in de zeilen. Projecten werden regelmatig getorpedeerd door de Afdeling Bestuursrechtspraak van de Raad van State, de Afdeling Advisering waarschuwde de regering tegen het gebruik van bestuurlijk lapmiddelen en de regering werd gedwongen veel meer geld uit te trekken voor de verbetering van luchtkwaliteit. Het volksgezondheidskamp echter kreeg een steeds geduchtere tegenstander in het infrastructuurkamp, een discourscoalitie van lobbygroepen en politieke partijen die wezen op het belang van economische ontwikkeling. Het kamp lanceerde de counter-claim dat het echte probleem met luchtkwaliteit niet lag in de sfeer van de volksgezondheid maar in de regulering dit het land op slot had gegooid voor verdere ontwikkeling. Zij brachten een verhaallijn naar voren dat wees op miljarden euro's aan schade en het op de tocht staan van wellicht 100.000 banen.

De fijnstofstrijd kon in de politieke arena alleen tot een einde worden gebracht door de consensus rond zwakke ecologische modernisering nieuw leven in te blazen. Het instrument van de programmatische aanpak was gebaseerd op consensus, sterk technisch van aard en leunde op een managementaanpak. Met deze aanpak vervat in het zogeheten Nationaal Samenwerkingsprogramma Luchtkwaliteit slaagde de regering er ten eerste in om bij de Europese Commissie enige uitstel voor het voldoen aan de normen te verkrijgen. Ten tweede werd het door de Raad van State uiteindelijk geaccepteerd als een instrument dat zowel ontwikkeling mogelijk maakte als de luchtkwaliteit veilig stelde.

Deze studie laat, voor zover ik weet voor de eerste maal, een zaak zien waarin de Raad van State de regering jaren van twee kanten in de greep kon houden, daarbij gebruik makend van zijn dubbele positie als bestuursrechter enerzijds en adviesorgaan anderzijds. De Afdeling Rechtspraak bleef besluiten vernietigen terwijl de

Afdeling Advisering de regering eraan bleef herinneren dat elke aanpassing van de wetgeving conform de Europese regels moest zijn.

Een derde manier om de fijnstofstrijd te analyseren is door te kijken naar de mate waarin een legaliteit gebaseerd op risico en compensatie en een legaliteit gebaseerd op voorzorg aanwezig waren in de politieke en juridische afhandeling van de strijd. Gedurende de strijd blokkeerden strikte standaarden gebaseerd op onzekere wetenschappelijke kennis belangrijke economische projecten in Nederland. Alleen door een balans te vinden tussen milieubelangen en economische belangen kon de strijd worden beslecht. De snelle afkondiging van de Europese normen, het succes van de verhaallijn rond de status van luchtkwaliteit als recht en de gehechtheid aan de koppeling die de Raad van State liet zien, zijn indicaties dat een voorzorglegaliteit inderdaad een rol heeft gespeeld. Echter, de sterke reactie tegen de kortstondige juridische prioritering van milieubelangen in Nederland, maar ook in Europa laat zien dat de voorzorglegaliteit nog een lange weg te gaan heeft voordat kan worden gesteld dat zij zich een dominante positie heeft verworven in de Nederlandse rechtsorde.

Summary: the social construction of the Dutch air quality clash

In April 2007, a short time before I started this dissertation, I returned to the Netherlands after having spent six years in Istanbul, Turkey. After my return, I quickly picked up a new word that was being used daily in Dutch: 'fijnstof', meaning 'particulate matter' in English. The precise nature of particulate matter remained unclear, but some things were certain: it was a form of air pollution, it was dangerous, and it was emitted primarily by cars. Mostly present in busy streets, it was the number one public health hazard. This 'new' form of air pollution intrigued me, because I had just come back from a metropolis that was home to over 10 million people, featuring perpetually congested motorways and coal-fired steam boats. In Istanbul, however, air pollution was rarely discussed, whereas in Amsterdam it was a topic of fierce debate.

In this dissertation, the social problem of air pollution by particulate matter (PM10) is under scrutiny. The question concerns why air quality became a major public issue in the Netherlands in the period between September 2004 and April 2010. Those years have been researched extensively, because in 2004 the highest Dutch administrative Court, the Council of State,[1] started to terminate the development of high profile infrastructure projects due to exceedance of the standards for particulate matter. In 2009 the European Commission granted the Netherlands postponement from the European air quality standards in order to implement its 'programmatic approach'. This approach was a Dutch Governments' proposed solution to solve the impasse. Finally, on the 31st of March 2010 the Council of State Administrative Jurisdiction Division approved of the Government's solution to the impasse, the so called programmatic approach.

During these years two camps of actors emerged that each proposed a different definition of the problem and a different solution. From the perspective of public health, various institutions pointed out the risks posed by particulate matter. It was argued that people living close to highways were particularly susceptible to cardiovascular and respiratory disease, and that particulate matter was a significant factor in the occurrence of premature mortality found in many urban areas. Other actors, however,

1. The Council of State is an organisation that functions as both the highest Dutch administrative court and as an important policy adviser that must be consulted before legislation is enacted.

pointed out that the legislation by which air quality was regulated was an unjustifiable economic burden, with severe consequences for employment in terms of the Dutch competitive position. Especially in 2005, 2006 and 2007 these two points of view collided and fought for dominance in the political arena. Therefore, I refer to this situation as the 'air quality clash'. In essence, the clash is one of interests: namely, which should prevail – public health or the economy? Over the years, however, the clash became more complex, and questions have arisen about the increasing influence of the European Union on national law and policy, the influence that the Dutch Council of State may wield in the political arena, and to what extent the population should be expected to tolerate health risks.

A number of aspects have been examined further in this socio-legal study. Firstly, the role of 'Europeanisation' in the emergence of the air quality clash is under investigation. Strict European rules have found their way into Dutch national legislation, and these rules have caused the air quality standards to be significantly tightened. On the basis of these standards, the Council of State has annulled a number of prestigious high-profile construction projects. Secondly, choices made by the Dutch Government are under scrutiny. These choices have been important in the emergence of the air quality clash and I argue that they are informed to a large extent by the philosophy underlying Dutch environmental policy. Thirdly, the air quality clash has been related to a supposed transition within the Dutch legal order. The question is whether the legal order is in transition from what is termed a legality of risk and compensation to a legality of precaution. The hypothesis is that whereas compensation of victims because of damages was considered previously to be imperative, the current prevention of damages altogether is considered crucial, even if prevention were to lead to very high costs.

Methodology

A social constructivist methodology was chosen for this investigation. In this perspective, social problems are not assumed to be objectively given states, but are the result of an intersubjective process of signification by which certain conditions become labelled as social problems. Certain people experience these conditions as problematic and if these people convince a large enough number of others that a particular state of affairs is indeed undesirable, the situation is then considered to be a 'social problem'. Social constructivism has proven to be an effective methodological perspective for studying the social problems, as it shifts from the question of how people respond to the emergence of problems to the process by which certain situations come to be defined as problematic. In the social sciences, it is now one of the most widely used frames of reference for the purpose of analysing the emergence of social problems.

In this dissertation, I have used social constructivist concepts employed in two different types of social constructivism. I refer to the first type as being actor-centred and to the second type as being discourse analytic. Actor-centred social constructivism assumes that certain groups of people have a particular interest in defining a situation as problematic. The unemployed, for instance, are keen to have unemployment accepted as a major social issue. Minority groups, for example, will benefit if society begins to view racism as a troubling social condition. These actors, referred to as 'claims makers', will attempt to convince others of their definition of the problem. Through various channels, they will try to persuade the public to accept the claim that a certain situation is problematic. But of course claims themselves tend to elicit counter-claims. The representation of the situation by claims makers will not be shared immediately by everybody, and groups that have an interest in maintaining the status quo might oppose it. The claim that abortion should be legalised, for instance, immediately met with the counter-claim that abortion in fact amounted to murder. When claims and counter-claims come into conflict, heated societal discussions erupt.

Claims makers and counter-claims makers exchange their arguments in writing and in front of cameras in talk shows, at public rallies, in Parliament, and sometimes in court. All these locations in which arguments are exchanged are referred to as 'arenas'. If a claim meets with broad popular support, policy makers may start to propose laws aimed at resolving the specific social problem, and these proposals could lead eventually to new regulation. Nevertheless, some may consider these new rules to be problematic, in which case the game begins anew.

The social constructivist sees the social order as the outcome of a continuous struggle between social groups making their claims. If these groups manage to have their assertions accepted by a substantial number of people, these claims will become institutionalised and often lead to new regulation. Actors are seen as players in a far-reaching game in which their claims-making moves are directed towards obtaining a favourable position from which to defend their interests.

In the last 20 years, a new social constructivist strategy has come to the fore, and focuses on the arguments on which actors make and defend their claims. In the social sciences, this shift in focus from actors towards arguments and discursive strategies is referred to as the 'discursive turn'. Discourse analysts point out that a claim can only be accepted if it fits within the framework of current public attitudes and opinions. Arguments for certain claims have to be plausible in light of other beliefs and opinions held among groups in society. The term 'discourse' refers to the way in which the dominant point view in a certain field of social interaction manifests itself in language and argumentation, and in the practices by which it is institutionalised.

Claims must be compatible with the discourse; therefore, they will never be brought forward in a neutral and objective fashion, but will be embedded within a certain narrative. A problematic situation is presented as a specific undesirable condition with a certain history, causes, and effects, and its own specific solution. In this dissertation, I refer to such a narrative as a storyline. These storylines are used by actors, but can also be taken up by other actors in the course of the debates. In the event that different actors embrace the same storylines, we speak of a discourse coalition. Storylines must be compatible with the reigning discourse, but discussions also change shape under the influence of newly emerging storylines. This dynamic interplay between storylines and the larger discursive context makes this social constructivist method appropriate to investigate the constellation of different points of view that are held within society, or by groups within society. These societal convictions are sublimated within storylines.

Ideal types

The clash concerning air quality is not only related to the individual behaviour of certain actors but also to the Dutch policy philosophy concerning environmental issues and structural changes in the legal order. To that end, it is related to two ideal types: ideal typical ecological modernisation and the ideal typical legality of precaution. In the scope of this study, ecological modernisation is considered to be a discourse in the field of environmental policy that made gains in the 1980s in the Netherlands and Europe, and became the dominant discourse in the 1990s. In light of ecological modernisation, the question is posed whether the extent to which Dutch policy choices themselves have contributed to the emergence of the air quality clash. In the table below, an ideal typical reconstruction is presented, consisting of ecological modernisation and its typical characteristics. This environmental discourse is contrasted with an earlier debate dominant in the 1970s, which I named 'limits to growth'. It obtained its characteristic elements from the report for the Club of Rome from 1971, with the same name. In addition, the discourse on ecological modernisation is divided into a strong and a weak variant. On the one hand, strong ecological modernisation demands far-reaching changes in our economic structure and behaviour vis-à-vis the environment. On the other hand, weak ecological modernisation is easily compatible with current capitalist economic relations. The two types of ecological modernisation have been outlined together with the 'limits to growth' discourse in table 12.

The term legality, coined by Dutch sociologists of law André Hoekema and Niels Van Manen and inspired by the work of Max Weber, is used to define a certain set of assumptions prevalent in the legal order in a certain time. According to sociology of law the legal order is not a stable set of rules, but a representation of the changing convictions held in society at a given time. Throughout legal history different sets of convictions have dominated the legal order, without however totally replacing earlier convictions. The legality of precaution is my identification of a new type of legality in which the necessity of the prevention of damage is the paramount imperative. My socio-legal hypothesis is that in environmental law and policy, this type of legality

Table 12

Policy discourse	Limits to growth	Weak ecological modernisation	Strong ecological modernisation
Relationship between economy and ecology	Inimical: economic growth threatens the environment.	Peaceful coexistence: Economy and ecology may progress side by side.	Positively intertwined: Ecological progress will lead to economic growth.
Role of science	Natural science and ecological science are used to assess the state of environmental degradation. Natural scientific considerations should steer policy.	Science provides the data to discuss ecological problems in economic, natural scientific and managerial terms.	Science is applied to take stock of environmental threats, and economic problems are discussed by including indicators of environmental performance. Resource use, depletion, and pollution become an important concern in economic management.
Consensus building	Market parties should be forced to commit to the goals of environmental policy through top-down regulation.	Negotiations between the Government, environmental pressure groups and industry on the topic of environmental regulation.	Broad mechanisms of participation for industry and the environmental movement in policy making.
Preventative approach	Ecological disturbance should be reduced to a minimum. This implies rigid application of the precautionary principle and an antagonistic view towards economic growth.	Pollution should be prevented by cleaner processes of production and if necessary application of the precautionary principle.	Targeting the polluting individual, enterprise and consumer him or herself. Broad application of the precautionary principle and internalization of it by addressees.
Responsibilisation	Education and legislation should be combined to raise the right environmental mentality.	Awareness raising through media campaigns, supplying information and subsidising environmental pressure groups.	Active involvement of citizens through education, granting rights to review corporate registries and access to justice, alliance between environmentalists and Government.
Comprehensive policies	The earth is considered one interdependent ecosystem. Ideally, policy is integrated and formulated holistically. In practice though, only piecemeal regulation could be realised in the 1970s.	Targeting multiple forms of pollution in multiple environmental media with the same policy.	Targeting multiple forms of pollution as well as establishing behaviour change with regard to polluters within the same policy.

is replacing an older type, the legality of risk and compensation. In this type of legality compensation of victims for damages is the prime imperative.

This shift from compensation to precaution has repercussions for different fundamental assumptions of the Dutch legal order. The core element of this new type of legality is the application of the precautionary principle, which holds that uncertainty as to the scope and nature of environmental damage cannot be used as an excuse not to take protective measures. In table 13, we may find an equally ideal typical representation of the legality of precaution, contrasted with the legality of risk and compensation.

Research design and research material

This research was conducted by analysing parliamentary documents that contained the phrase 'air quality'. In particular, the parliamentary material between May 2004 and December 2007 has been investigated, but documents from earlier and later periods have also been examined. I have chosen to focus on the aforementioned period because that is when the air quality clash took place in the Dutch parliamentary arena. However, because the historical roots of the clash lay earlier, a longer period had to be investigated. Dutch environmental policy came of age in the

Table 13

	Legality of risk and compensation	**Legality of precaution**
Core aspect of the legal order	Compensation of damage by spreading of risk is the prime imperative; hence, strict liability and public and private insurance schemes are essential. Prevention of damage if it is cost-effective to do so.	Prevention of damage is the prime imperative; hence, application of the Precautionary Principle is essential. Prevention extends to possible threats of an uncertain magnitude.
Category of damage considered of most concern	Short- to medium-term damage of a relatively certain magnitude and probability.	Long-term damage of the potentially catastrophic kind but of an uncertain probability.
Moral reaction to damage	Damage cannot be prevented totally, but compensation is in order when it occurs.	Damage is disgrace and should have been prevented.
Perception of the victim	Rights-bearing member of a risk collective.	Civilian damaged by negligence of responsible authority, agency, or enterprise.
Level of proof and safety guarantees required in decision making	Decision making procedures should rely on scientific knowledge and cost benefit assessment to determine optimal solutions.	Decision making procedures should be based on proof that the chance of harm is negligible, and proof is required from the actor undertaking the risky activity.
Stance on the possibility of social engineering by law	Society may be improved by piecemeal policies and an instrumental use of legislation.	Society may be improved by comprehensive policy making (i.e. long-term holistic planning) and societal awareness raising.
View on public participation	Civilians are seen as laypersons with little knowledge of risks or expert systems. Policy needs to be determined by experts.	Imperative because the citizen is an 'experiential expert', and policy and law need to provide opportunities for citizens to present their views.
Value most worthy of protection	Economic development: growth and expansion of construction, mobility, and innovation.	Ecological harmony: balance of eco-systems, public health, and safety from possible threats.

1980s and 1990s and in order to understand the context of the clash, the whole period from 1972 to 2010, has been investigated, by examining the literature and major policy documents such as the National Environmental Policy Plan.

The problem of air quality obviously had a significant 'European' component, since the air quality regulation was of European origin. This made it necessary to investigate EU documents as well. Moreover, parliamentary documents do not provide a sufficiently complete account. Though a crucial source, because every social problem is eventually discussed in Parliament, they provide only limited information about a more encompassing social reality. Therefore, a large number of scientific reports and newspaper articles, as well as extensive 'grey literature', has been examined in addition to parliamentary documents. Interviews with experts, politicians, and campaigners have also been conducted, and have helped considerably to assess the information gained from the documents in a broader context.

As mentioned, the investigation concerned an extensive period of time. As regards the years of the clash, Dutch parliamentary documents have been crucial, while European documents relating to the 1990s have been most important. I studied the preceding years by reviewing the literature as well as the most significant Dutch parliamentary memorandums in the sphere of air quality.

Short history of the air quality clash

The air quality clash began when the Council of State Administrative Jurisdiction Division, began to reject permits granted for road expansions and other infrastructural works due to a conflict with the standards laid down in the Air Quality Order 2001. These court orders created an economic setback, because a great many projects could not be completed. In particular, the programme to expand roads on a large scale was hit heavily. Spokespersons of the Ministry of Transport and Water Management (V&W) complained that they 'could not put a single spade in the ground' because of the air quality issue. However, the verdicts were considered to be a victory for the pressure groups and political parties that had pointed out the public health risks associated with poor air quality.

These risks were first brought to the public attention in the Netherlands by the GGD, the municipal health service of the Rotterdam area. American epidemiological investigations in the 1990s had already indicated the prevalence of premature mortality in areas affected by bad air quality, and Dutch studies revealed that children living along busy motorways were more susceptible to ailments involving the respiratory system. In 1999, these findings led to unrest in the area of Overschie, a densely populated residential area close to Rotterdam, which was divided into two by the busy A10 thoroughfare. In an alarming report, the GGD compared the air quality in

Overschie to the passive smoking of 17 cigarettes a day. The risks of passive smoking are in fact comparatively very small, but the claim proved strong enough to attract political attention. Jan Pronk, Minister of Housing, Spatial Planning, and the Environment (VROM) at the time, decided to conduct a test involving a reduction of the speed limit on the A10 near Overschie, because it was known that cars driving at a slower speed emitted less PM10. The GGD's claim of bad air quality near Overschie was taken up further by parties on the left of the political spectrum, and a storyline evolved around the harmful effects that an increase in automobility and the construction of highways had on public health. Epidemiological research was cited as evidence for the claim, and in 2001 the Air Quality Order appeared, containing the European standards for clean air. The standards for particulate matter would become valid from 2005 onwards.

In 2004, the Council of State Administrative Jurisdiction Division established a consistent line of precedents regarding the standards for PM10 and nitrogen dioxide. However, insufficient research into the effects of the decision on air quality would lead to the annulment of the administrative decisions, irrespective of whether people would be exposed to pollution. During the air quality clash, the Council of State continued to terminate projects, and two camps of actors emerged in 2005, each with its own typical perspective regarding air quality. The term 'camp' denotes a set of actors that put forward the same claims and use the same storyline to support them. Therefore, we may also speak of discourse coalitions. The first camp consisted of scientific institutes, pressure groups, and political parties that claimed poor air quality was harmful to public health, and demanded that structural measures be taken to improve it. I call this discourse coalition the pro-health camp. This camp took the offensive especially during the first year after the Council of State started to terminate infrastructure projects, and it demanded fewer roads as well as more funds for air quality measures. The pressure group 'Milieudefensie' in particular demanded a general reduction of the speed limit.

In response to the termination of projects, an important counter-claim emerged. Air quality legislation itself was a problem, according to an alliance of various pressure groups representing mostly the construction sector and a group of political parties, among which were the CDA, the VVD, and the LPF. The CDA and the VVD also held the most seats in the Dutch coalition government. This camp argued that the standards were unfeasible for the Netherlands, and that their economic consequences were not acceptable. According to this camp, air quality regulation and its interpretation by the Dutch judiciary had unjustifiably 'blocked' the Netherlands. The pro-infrastructure camp offered a different solution: namely, the release of the typical Dutch link between environmental quality standards and administrative decisions in the field of spatial planning. In the Netherlands, it was possible for the court to terminate administrative decisions that might lead to pollution in excess of the standards in a certain area. In other countries, however, these legal repercussions were not as severe.

A series of debates and new legislation followed in and after 2005, although a structural solution to the clash was not found until the end of 2007. Ultimately, it was decided to release the link gradually, and to transfer to a system now known as the 'programmatic approach'. All spatial projects that had significant consequences for air quality were required to be part of a plan that included those projects as well as a list of beneficial measures that needed to be adopted. Ultimately, the sum of the positive measures needed to top the negative consequences of the projects, and air quality would have to improve structurally, ensuring that the standards would be reached at a certain point in the future. The law that made this approach possible was drafted at the end of 2007, but it was not until 2009 that this elaborate plan involving projects and measures would see the light of day.

The plan, however, could not bring immediate relief, as it was aimed at long-term structural improvement, and it was still undecided whether the Dutch administrative judge and especially the European Commission would accept this solution. The air quality regulation was European in origin, and the Netherlands needed a postponement of the standards for its own plan to be effective. It was possible to obtain such a postponement, because the possibility of appealing for it had been included in a new European Air Quality Directive enacted in 2008. Ultimately, in 2009 the European Commission indeed accepted the Dutch plan as a ground for postponement. In March 2010, the Dutch Council of State also accepted the plan as a sufficient guarantee for projects having possible negative consequences for air quality.

Nevertheless, not everyone was happy with the eventual solution, and the environmental movement and certain ecological political parties in particular considered the air quality order to be a trick to rescue the infrastructural projects, despite their possible harmful consequences. In Dutch environmental spatial law, however, this new programmatic approach led to new legal innovations. Several new laws in the field of spatial planning now contain elements of the programmatic approach.

The air quality clash in light of Europeanisation, ecological modernisation, and precaution

The political history of air quality in the Netherlands and Europe yielded interesting socio-legal insights in three different fields: Europeanisation, ecological modernisation, and the legality of precaution. The most important conclusions in all three fields will be discussed below in more detail.

Europeanisation

Europeanisation has not been a widely debated topic among Dutch sociologists of law, but for environmental law in particular it is a most significant tendency.

Europeanisation is a term describing the growing influence of European law and regulation on the national law and policy of Member States. Insight into and knowledge of the way this regulation is formed is therefore highly important, but the interplay between national and European regulation turns out to be extremely complex.

In the case of air quality, European standards have caused national administrative problems because they turned out to be unfeasible for the Netherlands. In 1999, the standards were decreed in the European air quality Daughter Directive, and were subsequently incorporated into Dutch law by way of the Air Quality Order 2001. Since these standards were of European origin, the Dutch government liked to hide behind the excuse that it was taken by surprise, and that 'Europe' had promulgated poor standards. However, this study made clear that the Netherlands had been involved in air quality regulation from very early onwards, and it was not averse to strict standards.

Plans for a new set of Air Quality Directives were revealed in 1994, with this new integral approach to air quality management being based on the 5th European Action Programme. The strategic framework would be laid down in the first of what were called Framework Directives, while the standards themselves would be set out in a subsequent Directive, referred to as the Daughter Directive. The standards would be based on those recommended by the WHO in its air quality guideline of 1987, and which would be revised in the 1990s. The WHO was considered to play a key role, because as an independent organisation it would be able to supply the scientific legitimacy to compel all Member States to accept the new regulatory proposals. Strategically, the standards would be combined with provisions on the necessity of providing information in order to heighten environmental awareness, and the standards needed to be strict enough to force Member States to develop innovative strategies to meet them.

The Dutch contributed to the involvement of the WHO as well as to the strategic angle of the European air quality policy. In regard to the involvement of the WHO, Dutch civil servants had contacted this organisation early in the 1980s to ask it to draft air quality guidelines. The Netherlands intended to set up a system of air quality standards itself, but did not have sufficient scientific expertise. In addition, it was looking for an institution with the appropriate credentials to convince the opposition in the Netherlands of the necessity of such standards, as well as to persuade neighbouring countries to adopt a similar strategy. This last point has to do with considerations of competition. If the Netherlands were to force stricter measures on its own industry, other countries would then have a competitive advantage. The same idea formed the basis of the EU proposal to adopt a uniform set of air quality standards. It would ensure the emergence of an 'equal environmental playing field'. Strict standards were considered crucial in order to dissuade other countries from adopting even more rigid ones.

The Dutch efforts found a willing audience in the WHO, and the Netherlands paid for research into air quality recommendations. By the time results became available in

1987, however, Dutch enthusiasm regarding air quality standards had waned, although In the EU they had started to become a benchmark. The target that European air should meet the WHO standards was taken up in the previously mentioned 5th EAP. When the eventual regulation was under negotiation in Europe, the Netherlands did not relinquish its progressive attitude, and supported strict regulation.

As regards the strategic angle of the proposal, the Netherlands contributed as well, and the switch to ecological modernisation in the 1980s was a significant factor. Dutch politicians made their mark on European environmental policy in the early 1990s, as the 5th EAP was influenced by the then innovative Dutch National Environmental Policy Plan, and was strongly eco-modernist in character. Key considerations of ecological modernisation involved supplying information, stimulating environmental innovations, and mobilising the public to support environmental goals. The air quality regulation was a result of the plans laid down in the 5th EAP because this plan included the target that the air quality in the EU would match WHO recommendations. Like 5th EAP the Air Quality Directives contained the typical eco-modernist goals of involving the public, stimulating innovative policies and informing the public about remaining air quality problems.

Even during the time of the negotiations on the Air Quality Directives, the Dutch generally adopted an ambitious pro-environment stance. Strict rules were in Dutch interests in general, as the Netherlands had a strong environmental movement and, as a small country, was dependent on the environmental policy of other nations. It is not surprising, therefore, that the Netherlands lobbied for an international regime for air quality.

The Netherlands however ended up getting a lot more than it had bargained for. The standards were too strict and the 'Dutch' characteristics like the rights to information, the stimulation of innovations and public mobilisation exacerbated the problematic character of the new European regulation. The question is why a Directive initially favourable to Dutch interests ended up becoming a thorn in the side. In the European arena files have a very long lead time, and during the time it takes for such an initiative to yield results, national policy may already have taken a different turn. Another reason is that in the European arena many actors and ideas became involved. In order to guarantee a level playing field, European institutions aimed for strict standards that were valid on the territory of the whole of the EU. The Dutch approach was based on the idea that the standards should only be valid in places where people might be exposed. In addition, the process of drafting the air quality strategy was steered by a UK civil servant, and the eventual European plan of action contained many elements that were present in the UK strategy. In the preceding years, the UK had unveiled an air quality action plan in response to a national social problem that had made headlines: namely, childhood asthma. Automobility was held responsible, and the UK developed its strategy in 1997, a year before a common position was reached on the European strategy. This strategy began to function as a benchmark for the EU as well. Moreover, the idea of buttressing a regulatory proposal with a cost-benefit analysis became fashionable, and the assessment supporting the air quality strategy predicted that the benefits would significantly outweigh the costs. The study at hand reveals that assumptions underpinning the cost-benefit analysis were faulty, and severely undercut this analysis. The costs of implementing this new policy proved to be far higher than expected. By then the Dutch Government started to raise objections regarding the feasibility of the proposed standards, but by then it could not influence the process anymore and a policy proposal was adopted that included Dutch ideas but applied them in a way that proved to be problematic.

It turns out to be tempting for Member States to play in the European arena, because if European policy closely matches national policy, few costs need to be incurred in adjusting national policy to European rules. In addition, when a Member State enacts an ambitious environmental regulation, it is rational to make 'Europe' propose a similar regulation as well, in order not to burden one's own industry with a competitive disadvantage. This leads to a situation in which each Member State will try to have its own favourite regulatory strategies adopted by the EU. Moreover, European institutions such as the European Commission and the European Parliament have their own reasons for becoming actively involved in European environmental policy.

The Commission, for instance, would like the EU to profile itself as the safe keeper of its citizens' environmental interests, and demands that the EU strives for a high level of protection. In this manner, the Commission is able to defend the 'raison d'etre' of this supranational body politic.

The air quality regulation became a patchwork of different elements from diverse regulatory systems and philosophies. The notion of 'bricolage' aptly describes European policy, as it is a gathering of different ideas and parts of various national regulatory systems. In the field of air quality, it had numerous diverse effects, many unintended by the Netherlands. This bricolage nature makes Europeanisation an ultimately unpredictable process.

The bricolage nature of European policy has consequences for our assessment of the phenomenon of Europeanisation. This research demonstrates that Europeanisation is in fact a two-way street. Indeed, European law is significant for the national legal order, but Member States such as the Netherlands also try to influence the European legal order by 'exporting' their own policy measure and concepts to the European arena. Member States make sure that they have people in crucial positions in European committees, other discussion forums, and European regulatory institutions. These representatives suggest the policy solutions and philosophies of the Member States that they represent. However, because many Member States, European institutions and lobby groups have a stake in this arena the member states that exported proposals to the European arena get them fed back in a way that they had never intended. Europeanisation has the character of a constant feedback loop from the EU arena back to the member states and vice versa in which the ideas and proposals get mangled, reconfigured and translated beyond recognition.

Dutch policy choices and ecological modernisation

Ecological modernisation was embraced at the beginning of the 1980s by the Dutch Government as the reigning discourse involving environmental problems. This entailed a discursive change in regard to the discussion dominant in the 1970s – the limits-to-growth discourse. In this pessimistic debate, the Government had the onus of setting strict limits to the emission of pollution by enacting legislation. The Ministry of the Environment, however, did not have nearly enough power to realise this restrictive regulation, and environmental regulation became bogged down in a bureaucratically complex system of permits. Nevertheless, in the field of spatial planning, the Government did accept a system that limited the extent of the pollution. The link between spatial administrative decisions and quality standards meant that if environmental quality was not sufficient in a certain area, the administration could not permit further polluting activities. This arrangement resulted in quality standards having a direct influence on individual decisions made by administrative bodies.

When the Ministry of Housing, Spatial Planning, and Environment (VROM) was founded, the environmental department was merged with those of housing and spatial planning, and modern managerial ideas began to permeate this new Ministry. The first VROM Minister was Pieter Winsemius of the VVD. He modernised the Ministry and exchanged the pessimistic limits-to-growth discourse for the optimistic dialogue on ecological modernisation, according to which, environmental problems could be dealt with by further modernisation in an environmentally friendly direction. Winsemius initiated a campaign to bring the 'target groups' on board, and to sway them with his approach to shared environmental responsibility. He was helped to a great extent by the considerable amount of social attention being paid to the problem of acid rain. A clean environment was presented as a necessary resource for further production, environmental alternatives were positioned as profitable, and pollution was framed as a by-product of inefficiency. Environmental degradation was compared to a budget deficit that would have to be repaid by future generations, and this new discourse resulted in a strong 'economisation' of environmental thought.

Winsemius demanded that Dutch industry and enterprises do their part, but he would also make them stakeholders in the drafting of environmental policy. He concluded covenants with industrial sectors, in which they had a large say in environmental policy in exchange for their commitment. Aimed at consensus, prevention, and management, this policy reached its summit with the integral National Environmental Policy Plan (NEPP) issued in 1989. In subsequent years, this plan was exported successfully to other countries and to the European Union.

This policy proved to be a double-edged sword, however, because it promised a positive interdependency between the economy and environmental policy. This meant that far-reaching sacrifices could not be demanded from the economically powerful target groups, and important sectors like the petrochemical industry and transport were spared. Moreover, the Government aimed for an international approach to environmental problems that implicitly relegated domestic measures in importance. In addition, the incorporation of the environmental movement in the traditional consensus building structures of the Dutch corporate democracy was not always successful.

As regards air quality, this discourse brought little that was new to the table after the 1980s. Even when reports started to come in that particulate matter might be more harmful than expected, the Government waited for Europe to act. When the Dutch Government was eventually forced to take steps, it did not realise what legal consequences the air quality standards might have. In hindsight, it can be seen that the attitude of the Dutch administration was naïve, and was a consequence of an implicit conviction that standards that did not take economic feasibility sufficiently into account could be considered legally binding. The Dutch Government was firmly convinced that after evaluation the standards would be adjusted downwards by the European Commission. The implicit assumption was that if one did all that was reasonable, one could not be asked to do more.

The Dutch Government made a strong case for ecological principles and a strategic long-term approach to environmental problems, but was reluctant to take costly measures. During the 1990s, ecological modernisation weakened gradually, and more standard economic considerations became increasingly dominant. Whereas the strong variation that was leading for a short time in the early 1990s called for a determined rearrangement of the Dutch economy and consumer behaviour, the weak variation was essentially a conservative strategy.

The discourse suffered its most significant setback in the area of transport and mobility. In the early 1990s an attempt was made at the ecological modernisation of transport, but here determined opposition from motorists caused the attempt to fail. The eco-modernistic Second Structural Scheme for Transport and Traffic contained proposals for road pricing, stimulation of public transport, a strict parking policy and called on motorists to make the switch from using the car to environmental friendly modes of transportation. However, the motorist had no intention of leaving his or her car in order to use cleaner methods of transportation, and certainly did not intend to pay for use of the road. The policy became seen as inimical to the interests of ordinary car users and gradually the eco-modernist ambitions were released in the course of the 1990s.

The end of ecological modernisation in this domain arrived with the inauguration of the first Cabinet led by Prime Minister Balkenende in 2002. The policy of the Balkenende Cabinets, especially the first from 2002 to 2003, was geared at stimulation of economic activity including the facilitation of transport and mobility. Subsidies for cleaner cars were abolished, and motorists were accommodated, by, among other things, the adoption of an Emergency Law on Road Expansion. Environmental interests became thoroughly marginalised by the Balkenende Cabinets.

Air quality though proved to be a good topic on which to fight back. Already the situation in Overschie had attracted some attention, but the breakthrough came when the Council of State started to annul the infrastructural projects cherished by the Government. Suddenly the opposition and other proponents of environmental interests had a powerful weapon at their disposal. After 2005 air quality became a divisive subject in Dutch environmental discourse, and protagonists soon returned to their old positions. The broad consensus concerning an approach that would bring about ecological protection and economic welfare broke down, with the pro-health camp arguing that the Government was blind to the effects of its pro-asphalt policies for public health, and the pro-infrastructure camp claiming that environmental standards halted economic progress. The Government, meanwhile, pleaded that it was doing 'everything feasible'. This was an implicit appeal to the weak ecological modernistic consensus in which one chooses environmental protection to the extent that it is economically feasible and reasonable. Nevertheless, the appeal failed to bring the parties together.

Ultimately, ecological modernisation though still proved to be the key to a solution. The programmatic approach which eventually ended the deadlock bore the hallmarks of weak ecological modernisation, since it considered infrastructure and environment interests to be able to coexist peacefully if managed correctly. The programmatic approach stimulated scientific research, and the involvement of stakeholders and lower administrative bodies. It amounted to a preventative approach, but not one that would halt infrastructure development.

With this approach, the Netherlands returned to a weak variation of ecological modernisation, a discourse in which difficult choices are avoided. Apparently, the Dutch cannot do away with environmental principles, but at the same time are not willing to make concrete sacrifices for them, at least not when mobility is concerned. The discourse of ecological modernisation successfully defused the seemingly insurmountable dichotomy between ecology and economy in the air quality case but it did not have the rejuvenating ambition anymore it had in the 1980s, and currently resembles a forced marriage between ecology and economy.

When we review the domain of air quality and ecological modernisation, the extent to which the Dutch eco-modernist discourse is essentially conservative becomes obvious. Principles are strongly espoused, but, in practice, compromises are made, and the adage that no one should be asked to do too much is adhered to. This chasm between principles and practical policy leads to problems when pressure groups begin to resist the Government's conservative policies on the basis of these idealistic principles. In the air quality clash, the Government was reprimanded by the judiciary, it pointed out that the Government had made binding agreements in Europe. The Government's response to the environmental movement was to rescind subsidies for groups like Milieudefensie. This indicates that Dutch ecological modernisation is mostly a discourse concerning accommodation and incorporation rather than an ambitious environmental dialogue. However, the rise of the European Commission and the judiciary as active players in the environmental arena means that the Dutch Government may be confronted unexpectedly with the principles it has preached. As the air quality clash has shown, this may result in environmental considerations replacing economic ones, albeit temporarily.

The air quality clash and precaution

The hypothesis formulated at the start of this inquiry was that the air quality clash was an indication of change within the legal order, making it less concerned with compensation and more with the prevention of damage. The analysis of the air quality clash indicated some support for this hypothesis, but one could not state unequivocally that a precautionary approach is rising in this part of the legal order. The case also demonstrated that when precaution tended to become dominant, countervailing forces arose that undercut precaution as the new fundament of the legal order. Different actors propagated or argued against precaution, and the question as to whether it is gaining ground can only be answered by examining the attitudes of different actors towards this approach.

The Dutch Government was generally not susceptible to embracing a precautionary legality. The policy discourse on ecological modernisation actually ensured that the legality of risk and compensation remained dominant. The reason is that the Dutch Government always prioritised cost-effectiveness in its environmental policy, and this is a crucial prerequisite within the weak ecological modernistic discourse. The precautionary principle is highlighted, but if costs became too high, the Dutch Government did not implement a strictly precautionary approach. When the Air Quality Directives had to be implemented in Dutch law, Dutch Minister Jan Pronk wrote a letter to European Commissioner Margot Wallström, arguing that strict implementation would not be cost-effective. Pronk was and is known as a supporter of the precautionary principle, but even he assumed that an unequal cost-benefit ratio would be a good reason not to fully implement the Directive.

Up to the year 2000, European policy makers were more inclined to take a precautionary stance, and the European Air Quality Directives from the 1990s displayed signs of the legality of precaution. They were implemented with limited scientific support, they were broad in scope, and they involved the citizens explicitly for the purpose of maintaining pressure on the Member States to reach and adhere to the required standards. The provision of information was aimed especially at creating 'pressure from below' on Member States to make sure that they implemented the provisions correctly.

The Council of State was also more amenable to a precautionary approach than the Dutch Government. Even though indications often remained implicit, there were signs that a precautionary legality gained ground. The Council of State demanded much more research into the consequences of air quality-related decisions than it had in the past. In addition, the Council of State acted as a champion of the strict link between air quality standards and administrative decisions.

The Council of State is not only the highest administrative judge but is also the highest advisory council of the State. In both capacities, it was critical of any policy that would jeopardise environmental legal guarantees: such as the Urgent Law on Road Expansion. Moreover, the Council of State Advisory Section criticised the Government, claiming that policy was one-sided in the orientation of economic growth to the expense of sustainability.

In this file, the Council of State proved to be a tenacious opponent of the Government. The position of the Council of State was reinforced because the standards were European in origin. European law is directly applicable in Dutch law, and the Council made the argument that laws may be changed in 'The Hague', but that did not mean that European standards could be set aside in litigation if they were appealed to by citizens. The Council of State pointed out that the Dutch Government had bound itself to agreements made in Brussels. The storyline that the Government had bound itself to international legal rules and that therefore it cannot solve the situation by changing national policy is very important when it comes to precaution. It indicated that the judiciary would not hesitate to overrule national policy when it did not conform to European agreements. Since 'Europe' tended to be more precautionary than the Dutch Government this legal route became an avenue through which precautionary considerations entered the Dutch legal order.

As can be expected, the pro-health camp used the most precaution-based arguments, the most innovative of which was the use of an explicitly legal storyline in the air quality clash. Milieudefensie in particular used the verdicts of the Council of State, and began to campaign using the narrative that the rights of Dutch citizens were being violated, because the administration refused to act in relation to the poor quality of air in dirty streets. In court, Milieudefensie demanded that the speed limit be lowered on tangential roads, and that the dirtiest streets be closed to traffic. Milieudefensie lost those court cases, but managed to make its point: clean air is a human right. With this argument, Milieudefensie also urged citizens to take the administration to court. It played a background role itself, while local pressure groups argued that air quality was being jeopardised by a certain project. In order to facilitate the litigants, Milieudefensie developed a 'do it yourself' package that detailed step by step the way to initiate proceedings.

The success of this legal strategy was one of the most stimulating socio-legal conclusions, because before the onset of the air quality clash, the argument that a clean environment is a human right was made in academia, but never by actors in a concrete environmental conflict. During the air quality clash it managed to mobilise a significant amount of people, at least significant enough to cause a surge in environmental litigation.

It is not surprising that the pro-infrastructure camp was a staunch opponent of a precautionary approach. This camp used the standard argument that legislation had negative economic consequences, and that the issue of air quality should not block the Netherlands in terms of economic development. Indeed, the strict interpretation by the Council of State threatened the prevalence of environmental interests. The storyline presented by the pro-infrastructure camp gained more adherence as the air quality clash dragged on. In general, the Government adopted a similar stance, and air quality gradually became seen as a problem of inflexible regulation. The programmatic approach was indeed more flexible than the old link, and in that sense the pro-infrastructure camp obtained what it had demanded. However, the programmatic approach also contained measures for clean air that earlier Cabinets had never intended to take, and in this sense it was a compromise. In terms of the legality of precaution, however, the plan should be seen as belonging to the legality of risk and compensation, with its detailed balance between measures and harmful projects.

It is remarkable that the rise of a precautionary legality was also halted in Europe. In this arena, lobby groups comprising project developers and other economically important sectors successfully petitioned for flexibility. Indeed, the Air Quality Directive of 2008 granted more flexibility, as it foresaw the possibility of postponing the standards, and the Netherlands was quick to make use of these provisions.

The conclusion, therefore, is that although air quality regulation was indeed a product of a developing legality of precaution, this legality did not yet gained the upper hand. It had not emerged without friction, and the countervailing powers that were mobilised against it were powerful on a national as well as a European level.

The air quality clash: summary conclusions

How and why did air quality emerge suddenly as a problematic issue in the Netherlands? This study shows that EU air quality regulation strongly contributed to the outbreak and continuation of the Dutch air quality clash. It also shows that the interplay between the EU regulation and Dutch attempts to influence it are highly important. From a long term historical perspective we may even say that the Dutch air quality clash is the result of Dutch influence at the EU level at an early stage. The strict EU norms would not have come about if the Dutch had not involved the WHO in order to promote scientifically underpinned norms. That the norms are as strict as they are cannot be solely attributed to Dutch influence. It is also the result of the influence exerted by another member state, the United Kingdom.

Another characteristic of Dutch environmental law played a crucial role as well. This is 'the link' between environmental law and permits for infrastructure development. Because of this link infrastructure could not be developed in situations where environmental norms are already exceeded. Because the EU standards were strict, they were transgressed in large parts of the Netherlands. Hence, the development of infrastructure came to a halt in major parts of the country. The fact that the Council of State Administrative Jurisdiction Division operated in quite a formalistic way meant that there was no room to consider the social impact of cancelling permits. However the Directives and the link did not give it much discretionary room to consider other options than to cancel permits because of noncompliance with the air quality regulation.

The clash can be further understood as an episode in Dutch politics where the consensus of weak ecological modernisation was dissolved. This happened first because the Balkenende cabinets from 2001 broke away from the consensus by prioritising infrastructure development. This political change infuriated the environmental movement, which then made the strategic move to turn to the judiciary and requested a stop to the one-sided developmental policies. As stated above, the legal conditions favoured their attempts and during the clash they found an ally in the Council of State. This even led to the storyline of clean air as a human right.

The fact that the pro-health camp found an ally in the Council of State prevented the Dutch Government from quickly finding an administrative solution to the situation where infrastructure permits were cancelled on an unprecedented scale. It was forced to propose a new air quality law to Parliament which contained a novel regulatory instrument, the programmatic approach.

In Parliament the pro-health camp initially had everything going for it. The Council of State Administrative Jurisdiction Division kept annulling projects, the Government was forced to make ever more funds available for air quality and the Council of State Advisory Division issued strongly negative opinions on the Governments' attempts to quickly patch up the Dutch air quality regulation. However, the pro health camp became confronted with an increasingly strong pro-infrastructure camp that made the counter-claim that the real problem with air quality was not its effects on public health, but that the regulation of it had blocked the country to further development. The storyline that it put forward was that the regulation itself was to blame and that billions of dollars and as much as 100.000 jobs were at stake.

The clash could only come to an end in the political arena by rekindling the eco-modernist consensus in the Air quality law of 2007. The consensual, managerial, technical and bureaucratic instrument that was developed out of this law, the NSL, first led the EU Commission to extend the Dutch some postponement of the standards. Second, it was accepted by the Council of State as an instrument that facilitated both air quality improvement and infrastructure development. This research shows – to my knowledge for the first time – a case in which the Council of State as a two pronged institute was able to keep policy formation deadlocked for several years. One prong – the Administrative Jurisdiction Division – kept annulling permits with reference to EU air quality standards while the other prong – the Advisory Division – kept reminding the government that policy adaptation would have to be in line with EU regulation. In the air quality clash, the Council of State acted as a fellow policy maker.

A third way to understand the clash is to look at the level to which the legality of risk and compensation and the legality of precaution can be found in the handling of the air quality clash in the legal and the political arena. During the clash strict standards based on uncertain scientific knowledge blocked important developments in the Netherlands. Only by balancing the health and environment interests with the interests of development and mobility could the clash be ended.

The swift promulgation of standards in the European Union, the quality of the clean-air-is-a-right storyline as a mobilisation device and the attachment of the Council of State Administrative Jurisdiction Division to the link are indications that a precautionary legality was cautiously on the rise. However, the strong backlash against this temporary legal preference of environmental interests over economic ones, in the Netherlands as well as in Europe, pointed out that the legality of precaution had a long way to go still before it could be considered to reach a dominant position within the Dutch legal order.

Appendix 1
List of respondents interviewed

Wijnand Duyvendak (Former Dutch politician in the Green Left political party) 23-04, 2012

Hans Eerens (RIVM) 09-06, 2009

Marjan van Giezen (VROM) 25-06 2009

Dick van den Hout (TNO, consultant to the EU Commission) 29-10, 2009

Benno Jimmink (RIVM) (e mail correspondence) 31-05, 2013

Klaas Krijgsheld (VROM) 27-01 2010

Niels Koeman (Legal Scholar) 29-12 2011

Rob Maas (RIVM) 09-06, 2009

Petra Mak (VROM) 27-01 2010

Richard Mills (former head of air and environmental quality division of the UK environmental Ministry DEFRA) 14-07 2011

Jan Pronk (Former Minister of VROM, from 1998 to August 2002) 03-12, 2012

Joris Wijnhoven (Former campaign leader of Milieudefensie) 28-04 2010

Martin Williams (Scientist, part of the UK negotiation team during the EU negotiations on Directive 99/30) 23-08, 2011

Kees Zoeteman (Former deputy head of the Environmental Directorate of VROM, 17-08 2011, 05-09, 2012.

Appendix 2
Cabinets and Ministers

Cabinet	Cabinet factions	Prime Minister	Main Govt. representative responsible for environment (VOMIL / VROM)	Minister of V&W	Duration	Key event relevant to the air quality clash
Biesheuvel I	KVP VVD ARP CHU DS'70	Barend Biesheuvel (ARP)	Min. Louis Stuyt (KVP)	Min. Willem Drees jr. (DS.70)	July 1971 – Aug. 1972	Release of the Urgent Memorandum
Biesheuvel II	KVP VVD ARP CHU	Barend Biesheuvel	Min. Louis Stuyt	Berend Jan Udink (CHU)	Aug. 1972 – May 1973	none
Den Uyl	PvdA, KVP, ARP, PPR, D66	Joop Den Uyl (PvdA)	Min. Irene Vorrink (PvdA)	Tjerk Westerterp (KVP)	May 1973 – Dec. 1977	Memorandum on Ambient Environmental Standards (1976)
Van Agt I	CDA, VVD	Dries Van Agt (CDA)	Min. Leendert Ginjaar (VVD)	Danny Tuinman (VVD)	Dec. 1977– Sept. 1981	Emergence of the Noise Abatement Act Stb. 1979, 99
Van Agt II	CDA, PvdA, D66	Dries Van Agt	Scrt. of State Ineke Lambers-Hacquebard (D66)	Min. Jaap van der Doef (PvdA)	Sept. 1982 – May 1982	Proposal to establish a system of air quality standards
Van Agt III	CDA, D66	Dries Van Agt	Scrt. of State Ineke Lambers-Hacquebard	Min. Henk Zeevalking (D66)	May 1982 – Nov. 1982	none
Lubbers I	CDA VVD	Ruud Lubbers (CDA)	Min. Pieter Winsemius (VVD)	Min. Neelie Smit Kroes (VVD)	Nov. 1982 – July 1986	Foundation of VROM Ministry
Lubbers II	CDA VVD	Ruud Lubbers	Min. Ed Nijpels (VVD)	Min. Neelie Smit Kroes	July 1986 – Nov. 1989	Emergence of the NMP
Lubbers III	CDA PvdA	Ruud Lubbers	Min. Hans Alders (PvdA)	Min. Hanja Maij Weggen (CDA)	Nov. 1989 – Aug. 1994	Release of worrying PM medical research

Continued

Cabinet	Cabinet factions	Prime Minister	Main Govt. representative responsible for environment (VOMIL / VROM)	Minister of V&W	Duration	Key event relevant to the air quality clash
Kok I (Purple Cabinet)	PvdA VVD D66	Wim Kok (PvdA)	Min. Margreeth de Boer (PvdA)	Min. Annemarie Jorritsma (VVD)	Aug. 1994 – Aug. 1998	Proposal of and common position on EU Air Quality Directives
Kok II (Purple Cabinet)	PvdA VVD D66	Wim Kok	Min. Jan Pronk (PvdA)	Min. Tineke Netelenbos (PvdA)	Aug. 1998 – July 2002	Implementation of EU Air Quality Directives in AQO 2001
Balkenende I	CDA VVD LPF	Jan Peter Balkenende (CDA)	Scrt. of State Pieter van Geel (CDA)	Min. Roelf de Boer (LPF)	July 2002 – May 2003	Budget cuts in VROM Ministry Emergency Law on Road Expansion
Balkenende II	CDA VVD D66	Jan Peter Balkenende	Scrt. of State Pieter van Geel	Min. Karla Peijs (CDA)	May 2003 – July 2006	Emergence of the air quality clash in NL.
Balkenende III	CDA VVD D66	Jan Peter Balkenende	Scrt. of State Pieter van Geel	Min. Karla Peijs	July 2006 – Feb. 2007	Oct 2006: Air Quality Law adopted by Second Chamber
Balkenende IV	CDA PvdA CU	Jan Peter Balkende	Min. Jacqueline Cramer (PvdA)	Min. Camiel Eurlings (CDA)	Feb. 2007 – Oct. 2010	Verdict A4 Burgerveen Leiden: Adoption of Air Quality Law in Parliament and release of the NSL

Appendix 3
Dutch Political parties (in the air quality clash)

CDA (Christendemocratisch Appèl), Christian Democratic appeal: the biggest confessional political party in the Netherlands. The CDA is the result of a merger of various confessional political parties, namely the KVP, ARP and CHU. In the Cabinet during the years of the air quality clash. Centrist.

CU (ChristenUnie) Smaller Christian party, slightly to the left of the CDA in economic and environmental issues

D66 (Democraten '66) Social liberal party founded in 1966, progressive on social issues and the environment, leaning towards free market liberalism in economic issues.

GROENLINKS

GL (GroenLinks) the 'Green Left', is a smaller, ecological party, progressive in social economic issues and the most environmentalist minded of the Dutch political parties.

LPF (Lijst Pim Fortuijn) the 'List Pim Fortuijn' is a party founded by and inspired by the ideas of its late political leader Pim Fortuijn who was assassinated in 2002. Its ideas may best be described as populist. It is the least environmentally sensitive party before and during the air quality clash.

PvdA (Partij van de Arbeid) The Dutch labour party is the biggest social democratic party in the Netherlands. It was the biggest party in the Government during the years of the Kok Cabinets from 1994 to 2002. Entered the Cabinet again during the later years of the air quality clash is 2007.

SP (Socialistische Partij) The Socialist Party is a smaller party to the left of the PvdA. Despite its respectable size it has as of yet not participated in any Cabinet. It profiles itself mainly on economic issues, but takes the environmental side often as well.

VVD (Volkspartij voor Vrijheid en Democratie) People's Party for Freedom en Democracy. The VVD is a liberal / conservative party that endorses liberal economic policy. It was never known for its environmentalist credentials, but has delivered a highly important Ministers of VROM such as Pieter Winsemius and Ed Nijpels. They are responsible for the introduction of ecological modernisation is the Netherlands. During the air quality clash it belongs to the pro-infrastructure camp. It is frequently in Government.

Appendix 4
European Political parties

ALDE (Alliance of Liberals and Democrats for Europe) The ALDE is an alliance of political parties advocating free markets and the protection of civil liberties. The Dutch political parties VVD and D66 are both part of the ALDE group. Its political position is centrist.

EPP (European People's Party) The EPP is formed by Christian democratic political parties in Europe. It represents the Christian democratic wing of the European Parliament. It is generally considered conservative. The CDA is a member of the EPP.

GUE / NGL European United Left/Nordic Green Left A group of political parties with a socialist or (post) communist orientation. The Dutch political party SP is a member.

S&D (Progressive Alliance of Socialists and Democrats) The block consisting of social democratic parties in Europe. It is the second largest political party in Europe after the EPP. The Dutch PvdA is a member of S&D.

The Greens / EFA (The Greens / European Free Alliance) An alliance of ecological political parties, mostly on the left of the political spectrum. The Dutch political party GroenLinks is a member.

Appendix 5
List of acronyms

AbRVS	(Afdeling Bestuursrecht Raad van State) Council of State Administrative Jurisdiction Division
Aedes	(Vereniging van Woningcorporaties) Association of Housing Corporations
AEA T	AEA Technology (global sustainability consultancy firm)
ACS	American Cancer Society
ALARA	As Low As can Reasonably be Achieved
ALDE	Alliance of Liberals and Democrats for Europe, see annex 4 European Political Parties
AMvB	(Algemene Maatregel van Bestuur) General Administrative Order
ANWB	(Algemene Nederlandse Wielrijders Bond) Dutch Motorist Association
ARP	(Antirevolutionaire Partij), Anti Revolutionary Party, Dutch Christian Political party, merged in the CDA in 1980.
AVV	(Adviesdienst Verkeer en Vervoer) Advisory Service Traffic and Transport
AQO	(Besluit Luchtkwaliteit) Air Quality Order
Bouwend Nederland	Dutch Builders Association
BS	Black Smoke
CAFE	Clean Air For Europe
CDA	(Christen Democratisch Appèl), Christian Democratic Appeal, Dutch Christian democratic party (see annex 3 Dutch political parties)
CHU	(Christelijk Historische Unie) Christian Historical Union, merged in the CDA in 1980
CO	Carbon Monoxide
CO_2	Carbon dioxide
COMEAP	Committee on the Medical Effects of Air Pollutants
CPB	(Centraal Plan Bureau) Central Bureau of Statistics
CRMH	(Centrale Raad voor Milieuhygiëne) Central Council for Environmental Hygiene)
D66	(Democraten '66) Democrats 66, see political parties

Continued

DETR	Department of the Environment, Transport and the Regions (UK Ministry for the Environment and Transport)
DG	Directorate General
DGMH	(Directoraat Generaal Milieuhygiëne) Directorate General for Environmental Hygiene
DOE	Department of the Environment (UK Environment Ministry)
DOT	Department of Transport (UK Transport Ministry)
DROM	(Deregulering Ruimtelijke Ordening en Milieu) Deregulation Spatial Planning and Environment
DS'70	(Democratisch Socialisten '70) Democratic Socialists, Dutch political party
EAP	Environmental Action Programme (EU)
ECN	(Energieonderzoek Centrum Nederland) Dutch Energy Research Center
ECU	European Currency Unit, the precursor of the Euro
EEA	European Environmental Agency
EEB	European Environmental Bureau
EM	Explanatory Memorandum
ENVI	Environmental Committee
EP	European Parliament
EPA	Environmental Protection Agency (US)
EPAQS	Expert Panel on Air Quality Standards
EPP	European Peoples Party (see annex 3 European Political parties)
EU	European Union
EUROPIA	European Petroleum Industries Association
EVO	(Eigen Vervoerders Organisatie) Organisation of Enterprises in Logistics
EZ	(Economische Zaken) Ministry of Economic Affairs
FES	(Fonds Economische Structuurversterking) Structural Economic Strengthening Fund
GAOSD	(AMvB Gevoelige Bestemmingen) General Administrative Order on Sensitive Destinations
GGD	(Gemeentelijke Gezondheidsdienst) Municipal Health Service
GL	(GroenLinks) Green Left (see annex 3 Dutch Political parties)
GUE/NGL	European United Left/Nordic Green Left (see annex 4 European Political parties)
HEI	Health Effects Institute
IMP	(Indicatief Meerjarenplan) Indicative Multi-Year Plan
IMP M	(Indicatief Meerjarenplan Milieubeheer) Indicative Multi-Year Plan on Environmental Management
KEMA	(Keuringsinstituut Electrische Materialen) Certification Institute Electrical Materials

Continued

KVP	(Katholieke Volkspartij) Catholic Peoples Party, Dutch political party mergerd in the CDA in 1980.
KWS 2000	(KoolWaterStoffen 2000 convenant) Hydrocarbons 2000 covenant
LPF	(Lijst Pim Fortuyn) Pim Fortuyn List. Dutch political party on the right of the political spectrum. Named after its assassinated leader Pim Fortuyn.
LRTAP Convention	Convention on Long Range Transboundary Air Pollution
MAAPE	Medical Aspects of Air Pollution Episodes
MEP	Member of the European Parliament
MIT	(Meerjarenplan Insfrastructuur en Transport) Multi Year Programme Infrastructure and Transport
MKB-Nederland	(Midden-en Klein bedrijf Nederland) Dutch association of small and medium sized enterprises.
MNP	(Milieu en Natuur Plan Bureau) Environment and Nature Planning Bureau
MP	Member of Parliament
Mug/m^3	Micrograms per cubic meter
NGO	Non-Governmental Organisation
NH_3	Ammonium
NMP	(Nationaal Milieubeleidsplan) National Environmental Policy Plan, in English literature known as the NEPP
OECD	Organisation of Economic Cooperation and Development
NEPROM	(Vereniging voor Nederlandse Projectontwikkelingsmaatschappijen) Dutch Project Developers Association
NO_2	Nitrogen dioxide
NO_x	Various oxides of nitrogen
NSL	(Nationaal Samenwerkingsplan Luchtkwaliteit) National Cooperation Plan Air Quality
NVB	(Nederlandse Vereniging van Bouwondernemers) Dutch Builders Association
NVM	(Nederlandse Vereniging van Makelaars) Dutch Association of Real Estate Agents
NVVP	(Nationaal Verkeer en Vervoersplan) National Traffic and Transport Plan
PAH	PolyAromatyc Hydrocarbon.
PBL	(Planbureau voor de Leefomgeving) Planning Bureau for the Living Environment
PIM	(Project Integratie Milieubeleid), Project Integration Environmental Policy
PPR	(Politieke Partij Radicalen) Radical Political Party, Dutch political party merged in GroenLinks in 1990

Continued

OECD	Organization for Economic Co-operation and Development
PM10	Particulate Matter 10 micron
PM2.5	Particulate Matter 2.5 micron
PMR	Project Mainport Rotterdam
PvdA	(Partij van de Arbeid), Labour Party, Dutch social democratic party (see annex 3 Dutch Political parties)
QUARG	Quality of Urban Air Review Group
RARO	(Raad voor de Ruimtelijke Ordening) Council for Spatial Planning
RIVM	(Rijksinstituut voor Volksgezondheid en Milieuhygiëne), National Institute for Public Health and Environmental Hygiene
ROS	Eeactive Oxygen Species
S&D	Progressive Alliance of Socialists and Democrats (see annex 4 European political parties)
SG	Steering Group (CAFE)
SER	(Sociaal Economische Raad) Social Economic Council
SNM	(Stichting Natuur en Milieu) Foundation for Nature and Environment
SO_2	(Zwaveldioxide) sulfur dioxide
SP	(Socialistische Partij) Socialist Party, Dutch party on the left of the political spectrum (see annex 3 Dutch Political parties)
Stb.	(Staatsblad) Law Gazette
SVV2	(Tweede Structuurschema Verkeer en Vervoer) Second Structural Scheme Traffic and Transport
SWAB	(Samen Werken Aan Bereikbaarheid) Memorandum on mutual cooperation on accessibility
TAG	Technical Analysis Group (CAFE)
The Greens / EFA	The Greens / European Free Alliance (see annex 4 European Political Parties)
TLN	(Transport en Logistiek Nederland) Transport and Logistics Netherlands, Dutch organisation of transporters.
TNO	(Nederlandse Organisatie voor Toegepast Natuurwetenschappelijk Onderzoek) Dutch Organisation for Applied Scientific Reseach.
TOPS	(Tijdelijk Onderhandelingsplatform Schiphol) Temporary Negotiation Platform Schiphol
TSP	Total Suspended Particulates
UNECE	United Nations Economic Commission for Europe
UNICE	Union of Industrial and Employers' Confederations of Europe
US EPA	US Environmental Protection Agency
V&W	(Ministerie Verkeer en Waterstaat) Ministry of Transport and Water Management

Continued

VBO	(Vereniging Bemiddelaars Onroerende zaken) Real Estate Agents' Association
VNG	(Vereniging Nederlandse Gemeenten) Association of Dutch Municipalities
VNO-NCW	(Verbond van Nederlandse Ondernemingen – Nederlands Christelijk Werkgeversverbond) Alliance of Dutch Enterprises – Dutch Christian Employers Union: Dutch Employers Union.
VCRMH	Voorlopige Centrale Raad voor de Milieuhygiëne) Preliminary Central Council for Environmental Hygiene.
VINO/VINEX	(Vierde Nota Ruimte Ordening, plus de Vierde Nota Ruimtelijke Ordening Extra) Fourth Memorandum Spatial Planning, plus the Fourth Memorandum Spatial Planning Extra
VOMIL	(Volksgezondheid en Milieu) Ministry of Public Health and the environment
VROM	(Ministerie van Volkshuisvesting, Ruimtelijke Ordening en Milieu), Ministry of Housing Spatial Planning and the Environment
VVD	Volkspartij voor Vrijheid en Democratie (see annex 3 Dutch Political parties).
WG Imp	Working Group on Implementation (CAFE)
WG PM	Working Group on Particulate Matter (CAFE)
WG TSP	Working Group on Target Setting and Policy Assessment (CAFE)
WHO	World Health Organisation

LITERATURE

Aarts, F.J.H., P.H.N.Vliet van, N.A.H. Janssen, H. Harssema, B. Brunekreef (1999). *Nader onderzoek naar effecten van verkeersgerelateerde luchtverontreiniging op de luchtwegen van kinderen wonend nabij snelwegen. EOH rapport nr 1999-486.* Wageningen: Environmental and Occupational Health Group.

Algemene Rekenkamer (2004). *Aandacht voor financiële gevolgen van Europees beleid.* Kamerstukken II 2004-2004, 29 751, nr. 2. Den Haag: Algemene Rekenkamer.

Andersen, M.S., & Massa, I. (2000). Ecological modernization, Origins, Dilemma's and Future Directions. *Journal of Environmental Policy and Planning Special Issue: Ecological Modernization*, 2(4), pp. 337-345.

Andeweg, R. I., R.B. Irwin (2005). *Governance and Politics of the Netherlands* (2nd edition). Houndmills / New York: Palgrave McMillan.

Andrews, T. (2012). What is Social Constructionism? *The Grounded Theory Review: An International Journal, 11 (1)*, pp. 39-46.

Annema, J. A., H. Booij, J.M. Hesse, A. Van der Meulen, W. Slooff (1994). *Basisdocument Fijnstof.* RIVM, Bilthoven.

Arentsen, M.J., J.Th.A. Bressers, P.J. Klok (1993). Van urgentienota naar NMP en Verder, In: *Raad voor het Milieubeheer, Twintig jaar Nederlands milieubeleid.* Den Haag: Delwel.

Arnoldussen, N.T. (2009). Precautionary Logic and a Policy of Moderation. *Erasmus Law Review, 2(2)*, pp. 259-285.

Arnoldussen N.T. (2011). The Limits of Numbers: Impact Assessments and Increased Uncertainty. In: J. Blad, M. Hildebrandt, N. Rozemond, M. Schuilenburg & P. Van Calster (eds.). *Governing Security under the Rule of Law.* OMV Reeks (2011), pp. 235-257. The Hague: Eleven International Publishing.

Arts, B.J.M., C. Dieperink, J.D. Liefferink (2002). The Netherlands: international innovator or self-interested negotiator? In: Driessen, P.P.J. & P. Glasbergen (eds.). *Greening society: the paradigm shift in Dutch environmental politics.* Dordrecht: Kluwer Academic Publisher.

AVV (2006). *Snelheidsverlaging en compact rijden op ringen van grote steden.* Den Haag: Adviesdienst Verkeer en Vervoer, Rijkswaterstaat.

Ayres et al. (2008). Evaluating the Toxicity of Airborne Particulate Matter and Nanoparticles by Measuring Oxidative Stress Potential – A Workshop Report and Consensus Statement. *Inhalation Toxicology*, 20(1), pp. 75-99.

Backes, C., T. Van Nieuwerburgh, R. Koelemeijer. (2005). Transformation of the first Daughter Directive on air quality in several EU Member States and its application in practice. *European Environmental Law Review*, 6, pp. 157-164.

Backes, C. W. (2006a). *Internationale Vergelijking EU richtlijnen Luchtkwaliteit.* Utrecht: Centrum voor Omgevingsrecht en Beleid/NILOS,

Backes, Ch.W. (2006b). Het dossier luchtkwaliteit: Anders verder! *Bouwrecht*, 2, pp. 88-102.

Beck, U. (1986). *Risikogesellschaft. Auf dem Weg in eine andere Moderne.* Frankfurt am Main, Germany: Suhrkamp.

Beck, U. (1994). The reinvention of politics. In: U. Beck, A. Giddens & S. Lash (eds.), *Reflexive Modernization, Politics, Tradition and Aesthetics in the modern social order.* Cambridge UK: Polity Press.

Beek, P. van, H. Stoelhorst, P. Havermars (2006). *Snelheidsverlaging voor luchtkwaliteit*, Contribution to the Colloquium Vervoersplanologisch Speurwerk 2006, 23-24 november 2006, Amsterdam.

Beelen, R. et al. (2008). Long-Term Effects of Traffic-Related Air Pollution on Mortality in a Dutch Cohort (NLCS-AIR Study). *Environmental Health Perspective*, 116(2), pp. 196-202.

Bennett, G. (1991). *Air Pollution Control in the European Community: the Implementation of the EC Directives in the Twelve Member States.* London: Graham & Trotman.

Berger, P.L. & T. Luckmann (1966). *The Social Construction of Reality: A Treatise in the Sociology of Knowledge.* Garden City, New York: Anchor Books.

Best, J. (2004). Theoretical Issues in the Study of Social Problems and Deviance. In: G. Ritzer (ed.). *Handbook of Social Problems A Comparative International Perspective.* London, UK, New Delhi, India: Sage Publications

Best, J. (2008). *Social Problems.* London, UK, New York, NY: WW Norton & Company.

Blühdorn, I. (2000). Ecological modernization and post ecologist politics. In: G. Spaargaren, A.P.J. Mol & F.H. Buttel (eds.). *Environment and Global Modernity.* London, UK: Sage Publications.

Blühdorn, I. & I. Welsh (2007). Eco-politics beyond the paradigm of sustainability: A conceptual framework and research agenda. *Environmental Politics*, 16(2), pp. 185-205.

Boender, K. (1985). *Milieuprotest in Rijnmond, sociologische analyse van milieusolidariteit onder elites en publiek.* PhD. thesis. Erasmus University Rotterdam.

Bomberg, E. & C. Burns (1999). The environment committee of the European Parliament: New powers, old problems. *Environmental Politics*, 8(4), pp. 174-179.

Borgers, H. (2012). *Duurzaam handelen. Een onderzoek naar een normatieve grondslag van het milieurecht.* Den Haag: SDU.

Borman (2003). Het ambacht: De voorhang van het besluit luchtkwaliteit: een wetgevingssoap in 25 afleveringen. *RegelMaat*, 2003, 6, pp. 244-254.

Börzel, T. (2002). Pace setting, foot dragging, and fence sitting, Member State's responses to environmental regulation. *Journal of Common Market Studies*, 40(2) pp. 193-214.

Bouwend Nederland et al. Press release 28 June 2011. (Website: www.ridder.punt.nl).

Bramwell, A. (1989). Ecology in the 20th Century, a History. London: Yale University Press.

Brandt, A.M. (1993). The Cigarette, risk, and American Culture. In: E.J. Burger (ed.). *Risk.* Michigan: The University of Michigan Press.

Bree, van L., F. Cassee. (2000). *Toxicity of Ambient Air PM10, A critical review of potentially causative PM properties and mechanisms associated with health effects.* Bilthoven: RIVM.

Breeman, G.E. & A. Timmermans (2008). *Politiek van de aandacht voor milieubeleid; een onderzoek naar maatschappelijke dynamiek, politieke agendavorming en prioriteiten in het Nederlandse milieubeleid*, WOT Rapport. Wageningen: Universiteit Wageningen.

Bressers, H. P., L. Plettenburg (1995). *Environmental Policy Capacity Building in the Netherlands.* Enschede: Centre For Clean Technology and Environmental Policies.

Brignell, J. (2004). *The Epidemiologists: Have They Got Scares for You!* London: Brignell Associates.

Brimblecombe, P. (1987). *The Big Smoke a history of air pollution in London since medieval times.* London: Methuen.

Broek, van den, J., Smit, R. (2005). Nederland hoeft niet op slot. *Forum*, 30-06, 2005.

Brokking, P. (2001). Moving with the Times: An Examination of Ten Years of Dutch Transport Planning. *Journal of Environmental Policy & Planning*, 3(1), pp. 31-42.

Brown, W. (1994). Dying From Too Much Dust. *New Scientist*, March 12, pp. 12-13.

Brunekreef, B., D.W. Dockery, F.E. Speizer, J.H. Ware, J.D. Spengler, B.G. Ferris (1989). *Home dampness and respiratory morbidity in children.* Am Rev Respir Dis. 140(5), pp. 1363-1367.

Brunekreef B, P.L. Kinney J.H. Ware, D. Dockery, F.E. Speizer, J.D. Spengler, B.G. Ferris Jr. (1991). Sensitive subgroups and normal variation in pulmonary function response to air pollution episodes. *Environ Health Perspect.* 1991 Jan. 90, pp. 189-193.

Brunekreef, B., N.A. Janssen, J. de Hartog, H. Harssema, M. Knape, & P. van Vliet (1997). Air pollution from truck traffic and lung function in children living near motorways. *Epidemiology*, 1997, 8, pp. 298-303.

Buijsman, E. (2003). *Er zij een meetnet, een geïllustreerde geschiedenis van het luchtmeetnet van het RIVM.* Bilthoven: RIVM.

Buijsman, E. (2007). *Een boekje open over fijnstof.* Houten: Tinsentiep.

Buijsman, E. (2009). Meten waar de Mensen zijn deel 5. *Lucht*, June 2009, pp. 12-16.

Burgess, A. (2004). *Cellular Phones, Public Fears, and a Culture of Precaution.* Cambridge: Cambridge University Press.

Buringh, E., A. Opperhuizen (2002a). *On health risks of ambient PM in the Netherlands.* Executive Summary. Bilthoven: RIVM.

Buringh, E., A. Opperhuizen (2002b). '*On health risks of ambient PM in the Netherlands.* Full Report, Bilthoven: RIVM.

Burns, C. (2005). The European Parliament, The European Union's Environmental Champion? In: A. Jordan (ed.). *Environmental Policy in the European Union, actors institutions and processes* (2nd edition). London: Earthscan.

Buttel, F.H. (2000a). Ecological Modernization as Social Theory. *Geoforum*, 31, pp. 57-65.

Buttel, F.H. (2000b). Classical theory and Contemporary Environmental Sociology. In: G. Spaargaren, A. Mol, F.H. Buttel (eds.). *Environment and Global Modernity*, London UK, Thousand Oaks CA: SAGE.

Carter, N. (2001). *The politics of the environment.* Cambridge: Cambridge University Press.

Cendra de Laragan, J. (2009). Regulatory dilemmas in EC environmental law: the ongoing conflicts between competitiveness and the environment. In: F. Caffagi, H. Muir Watt (eds.). *The regulatory function of European Private Law.* Cheltenham UK, Northampton USA: Edward Elgar.

Christoff, P. (2000). Ecological modernisation, ecological modernities. In: Young, S.C. (ed.). *The Emergence of Ecological Modernisation, Integrating the environment and the economy?* London: Routledge.

Clarenburg et al. (1995). *Deeltjesvormige luchtverontreiniging, inclusief toetsing basisdocument.* Den Haag: Gezondheidsraad.

Cohen, M. J. (2000). 'Ecological modernisation, environmental knowledge and national character: A preliminary analysis of the Netherlands'. *Environmental Politics*, 9(1), pp. 77-106.

Centraal Bureau voor de Statistiek (2010). *Bevolkingstrends 3e kwartaal 2010.* Den Haag: Centraal Bureau voor de Statistiek.

COMEAP (1995) *Non biological particles and health, Executive Summary.* London: HMSO.

Delmas, M., A. Terlaak (2002). Regulatory Commitment to Negotiated Agreements: Evidence for the United States, Germany, The Netherlands and France. *Journal of Comparative Policy Analysis*, 4, pp. 5-29.

Deth, van, J.W., Vis, J.C.P.M. (2006). *Regeren in Nederland*. Assen: Koninklijke Van Gorcum.

Dinkelman, G. (1995). *De Wisselwerking tussen problemen en oplossingen in het Nederlandse luchtverontreinigingsbeleid (1970-1994)*. Utrecht: Van Arkel.

Dockery, D.W. C.A. Pope X. Xu, J.D. Spengler J.H. Ware, M.E. Fay, B.G. Ferris, F.E., Speizer (1993). An association between air pollution and mortality in six US cities. *New England Journal of Medicine*, 329, pp. 1753-1759.

Drupsteen Th.G., Koeman, N.S.J. (eds.). (1996). *Tekst en Commentaar. Wet milieubeheer*. Deventer: Kluwer.

Dryzek, J.S. (2005). *The Politics of the Earth, Environmental Discourses*. Oxford: Oxford University Press.

ECORYS-Kolpron (2002). *Inventariseren/uitdiepen praktijkervaringen Stad & Milieu Eindrapportage*, Rotterdam, 30 oktober 2002.

Edwards, L. (1999). Air Pollution in European Cities, an Overview. In: J. Fenger, O. Hertel, F. Palmgren (eds.). *Urban Air Pollution, European Aspects*. Dordrecht: Kluwer.

Elsom, D. (1999). Development and implementation of strategic frameworks for air quality management in the UK and the European Community. *Journal of Environmental Planning and Management*. 42(1), pp. 103-121.

Elverding Commission (2008). Advies Commissie Versnelling Besluitvorming Infrastructurele Projecten: *'Sneller en Beter'*. April 2008.

EPA (2006). *Review of the national ambient air quality standards for particulate matter, policy assessment of scientific and technical information*. OAQPS Staff Paper, attachment A.

EVO (2005). Press release 22-11 2005.

Ewald, F. (1999). The return of a crafty Genius, An Outline of a Philosophy of Precaution. *Connecticut Insurance Law Journal*, 6(1), pp. 47-80 reprinted in: P. O'Malley (ed.). (2005). *Governing Risks*, Aldershot, UK, Burlington, USA: Ashgate.

Ewald, F. (2002). The return of Descartes's Malicious Demon: An Outline of a Philosophy of Precaution, In: T. Baker and J. Simon (eds.). *Embracing Risk, the Changing Culture of Insurance and responsibility*. Chicago: The University of Chicago Press.

Feinstein, A.R., and R.I. Horowitz (1982). Double standards, scientific methods, and epidemiologic research. *New England Journal of Medicine*, 307, pp. 1611-1617.

Feinstein, A.R. (1988a). Scientific standards in epidemiologic studies of the menace of daily life. *Science*, 242, pp. 1257-1263.

Feinstein, A.R. (1988b). Fraud, Distortion, Delusion, and Consensus: The Problems of Human and Natural Deception in Epidemiologic Science. *The American Journal of Medicine*, 84, pp. 475-478.

Feltz van der, G. (2006). Rechtspraak naar aanleiding van het Besluit luchtkwaliteit. In: S.T. Ramnewash – Oemrawasingh (ed.). *Regeling inzake luchtkwaliteit; Nederland op slot? Verslag van de 86ste ledenvergadering van de Vereniging voor Milieurecht op 2 december 2004*, 2006, 1.

Fisher, D.R. & W.R. Freudenburg (2001). Ecological Modernization and its Critics: Assessing the Past and Looking Towards the Future. *Society and Natural Resources, 14*, pp. 701-709.

Fleurke, F.M. and N.S.J. Koeman (2004). The Impact of the EU Air Quality Standards on the Planning and Authorisation of Large Scale Infrastructure Projects in the Netherlands. *Journal for European Environmental and Planning Law*, vol. 5, pp. 375-383.

Flynn, A., R. Bayliss (1996). Pollution regulation and ecological modernization: The formulation and implementation of best available techniques not entailing excessive costs. *International Planning Studies*, 1(3), pp. 311-329.

Folkert, R.J.M. et al. (2005). *Consequences for the Netherlands of the thematic strategy on air pollution*. Bilthoven: MNP.

Foster, J.B., B. Clark, R. York (2010). *The Ecological Rift: Capitalism's War on the Earth*. New York: Monthly Review Press.

Freestone, D. & E. Hey (1996). Origins and Development of the Precautionary Principle. In: D. Freestone & E. Hey (eds.). *The Precautionary Principle and International Law. The Challenge of Implementation*. The Hague: Kluwer

Furedi, F. (1997). *The Culture of Fear; Why Americans Are Afraid Of The Wrong Things*. New York: Basic Books.

Furedi, F. (2009). The precautionary culture and the rise of possibilistic risk assessment. *Erasmus Law Review 2(2)*, pp. 197-220.

Fürst, V. (2004). Umweltpolitisches Regieren im Wandel Zur Transformation von Staatlichkeit in weltmarktoffenen Staaten. Dissertation. Berlin: Freie Universität Berlin.

Geest, van der, H.J.A.M., T.E.P.A. Lam (2006). Wet Milieubeheer Year? Algemeen, plannen en milieukwaliteitseisen. In: Ch.W. Backes, P.C. Gilhuis, N.S.J. Koeman (eds.). *Milieurecht* (6th edition). Deventer: Kluwer.

Giddens, A. (1994). Living in a post traditional society. In: U, Beck A. Giddens & S. Lash (eds.). *Reflexive Modernization, Politics, Tradition and Aesthetics in the modern social order*. Cambridge UK: Polity Press.

Glasbergen, P., P.P.J. Driessen (2002). The paradigm shift in environmental politics. In: P.P.J. Driessen, P. Glasbergen (eds.). *Greening Society: The Paradigm shift in Dutch environmental politics*. Deventer: Kluwer.

Goldsmith, E. & R. Prescott-Allen. (1972) A Blueprint for Survival. *The Ecologist*, January 1972. Available online: http://www.edwardgoldsmith.org/books/a-blueprint-for-survival/.

Greenbaum, D. (2003). A Historical Perspective on the Regulation of Particles. *Journal of Toxicology and Environmental Health*, Part A: Current Issues, 66:16-19, pp. 1493-1498.

Greenpeace (2012). *Ranking Eastern Chinese Cities by their "Clean Air" Actions*. Website Greenpeace

Hafkamp, W. (1997). Covenants from instrument in environmental policy to implementation tool. In: C. Jeanenraud (ed.). *Environmental Policy Between Regulation and Market*. Neuchatel: Birkhauser.

Hajer, M.A. (1993). Discourse Coalitions and the Institutionalization of Practice, the Case of Acid Rain in Great Britain. In: F. Fisher & J. Forester (eds.). *The argumentative turn in policy analysis and planning*. Durham, London: Duke University Press.

Hajer, M.A. (1995). *The politics of environmental discourse, ecological modernization and the policy process*. Oxford: Clarendon Press.

Hajer, M.A. & Zonneveld, W. (2000). Spatial Planning in the Network Society-Rethinking the Principles of Planning in the Netherlands. *European Planning Studies*, 8(3), pp. 337-355

Hammingh, P., J.P. Beck, W.F. Blom, R.M.M. van den Brink, R.J.M. Folkert, K. Wieringa (2005). *Beoordeling van het Prinsjesdagpakket aanpak luchtkwaliteit 2005*. Bilthoven: MNP.

Hanekamp, J.C., R. Pieterman (2005). Misvattingen en mystificaties rond fijnstof. *Spil*, 6, pp. 15-23.

Hanekamp, J.C. (2015). *Utopia and Gospel: Unearthing the Good News in Precautionary Culture* (Dissertation Tilburg University), Tilburg.

Hanf, K., E. van de Gronden (1998). The Netherlands: Joint regulation and sustainable development, In: K. Hanf, A.I. Jansen (eds.). *Governance and Environmental Quality in Western Europe.* Essex: Addison Wesley Longman.

Haverland, M. (2009). How leader states influence EU policy-making: Analysing the Expert Strategy. *European Integration online Papers (EIoP)*, 13 (Website European Integration Online, last accessed 23-06, 2015).

Hegger C., R. Slob (1999). *Stad en Milieu Overschie, Luchtkwaliteit en Gezondheid.* Rotterdam: Gemeentelijke Gezondheidsdienst voor Rotterdam en Omstreken.

HEI (2000). *Reanalysis of the Harvard Six Cities Study and the American Cancer Society. Study of Particulate Air Pollution and Mortality.* A Special Report of the Institute's Particle Epidemiology Reanalysis Project. Cambridge Ma: Health Effects Institute.

Heijden van der, H.A. (2000). *Tussen Aanpassing en Verzet: Milieubeweging en Milieudiscours.* Amsterdam: AMBO.

Héritier, A., C. Knill, S. Mingers (1996). *Ringing the Changes in Europe, Regulatory Competition and Redefinition on the State. Britain, France, Germany.* New York: Walter de Gruyter.

Hey, C. (2000). *Towards Balancing Participation, A Report on Devolution, Technical Committees and the New Approach in EU Environmental Policies: The Cases of Standardisation, Chemicals Control, IPPC and Clean Air Policies in a Comparative Perspective.* Brussels: EEB.

Hey, C. (2005). EU environmental policies, a short history of the policy strategies. In S. Scheuer (ed.), *'EU environmental policy handbook'*, Brussels: EEB.

Hilgartner, S., C.L. Bosk (1988). The Rise and Fall of Social Problems, a Public Arenas Model. *The American Journal of Sociology*, 94(1), pp. 53-78.

Hill, A. B. (1965). The Environment and Disease: Association or Causation? *Proceedings of the Royal Society of Medicine*, 58, pp. 295-300.

Hippe, J., P. Lucardie, G. Voerman (2004). Kroniek 2002, Overzicht van de partijpolitieke gebeurtenissen van het jaar 2002. *Jaarboek 2002 Documentatiecentrum Nederlandse Politieke Partijen.* Groningen: DNPP.

Hiskes, R. (2012). Environmental Human Rights. In: T. Cushman (ed.). *Handbook of Human Rights*, Oxon, UK: Routledge.

Hoek, G. Brunekreef, B. Goldhohm, S., Fischer, P. and Van der Brandt, P.A. (2002). Association between mortality and indicators of traffic-related air pollution in the Netherlands, a cohort study. *The Lancet*, 360(9341), pp. 1203-1209.

Hoek, A.J.L. van der (1996). *Inventaris van de archieven van de Voorlopige Raad voor de Milieuhygiëne, 1974-1981, en de Centrale Raad voor de Milieuhygiëne (1978), 1981-1993.* Den Haag: Nationaal Archief.

Hoekema, A.J. & N.F. van Manen (2000). *Typen van Legaliteit.* Deventer: Kluwer.

Holzinger, K., Knill, C. and Sommerer, T. (2008). Environmental Policy Convergence? The Impact of International Harmonization, Transnational Communication and Regulatory Competition. *International Organization*, 60(4), pp. 553-587.

Hoogerbrugge, R., H.A.C. Denier van der Gon, M.C. Van Zanten & J. Matthijsen (2010). *Trends in fijnstof.* Bilthoven/Den Haag: Planbureau voor de Leefomgeving.

Hoop De, M. (2006). Luchtkwaliteit en Wegaanpassingen in het kader van de spoedwet wegverbreding. In: S.T. Ramnewash – Oemrawasingh (ed.). *Regeling inzake luchtkwaliteit; Nederland op slot? Verslag van de 86ste ledenvergadering van de Vereniging voor Milieurecht op 2 december 2004*, 2006, 1.

Hou, Y-J, Y-Y Zhao, B. Xiong, X-S Cui, N-H Kim et al. (2013). Mycotoxin-Containing Diet Causes Oxidative Stress in the Mouse. *PLoS ONE* 8(3).

Infomil (2000). KWS 2000 succesvol door aanpak en resultaten, *Informil nieuwsbrief*, 20, 5, 2000.

Jagt, van der, J.A.E. (2006). *Milieuconvenanten Gehandhaafd.* (Dissertation, Utrecht University), Utrecht.

Jamrozik, A. & L. Nocella (1998). *The Sociology of Social Problems, Theoretical Perspectives and Methods of Intervention.* Cambridge, UK: Cambridge University Press.

Jänicke, M. (2000). *Ecological Modernization: Innovation and Diffusion of Policy and Technology.* Report 00-08, Berlin.

Janse, D. (2008). Jurisprudentie Onderzoek BLK 2005 (master thesis at the Erasmus School of Law), Rotterdam.

Jasanoff, S. (1990). *The Fifth Branch: Science Advisers as Policymakers.* Harvard: Harvard University Press.

Jimmink, B.A., R.J.M. Folkert, R. Thomas, J.P. Beck, M.M. van Eerdt, H.E. Elzenga, K.W. van der Hoek, A. Hoen, C.J. Peek (2004). *The Dutch CAFE baseline: In or out of line? A review on the current information in the RAINS model.* Bilthoven: RIVM.

Jordan, A/, R. Würzel, A. Zito (2003). 'New' environmental policy instruments in Germany. In: A. Jordan, R. Würzel, A. Zito (eds.). *New Instruments of Environmental Governance?* London, UK / Portland, Or: Frank Cass Publishers.

Jörgens, H. (2004). 'Governance by diffusion: Implementing global norms through cross-national imitation and learning'. In: W. Lafferty (ed.). *Governance for sustainable development: the challenge of adapting form to function.* Massachussetts USA: Edward Elgar.

Judge, D. (1993). 'Predestined to save the earth', The Environmental Committee of the European Parliament. In: D. Judge (ed.). *A Green Dimension for the European Community: Political Issues and Processes.* London: Frank Cass and Company Limited.

Kabat, G.C. (2008). *Hyping Health Risks, Environmental Hazards in Daily Life and the Science of Epidemiology.* New York: Columbia University Press.

Klatte, E.R. (1997). The Evolution, Operation and Future of Environmental Management in the European Union. In: R. Baker, (ed.). *Environmental law and policy in the European Union and the United States.* Westport, Connecticut / London: Praeger.

Klink, P. van (2005). Schone lucht is een recht, *Grenzeloos*, 82, July/August 2005. (Website grenzeloos last accessed 23-06 2015)/

Knipschild, P. (2005). *Uit de Contramine.* Maastricht: Universiteit Maastricht.

Knill C. & Liefferink, D. (2007). *Environmental Politics in the European Union.* Manchester / New York: Manchester University Press.

Knol, A.B. & B.A.M. Staatsen (2005). *Trends in the environmental burden of disease in the Netherlands 1980-2020.* Bilthoven: RIVM.

Koeman, N.S.J. (2005a).Vuile lucht dwingt echt niet tot bouwstop. *Het Financieel Dagblad*, 26-04-2005.

Koeman, N.S.J. (2005b) Het Besluit Luchtkwaliteit; hoe nu verder. *Bouwrecht*, 6, 2005, pp. 504-509.

Koeman, N.S.J. (2006). Luchtkwaliteit en nieuwe projecten; op zoek naar Europeesrechtelijk aanvaardbare ontkoppelingsmogelijkheden. *Milieu en Recht*, 5, pp. 280-284.

Koeman, N.S.J. (2010). *Een wereld te winnen: Enkele beschouwingen over de toekomst van het omgevingsrecht.* Amsterdam: Universiteit van Amsterdam.

Kouridis, C., I Kioutsioukis, T. Papageorgiou, S. Mills, L. White, L. Ntziachristos (2011). *Uncertainty/Sensitivity analysis of the transport model TREMOVE.* Emisia SA report for the DG Climate. Thessaloniki: Emisia.

Krijgsheld, K.R. (1987). Richtlijnen voor de Luchtkwaliteit, een bijzondere Aktie van de Wereldgezondheid-organisatie. *Lucht en Omgeving, 4(3)*, pp. 74-78.

Kreis, I.A. & Cromwell, D.A. (2013). Study design for public health investigations: basic concepts and practice, illustrated with a case study on asthma. In: I.A. Kreis, A. Busby, G. Leonardi, J. Meara, V. Murray (eds.). *Essentials of environmental epidemiology for health protection, a handbook for field professionals.* Oxford: Oxford University Press.

Krewski et al. (2000). *Overview of the Reanalysis of the Harvard Six Cities Study and American Cancer Society Study of Particulate Air Pollution and Mortality.* Discussion paper for Health and Air Quality 2001 - Interpreting Science for Decision-making, Ottawa, January 28-29, 2000.

Krewski et al. (2000b). Reanalysis of the Harvard Six Cities Study and American Cancer Society Study of Particulate Air Pollution and Mortality. A special report of the Institute's Particle Reanalysis Project. Executive Summaries and Commentary. Cambridge MA: HEI.

Krewski et al. (2003). Overview of the Reanalysis of the Harvard Six Cities Study and American Cancer Society Study of Particulate Air Pollution and Mortality. *Journal of Toxicology and Environmental Health*, Part A, 66, pp. 1507-1551.

Krewski et al. (2009). *Extended Follow-Up and Spatial Analysis of the American Cancer. Society Study Linking Particulate Air Pollution and Mortality.* Boston MA: Health Effects Institute.

Kronsell, A. (2000). A "sustainable" impact on the EU? In: S. C. Young (ed.). *The emergence of ecological modernization: integrating the environment and the economy.* London: Routledge.

Langeweg F. (ed.). (1988). *Zorgen voor Morgen National milieuverkenning 1985-2010.* Bilthoven: RIVM / Samsom Tjeenk Willink.

Langhelle, O. (2000). Why ecological modernization and sustainable development should not be conflated. *Journal of Environmental Policy and Planning*, 2(4), pp. 303-322.

Laverman, W. (2005). Peijs: 'Voor infra geld zat, luchtkwaliteit, dát is het echte probleem', *Building Business*, april, pp. 16-19.

Le Fanu, J. (1999). *The Rise and Fall of Modern Medicine.* London: Little Brown.

Lenschow, A. (1997). *Transformation in European Environmental Governance*, European University Institute, Working Paper RSC No 97/61.

Lenschow, A. & A.R. Zito (1997). *Institutional Linkages across EC Economic and Environmental Policy Realms*, Paper prepared for the 5th Biennale ECSA International Conference, 29 May-1 June 1997, Seattle, Washington.

Leroy, P. & D.M. Wiering, M. (2007). Inhoudelijke ontwikkelingen, doelstellingen en uitgangspunten van het milieubeleid. In: P.P.J. Driessen, P, Leroy (eds.). *Milieubeleid, Analyse en Perspectief.* Bussum: Coutinho.

Liefferink, D. & K. Birkel (2011). The Netherlands: A case of "cost-free leadership". In: Wurzel & Connelly (eds.). *The European Union as a Leader in International Climate Change Politics.* London: Routledge.

Lijphard, A. (1968). *The Politics of Accommodation. Pluralism and Democracy in the Netherlands*, Berkeley: University of California Press.

Loseke, D. (2015). Introduction to Constructionist Futures: New Directions in Social Problems Theory. *Qualitative Sociology Review*, 11(2), pp. 6-14

MAAPE (1992). *Sulphur Dioxides, Acid Aerosols and Particulates.* London: HMSO.

MacNee, W. & K. Donaldson (2003). Mechanism of lung injury caused by PM10 and ultrafine particles with special reference to COPD, *European Respiratory Journal*, 21 suppl. 40, pp. 47s-51s.

Mantzos, L (2010). *PRIMES Energy System Model: Uncertainties and their Treatment.* Paper presented at the Joint TFIAM/EC4MACS Workshop on Uncertainty Treatment, in Laxenburg, Austria, November 3-4, 2010.

Marseille, B. (2011). Toegang tot de bestuursrechter in Nederland en Duitsland. *Ars Aequi,* March, 2011, pp. 238-245.

Mathijssen H., J. Visser (2006). *PM10 in Nederland, Concentraties, rekenmethodiek en onzekerheden.* Bilthoven: MNP.

Meadows, D.H., D.L. Meadows, J. Randers, W.W. Behrens III (1972). *The Limits to Growth, a Report for the Club of Rome's Project of the Predicament of Mankind.* New York: Universe Books.

Meulen, van der, A., P.J.A. Rombout, C.J. Prins, P.G.N. Kramers, G.J. van Esch, E. Heijna-Merkus, A.C. Besemer, H.J. Huldy, H.C.M. Mulder, J. Venselaar (1987). *Criteriadocument fijnstof.* Bilthoven: RIVM.

Meuwese, A. (2008). *Impact Assessment in EU Lawmaking.* The Hague: Kluwer Law International.

Michiels, F.C.M.A. (1998). *De Wet Milieubeheer.* Deventer: Tjeenk Willink.

Milieudefensie (2005). *Luchtalarm, maak de lucht weer schoon,* krant Milieudefensie, 28-04, 2005.

Milieudefensie (2006). *Letter of Milieudefensie to the Dutch Council of State, 19 April 2006,* appeal trajectory A4 Burgerveen, Leiden.

MNP (2000). *Milieubalans 2000.* Bilthoven: MNP.

MNP (2005). *Fijnstof Nader Bekeken, een overzicht van de stand van zaken in het dossier fijn stof.* Bilthoven: MNP / RIVM.

MNP (2006a). *Nieuwe Inzichten in de omvang van de fijnstof problematiek.* Bilthoven: MNP.

MNP (2006b). *Milieubalans 2006.* Bilthoven: MNP.

MNP / RIVM (2004). *Milieubalans 2004.* Bilthoven: MNP/RIVM.

Mol, A.P.J., & D.A. Sonnenfeld (eds.) (2000). *Ecological Modernisation Around the World. Perspectives and Critical Debates.* London/Portland OR: Frank Cass.

Mol, A. P. J. & G. Spaargaren (2000). Ecological modernisation theory in debate: A review. *Environmental Politics,* 9(1), pp. 17-49.

Mol, A. P. J. & M. Jänicke (2010). The Origins and Theoretical Foundations of Ecological Modernization Theory. In: A.P.J. Mol, D.A. Sonnenfeld, G. Spaargaren (eds.). *The Ecological Modernisation Reader.* London and New York: Routledge.

Moolgavkar, S. (2005). A review and critique of the EPA's rationale for a fine particle standard. *Regulatory Toxicology and Pharmacology,* 42(1), pp. 123-144.

NationalAudit Office (2010). *Tackling inequalities in life expectancy in areas with the worst health and deprivation, executive summary.* London: The Stationary Office.

Nilsson, M, K. Eckerberg, A. Persson (2007). *Environmental Policy Integration and Changes in Governance in Swedish Energy and Agriculture Policy over two Decades.* Berlin: Ecologic/ IEEP.

O'Byrne, D. (2011). *Introducing Sociological Theory.* London: Pearson.

OECD (1995). *Environmental Performance Review, The Netherlands.* Paris: OECD Publications Office.

OECD (2003). *Environmental Performance Review, The Netherlands.* Paris: OECD Publications Office.

Olsthoorn, X. et al. (1997). *Economic Evaluation of air quality targets for sulphur dioxide, nitrogen dioxide, fine suspended particulate matter and lead,* final Report. Brussels: European Commission

O'Malley, P. (2013). Uncertain Governance and Resilient Subjects in the Risk Society. *Oñati Socio-legal Series* [online], 3(2), pp. 180-195. http://ssrn.com/abstract=2221288.

Oosten, van, K. (2012). *Gebakken Lucht, Tien jaar luchtkwaliteitsbeleid in Nederland: over de versoepeling van de regelgeving en het schoonrekenen van de lucht.* Breda: Papieren Tijger.

Pallemaerts, M., Wilkinson, D., Bowyer, C., Brown, J., Farmer, A., Farmer, M., Herodes, M., Hjerp, P., Miller, C., Monkhouse, C., Skinner, I., ten Brink, P. and Adelle, C. (2006). *Drowning in Process? The Implementation of the EU's 6th Environmental Action Programme Report for the European Environmental Bureau*. London: IEEP

Peper, B. (1998). *Sociale Problemen en de Moderne Samenleving. Een Cultuursociologische Beschouwing*. (Diss. University of Amsterdam) Amsterdam: Het spinhuis.

Pettenger, M.E. (2007). The Netherlands Climate Change Policy, constructing themselves/constructing climate change. In: M.E. Pettenger (ed.). *The Social Construction of Climate Change: Power, Knowledge, Norms, Discourses*. Aldershot England, Burlington VT: Ashgate.

Phalen, R.F. (2002). *The Particulate Air Pollution Controversy: A Case Study and Lessons Learned*. Dordrecht: Kluwer Academic Publishers.

Phalen, R.F. (2004). The Particulate Air Pollution Controversy. *Nonlinearity Biology Toxicology Medicine*, 2(4), pp. 259-292.

Pieterman, R. (2008). *De voorzorgcultuur, streven naar veiligheid in een wereld vol risico en onzekerheid*. Den Haag: Boom Juridische Uitgevers.

Pieterman, R. (2010). Geschipper rond visserij. Het voorzorgbeginsel en de overlevingskansen van een Nederlandse traditie. In: N. Doornbos, W. van Rossum, N. Huls (eds.). *Rechtspraak van Buiten*. Deventer: Kluwer.

Pieterman, R. & N.T. Arnoldussen (2008). Het voorzorgbeginsel, over Ideologie en Onzekerheid. *Rechtstheorie en Rechtfilosofie*, 2008, 3, pp. 230-251.

Platform Anders Betalen voor Mobiliteit (2005). *Anders Betalen Voor Mobiliteit*. Rapport 26 May 2005. Platform anders Betalen voor Mobiliteit.

Pope III CA, Thun MJ, Namboodiri MM, Dockery DW, Evans JS, Speizer FE, Heath CW Jr. (1995). Particulate air pollution as a predictor of mortality in a prospective study of U.S. adults. *American Journal of Respiratory and Critical Care Medicine*. 151(3 Pt 1), pp. 669-74.

Pope III, C.A., D.W. Dockery (1999). Epidemiology of Particle Effects. In: S.T. Holgate, H.S. Koren, J.M. Samet, R.L. Maynard (eds.). *Air Pollution and Health*. London UK San Diego CA: Academic Press.

Pope III C.A. (2003). The Particulate Air Pollution Controversy: A Case Study and Lessons Learned, book review. *International Journal of Epidemiology*, 32(5), p. 890.

Posner, R.A. (1982) *Tort Law, Cases and Economic Analysis*. Boston, Toronto: Little, Brown & Company.

Powers S.K., M.J. Jackson (2007). Exercise-induced oxidative stress: cellular mechanisms and impact on muscle force production. *Physiology Review*, Oct, 88(4), pp. 1243-1276.

Priemus, H. (2006a). Fijnstofnormering: Patstelling tussen volksgezondheid en ruimtelijke ontwikkeling? *NovaTerra*, 6(1), pp. 4-11.

Priemus (2006b). Wat ging er mis met het fijnstofbeleid? *Internationale Spectator*, 60(6), 2006.

Prittwitz Von, V. (1990). *Die Katastophenparadox Elemente einer Theorie der Umweltpolitik*. Opladen: Leske + Budrich.

QUARG (1993a). *Urban Air Quality in the United Kingdom*. London: HMSO.

QUARG (1993b). *Diesel vehicles and urban air quality*, Second QUARG report. London: HMSO.

Raad voor Verkeer en Waterstaat (1998). *SVV-III, Nee, Tenzij*. Den Haag: Raad voor Verkeer en Waterstaat.

Raak van R, R., van der Brugge, S. van der Krogt, H. te Riele (2006). *The Jellyfish clash, Ontknoping van het fijnstofprobleem*. Rotterdam: Erasmus Universiteit Rotterdam, Drift.

Redclift M., E. Shove, B. Van Der Meulen, B. S. Raman (2000). *Social Environmental Research in the European Union*. Cheltenham UK, Northhampton USA: Edward Elgar.

RIVM (1998). Fijnstof in Nederland, een tussenbalans. Bilthoven: RIVM.

RIVM (2000). *Nationale Milieuverkenning 5*. Bilthoven: RIVM.

RIVM (2003). *Milieu en natuureffecten hoofdlijnenakkoord Balkenende-2*, rapport 500023002. Bilthoven: RIVM.

Roo, G. de (2003). *Environmental Planning in the Netherlands – Too Good to be True. From Command-and-Control Planning towards Shared Governance*. Farnham: Ashgate.

Rood, J. M. van Keulen, S. Nollen, S. Arts (2005). *Nederland en de tot standkoming van EU milieurichtlijnen*. Den Haag: CESP Clingendael.

Saedeleer De, N., G. Roller, M. Dross (2002). *Access to Justice in Environmental Matters* ENV.A.3/ETU/2002/0030 Final Report. (Website ENV.A.3Final report, last accessed 25-06 2015).

Schenkel, W. (1998). *From Clean Air to Climate Policy in the Netherlands and Switzerland – Same Problems, Different Strategies?* Bern: Peter Lang.

Schumacher, E.F. (1973). *Small is Beautyful, A study of economics as if people mattered*. London: Blond & Briggs

Schutte Postma, E.T. (2006). Luchtkwaliteit en EU, Europese achtergronden van het Besluit luchtkwaliteit. In: S.T. Ramnewash – Oemrawasingh (ed.). *Regeling inzake luchtkwaliteit; Nederland op slot? Verslag van de 86ste ledenvergadering van de Vereniging voor Milieurecht op 2 december 2004*, 2006, 1.

Schwitters, R.J.S. (2008). *Recht en Samenleving in Verandering, Een Inleiding in de Rechtssociologie*. Deventer: Kluwer

SER (2006). *Nederland en de totstandkoming van EU milieurichtlijnen*. Den Haag: Sociaal Economische Raad.

Sluis, van der C. (2010). NSL dan toch echt geaccepteerd. *Lucht*, 3, pp. 27-29.

Smaal, M. (2012). *Politieke strijd om de prijs van de automobiliteit. De geschiedenis van een langdurend discours: 1895-2010*. (Diss. Tilburg University), Delft: Eburon.

Smit, R. (2006) Hoe Nederland zichzelf in de problemen bracht met fijnstof, zendeling in Europa. *Forum*, 12-01, 2006.

Somsen, H. (2011). When Regulators Mean Business: Regulation in the Shadow of Environmental Armageddon. *Rechtsfilosofie & Rechtstheorie, 4*, pp. 47-57.

Spaans, L.A.J., F.C.M.A. Michiels. (2000). *De waarde(n) van milieukwaliteitsnormen*. Den Haag: Boom Juridische Uitgevers.

Spector, M. & J. I. Kitsuse. (1977). *Constructing Social Problems*. Menlo Park CA: Cummings.

Swedish Environmental Research Institute (2004). *Review of the RAINS Integrated Assessment Model Draft*. Final Report for Commission Comment 16th July 2004.

Taskforce Environment and Internal Market (1990). *1992, The Environmental Dimension*. Brussels: EU.

Tatenhove, van J.P.M. (1993). *Milieubeleid Onder Dak? : beleidsvoeringsprocessen in het Nederlandse milieubeleid in de periode 1970 - 1990: nader uitgewerkt voor de Gelderse Vallei*. (Diss. Universiteit van Wageningen), Wageningen.

Tatenhove, van J.P.M. & H. Goverde (2007). Institutionalisering van het Nederlandse milieubeleid. In: P.P.J. Driessen, P. Leroy (eds.). *Milieubeleid, Analyse en Perspectief*. Bussum: Coutinho.

Tatenhove, van J.P.M & P. Leroy (2010). Environment and Participation in a Context of Political Modernisation. In: A.P.J. Mol, D.A. Sonnenfeld, G. Spaargaren (eds.). *The Ecological Modernisation Reader*. London and New York: Routledge

Tieman, J.R.C., Van Dijk Barkmeijer, A.R.G, Pennarts, H.F.T. (2007). *Recht* op *schone lucht* : een *analyse* van *mogelijkheden tot ontkoppeling* van *ruimtelijke ordening* en *luchtkwaliteit*. Den Haag: Instituut voor Bouwnijverheid / Erasmus Universiteit Rotterdam.

TNO (2003). *Onderzoek naar de effecten van de 80km/u maatregel op de A 13 in Overschie, rapport nr. 2003/258*. Apeldoorn: TNO.

TNO (2012). *Surveys of views of stakeholders, experts and citizens on the review of the EU clean air policy Part 1: main results*. Utrecht: TNO.

Trouwborst A. (2009) Prevention, Precaution, Logic and Law, The Relationship Between the Precautionary Principle and the Preventative Principle in International Law and Associated Questions. *Erasmus Law Review*, 2(2), pp. 105-127

Tuinstra, W. (2007). Preparing for the European Thematic Strategy on Air Pollution: At the Interface Between Science and Policy. *Environmental Science & Policy*, 10, pp. 434–444.

UNICE (2002). Letter from Suzie Baverstock to Andre Zuber, UNICE comments on target setting and policy development WG TG SA, 16 December 2002.

UNICE (2004). Letter from Suzie Baverstock to Andre Zuber, UNICE COMMENTS ON TARGET SETTING AND POLICY ASSESSMENT WG DEVELOPMENTS FOLLOWING ITS MEETING ON FEBRUARY 17, 2004.

UNICE (2005). Letter from Philippe de Buck to Jose Manuel Barroso. UNICE's concerns about the Thematic Strategy on Air Pollution (CAFE) 7 June 2005.

Vanhala, l. (2012). Legal Opportunity Structures and the Paradox of Legal Mobilization by the Environmental Movement in the UK. *Law and Society Review*, 46(3) pp. 523–556.

Velzen van, K., H.S.M.A. Diederen, H.W. ter Maat, L. van Bree (2000). *Schets van Knelpunten in de luchtkwaliteit in Nederland*. Bilthoven: RIVM.

Vliet, P. van, M. Knape, J. Hartog, N. Jannssen, H. Harssema, B. Brunekreef (1996). *Luchtverontreiniging, snelwegen, luchtwegen, Een onderzoek naar het verband tussen luchtveronreiniging afkomsitig van snelwegen en de respiratoire gezondheid van kinderen die wonen bij snelwegen*. Wageningen: Landbouw Universiteit Wageningen.

Vos, E. (2009). *50 Years of European Integration, 45 Years of Comitology*. Maastricht Faculty of Law Working Paper , No. 3, 2009.

Vries, B. de, and L. Francot Timmermans (2013) Eyes wide shut: on Risk, Rule of Law and Precaution. *Ratio Juris*, 26(2), pp. 282-301

Vrijssen, E. (2010). Anti-autoverkeersbeleid, Elke dag Zwarte Zaterdag. *Elsevier*, 30 Nov. 2010.

VROMraad (1999). *Mobiliteit met Beleid*. Den Haag: VROMraad.

VROMraad (2008). *Brussels Lof, Handreiking voor ontwikkeling en implementatie van Europees Recht en Beleid*. Den Haag: VROMraad.

VROMraad (2009) *Dynamiek in gebiedsgericht milieubeleid: Ontwikkelen door herschikken*. Den Haag: VROMraad.

Weale, A. (1992). *The New politics of pollution*. Manchester / New York: Manchester: University Press.

Weale A. (2005). Environmental Rules and Rule Making in the European Union. In: A. Jordan (ed.). *Environmental policy in the European Union: Actors, Institutions and Processes* (2nd edition). London UK/USA: Sterling.

Weale A., G. Pridham, M. Cini, D. Konstadakopulos, M. Porter, B. Flynn (2000). *Environmental Governance in Europe: An Ever Closer Ecological Union?* Oxford: Oxford University Press.

Weggeman, J. (2003). *Controversiele besluitvorming, opkomst en functioneren van groen polderoverleg*. (Diss. Erasmus University Rotterdam), Rotterdam.

Wenk J., Brenneisen P., C. Meewes, M. Wlaschek, T. Peters, R. Blaudschun, W. Ma, L. Kuhr, L. Schneider, K. Scharffetter-Kochanek (2001). UV-Induced Oxidative Stress and Photoaging. In: J. Thiele, & P. Elsner (eds.). *Oxidants and Antioxidants in Cutaneous Biology. Current Problems in Dermatology*, vol. 29, pp. 83-94. Basel: Karger.

Werf, van der E. (1997). Europese richtlijnen bieden geen directe oplossingen voor de praktijk. *Lucht*, 2.

Wettestad, J and A. Farmer (2001).*The EU Air Quality Framework Directive: EU Pitcher; International Catcher.* Lysaker and London: Fridtjof Nansen Institute / IEEP

Wettestad, J. and A. Farmer (2003). *The EU Air Quality Framework directive: Shaped and Saved by Interaction?* Brussels: Fridtjof Nansen Institute / IEEP.

Wettestad, J. (2006). The EU Air Quality Framework directive: Shaped and Saved by Interaction? In: S. Obertuer & T. Gehring (eds.). *Institutional Interaction in Global Environmental Governance: Synergy and Conflict among International and EU Policies.* Cambridge MA, London, UK: MIT Press.

WHO (2003). *Health Aspects of Air Pollution with Particulate Matter, Ozone and Nitrogen Dioxide Report on a WHO Working Group*, Bonn, Germany 13–15 January 2003.

Wilkinson, D., C. Monkhouse, M. Herodes, A. Farmer (2005). *For better or for worse, The Eu's Better regulation agenda and the Environment.* London: IEEP.

Winsemius, P. (1986). *Gast in eigen huis, Beschouwingen over milieumanagements.* Alphen a/d Rijn: Samsom H.D. Tjeenk Willink.

WRR (1998). *Ruimtelijke Ontwikkelingspolitiek.* Den Haag: SDU Uitgevers.

Würzel, R. (2002). *Environmental policy making in Britain, Germany and the European Union.* Manchester: Manchester University Press.

Würzel, R. (2003). *The EU Presidency, honest broker or driving seat.* Anglo German Foundation for the study of industrial society.

Yang, W., S.T. Omaye (2009). Air pollutants, oxidative stress and human health. *Mutation Research/Genetic Toxicology and Environmental Mutagenesis*, 674(1-2), pp. 45-54.

Young, S.C., (2000). Introduction: the origins and evolving nature of ecological modernisation. In: S.C. Young (ed.). *The Emergence of Ecological Modernisation, Integrating the environment and the economy?* London: Routledge.

Dutch Parliamentary Documents and Policy Documents

Concept Ministerial Decree on Air Quality (2005). Version 18-01 2005, Website Bigwobber, last accessed 23 06 2015.

Council of State (2002). Advies W09.02.0427/V.

Council of State (2005a). Advies W08.05.0081/V/A, 30 Maart 2005.

Council of State (2005b). Jaarverslag van de Raad van State over het jaar 2004.

Council of State (2005c). Advies W06.05.0257/IV, 12 September 2005.

Council of State (2005d). Advies W08.05.0176/V, 15 Juni 2005, Addition Staatscourant 12 July 2005, nr. 132.

Council of State (2005e). Advice of the Council of State, W08.05.0401/V, 11 November 2005, Kamerstukken II 2005-2006, 30 489, nr. 4.

Council of State (2006a). Jaarverslag van de Raad van State over het jaar 2005.

Council of State (2006b). Advice of the Council of State, W08.06.0164/V/A, 1 June 2006.

EZ (1976). Memorandum on Selective Economic Growth, Kamerstukken II 1975-1976, 13955 nr. 2.

EZ (2005). Nu Voor Later, Energierapport 2005. The Hague: Ministerie van Economische Zaken.

Geel van, P.L.B.A. (2004). Interpretatie Besluit Luchtkwaliteit, Letter to the Mayors and Aldermen and the provincial estates, 30 September 2004 LMV 2004091308.

Handelingen I 1999-2000, nr. 21, 21 March 2000, pp. 922-951.

Handelingen II 2000-2001 nr. 13, 17 October 2000, 889. Aan de orde is de behandeling van: het wetsvoorstel Vaststelling van de begroting van de uitgaven en de ontvangsten van het Ministerie van Volkshuisvesting, Ruimtelijke Ordening en Milieubeheer (XI) voor het jaar 2001 (27 400 XI).

Handelingen II 2000-2001 nr. 24, 16 November 2000, 1980 appendix. Schriftelijke antwoorden van de minister en de staatssecretaris van Verkeer en Waterstaat op vragen, gesteld in de eerste termijn van de behandeling van de begroting van Verkeer en Waterstaat voor het jaar 2001 (27 400-XII).

Handelingen II 2001-2002 nr. 25, 20 November 2001, 1802-1844. Aan de orde is de behandeling van: het wetsvoorstel Vaststelling van de begroting van de uitgaven en de ontvangsten van het Ministerie van Volkshuisvesting, Ruimtelijke Ordening en Milieubeheer (XI) voor het jaar 2002 (28000 XI).

Handelingen II 2002-2003 nr. 62, 15 April 2003, 3709-3711.

Handelingen II 2003-2004 nr. 24, 25 November 2003, 1669-1691.

Handelingen II 2004-2005 nr. 27, 24 November 2004, p. 1756. Aan de orde is de voortzetting van de behandeling van: het wetsvoorstel Vaststelling van de begrotingsstaten van het Ministerie van Verkeer en Waterstaat (XII) voor het jaar 2005 (29800 XII).

Handelingen II 2004-2005 nr. 34, 14 December 2004, 2248. Aan de orde zijn de stemmingen over moties, ingediend bij het debat over verkeersemissies, te weten: -de motie-Duyvendak c.s. over verbetering luchtkwaliteit door snelheidslimiet van 80 km/u (29667, nr. 4); -de motie-Van Velzen over een minimumafstand tot snelwegen (29667, nr. 6); -de motie-Van Velzen c.s. over een Europese heffing op kerosine (29667, nr. 7).

Handelingen II 2004-2005 nr. 82, 19 May 2005, pp. 4900-4920. Verantwoordingsdebat.

Handelingen II 2004-2005 nr. 96, 28 June 2005, pp. 5776- 5804 Debat over de Voorjaarsnota 2005.

Handelingen II 2004-2005 nr. 103, 7 September 2005, pp. 6263-6283. Debat over milieu, duurzaamheid en rentmeesterschap.

Handelingen II aanhangsel 2004-2005 nr. 148, pp. 313- 314.

Handelingen II 2005-2006 nr. 19, 9 November 2005, pp. 1139-1168. Aan de orde is de behandeling van het wetsvoorstel Vaststelling van de begrotingsstaten van het Ministerie van Volkshuisvesting, Ruimtelijke Ordening en Milieubeheer (XI) voor het jaar 2006 (30300-XI).

Handelingen II 2006-2007 nr. 14, 18 October 2006, pp 884-897. Aan de orde is de behandeling van: - het wetsvoorstel Wijziging van de Wet milieubeheer (luchtkwaliteitseisen) (30489), en van: - de motie-Spies over uitbreiding van de aanwijzingsbevoegdheid (30489, nr. 18).

Handelingen II anhangsel 2007-2008 nr. 2, 19 September 2007, pp. 5-41. Algemene Beschouwingen

Handelingen II 2007-2008 nr. 74, 9 April 2008, pp. 5157-5184.

Handelingen II 2007-2008 nr. 38, 19 December 2007, pp. 2983-3008

Handelingen II 2009-2010 nr. 24, 12 November 2009. pp. 2140-2204. Het wetsvoorstel Regels met betrekking tot versnelde ontwikkeling en verwezenlijking van ruimtelijke en infrastructurele projecten (Crisis- en herstelwet) (32127).

Kamerstukken I 2006-2007, 30 489 nr. J. Wijziging van de Wet milieubeheer (luchtkwaliteitseisen) NADERE MEMORIE VAN ANTWOORD. Ontvangen 24 april 2007.

Kamerstukken I 2007-2008, 30 489 nr. N. Wijziging van de Wet milieubeheer (luchtkwaliteitseisen) BRIEF VAN DE MINISTER VAN VOLKSHUISVESTING, RUIMTELIJKE ORDENING EN MILIEUBEHEER Den Haag, 26 september 2007.

Kamerstukken II 1988-1989, 21 163 nr. 3. Uitbreiding en wijziging van de Wet algemene bepalingen milieuhygiëne en daarmee samenhangende wijzigingen van andere wetten (milieubeleidsplanning en milieukwaliteitseisen; provinciale milieuverordening; totstandkoming algemene maatregelen van bestuur).

Kamerstukken II 1996- 1997, 25 005 nr. 3. BRIEF VAN DE MINISTER VAN VOLKSHUISVESTING, RUIMTELIJKE ORDENING EN MILIEUBEHEER, 17 September 1996.

Kamerstukken II 1997-1998, 21 501-08 nr. 72. VERSLAG VAN EEN ALGEMEEN OVERLEG. Vastgesteld 7 april 1998.

Kamerstukken II 1997-1998, 25 605 nr. 2. Milieuprogramma 1998-2001 ingediend 16-09 1997.

Kamerstukken II 1999/2000, 26 804 nr. 2. Milieuprogramma 2000–2003.

Kamerstukken II 2000-2001, 27 400 XI nr. 56. AO 14 februari 2001. Vaststelling van de begroting van de uitgaven en de ontvangsten van het Ministerie van Volkshuisvesting, Ruimtelijke Ordening en Milieubeheer (XI) voor het jaar 2001.

Kamerstukken II 2000-2001, 27 408 nr. 21. Meerjarenprogramma Infrastructuur en Transport 2001-2005 (MIT), Verslag van een notaoverleg

Kamerstukken II, 2000-2001, 27 408 nr. 26. Meerjarenprogramma Infrastructuur en Transport 2001-2005 (MIT) BRIEF VAN DE MINISTER VAN VERKEER EN WATERSTAAT Aan de Voorzitter van de Tweede Kamer der Staten-Generaal Den Haag, 20 april 2001

Kamerstukken II 2001-2002, 27 793 nr. 1/309.

Kamerstukken II 2001-2002, 27 793 nr. 2.

Kamerstukken II 2001-2002, 27 793 nr. 3.

Kamerstukken II 2001-2002, 27 793 nr. 8.

Kamerstukken II 2001-2002, 28000 A nr. 29. Vaststelling van de begroting van de uitgaven en de ontvangsten van het Infrastructuurfonds voor het jaar 2002 BRIEF VAN DE MINISTER VAN VERKEER EN WATERSTAAT Aan de Voorzitter van de Tweede Kamer der Staten-Generaal, Den Haag, 1 mei 2002.

Kamerstukken II 2001-2002, 28 003 nr. 2. Milieuprogramma 2002–2005.

Kamerstukken II 2001–2002, 28 089 nr. 3.

Kamerstukken II 2001-2002, 28 375 nr. 5. WERKEN AAN VERTROUWEN, EEN KWESTIE VAN AANPAKKEN STRATEGISCH AKKOORD VOOR KABINET CDA, LPF, VVD.

Kamerstukken II 2002–2003, 27 793 nr. 8. Letter of the Secretary of State for Housing, Spatial Planning and the Environment to the chairman of the Second Chamber of Parliament, 7 October 2002.

Kamerstukken II 2002-2003 28 600 A nr. 2 A Vaststelling van de begrotingsstaat van het Infrastructuurfonds voor het jaar 2003 MEMORIE VAN TOELICHTING.

Kamerstukken II 2002–2003, 28 600 XI nr. 73. Vaststelling van de begrotingsstaat van het Ministerie van Volkshuisvesting, Ruimtelijke Ordening en Milieubeheer (XI) voor het jaar 2003. VERSLAG VAN EEN NOTAOVERLEG , vastgesteld 29 november 2002.

Kamerstukken II 2002-2003, 28 663 nr. 2. Lijst van vragen en antwoorden, 19 November 2002.

Kamerstukken II 2002-2003, 28 679 nr. 3. Spoedwet wegverbreding Memorie van Toelichting.

Kamerstukken II 2002-2003 28 679 nr. 40. Regels ter bespoediging en vereenvoudiging van procedures met het oog op het zo spoedig mogelijk vergroten van de capaciteit van een aantal hoofdwegen door middel van een betere benutting en verbreding van die wegen (Spoedwet wegverbreding). Vastgesteld 18 december 2002. VERSLAG VAN EEN WETGEVINGSOVERLEG.

Kamerstukken II 2003-2004, 28 663 nr. 8. Milieubeleid 2002 – 2006, BRIEF VAN DE STAATSSECRETARIS VAN VOLKSHUISVESTING, RUIMTELIJKE ORDENING EN MILIEUBEHEER Aan de Voorzitter van de Tweede Kamer der Staten-Generaal.

Kamerstukken II 2003-2004 28 663 nr. 9. Milieubeleid 2002–2006 VERSLAG VAN EEN ALGEMEEN OVERLEG Vastgesteld 10 november 2003.

Kamerstukken II 2003-2004, 28 663 nr. 11, BRIEF VAN DE MINISTER VAN VERKEER EN WATERSTAAT, 17 November 2003.

Kamerstukken II 2004-2005, 28 663 nr. 26. BRIEF VAN DE MINISTER VAN VERKEER EN WATERSTAAT, 29 Oktober 2004.

Kamerstukken II 2004-2005 28 663 nr. 27. BRIEF VAN DE STAATSSECRETARIS VAN VOLKSHUISVESTING, RUIMTELIJKE ORDENING EN MILIEUBEHEER Den Haag, 28 oktober 2004.

Kamerstukken II 2004/2005, 28 663 nr. 30. Letter of the Secretary of State for Housing, Spatial Planning and the Environment to the chairman of the Second Chamber of Parliament, 26 November. 2004.

Kamerstukken II 2004-2005, 29 667 nr. 2. Beleidsnota Verkeersemissies BRIEF VAN DE STAATSSECRETARIS VAN VOLKSHUISVESTING, RUIMTELIJKE ORDENING EN MILIEUBEHEER Den Haag, 1 december 2004.

Kamerstukken II 2004-2005, 29 667 nr. 8. VERSLAG VAN EEN ALGEMEEN OVERLEG. Vastgesteld 22 december 2004 AO, 4 november 2004.

Kamerstukken II 2004-2005 29 667 nr. 12. BRIEF VAN DE STAATSSECRETARIS VAN VOLKSHUISVESTING, RUIMTELIJKE ORDENING EN MILIEUBEHEER, Den Haag, 20 april 2005.

Kamerstukken II 2004-2005 29 667 nr. 18. VERSLAG VAN EEN ALGEMEEN OVERLEG 26 april 2005. Vastgesteld 6 juni 2005.

Kamerstukken II 2004-2005, 29 667 nr. 21.Beleidsnota verkeersemissies BRIEF VAN DE STAATSSECRETARIS VAN VOLKSHUISVESTING, RUIMTELIJKE ORDENING EN MILIEUBEHEER Aan de Voorzitter van de Tweede Kamer der Staten-Generaal, Den Haag, 31 augustus 2005.

Kamerstukken II 2004-2005, 29 800 A nr . 43. Vaststelling van de begrotingsstaat van het Infrastructuurfonds voor het jaar 2005. VERSLAG VAN EEN NOTAOVERLEG.

Kamerstukken II 2004-2005 29 800 XI nr. 93. Vaststelling van de begrotingsstaten van het Ministerie van Volkshuisvesting, Ruimtelijke Ordening en Milieubeheer (XI) voor het jaar 2005. VERSLAG VAN EEN WETGEVINGSOVERLEG. Vastgesteld 13 december 2004.

Kamerstukken II 2004-2005, 30 175 nr. 2. Besluit Luchtkwaliteit 2005.

Kamerstukken II 2004-2005, 30 175 nr. 9. Besluit luchtkwaliteit 2005.

Kamerstukken II 2005- 2006, 21 501-08 nr. 215. Milieuraad, VERSLAG VAN EEN ALGEMEEN OVERLEG. Vastgesteld 31 januari 2006.

Kamerstukken II 2005-2006, 22 112 & 21 501-08 nr. 454. Nieuwe Commissievoorstellen en initiatieven van de lidstaten van de Europese Unie; Milieuraad.

Kamerstukken II 2005-2006, 28 679 nr. 50. BRIEF VAN DE MINISTER VAN VERKEER EN WATERSTAAT aan de Voorzitter van de Tweede Kamer der Staten-Generaal Den Haag, 8 november 2005.

Kamerstukken II 2005-2006, 29 644 nr. 12. Nota Mobiliteit Den Haag. BRIEF VAN DE MINISTERS VAN VERKEER EN WATERSTAAT EN VAN VOLKSHUISVESTING, RUIMTELIJKE ORDENING EN MILIEUBEHEER, Den Haag, 31 augustus 2005.

Kamerstukken II 2005-2006, 30 175 nr. 10. Besluit Luchtkwaliteit 2005, BRIEF VAN DE MINISTER EN DE STAATSSECRETARIS VAN VOLKSHUISVESTING, RUIMTELIJKE ORDENING EN MILIEUBEHEER Aan de Voorzitter van de Tweede Kamer der Staten-Generaal, Den Haag, 20 september 2005.

Kamerstukken II 2005–2006, 30 175 nr. 15. LIJST VAN VRAGEN EN ANTWOORDEN, Vastgesteld 1 februari 2006.

Kamerstukken II 2005-2006, 30 175 nr. 16. BRIEF VAN DE STAATSSECRETARIS VAN VOLKSHUISVESTING, RUIMTELIJKE ORDENING EN MILIEUBEHEER, 10 Maart 2006.

Kamerstukken II 2005-2006, 30 175, nr. 17. Besluit luchtkwaliteit 2005, VERSLAG VAN EEN ALGEMEEN OVERLEG. Vastgesteld 16 maart 2006.

Kamerstukken II 2005-2006, 30 300, nr. 3. NOTA OVER DE TOESTAND VAN 'S RIJKS FINANCIËN, advies en nader rapport.

Kamerstukken II 2005-2006 30 300 XI nr. 2. Vaststelling van de begrotingsstaten van het Ministerie van Volkshuisvesting, Ruimtelijke Ordening en Milieubeheer (XI) voor het jaar 2006 MEMORIE VAN TOELICHTING.

Kamerstukken II 2005- 2006 30 300 XI nr. 10. Vaststelling van de begrotingsstaten van het Ministerie van Volkshuisvesting, Ruimtelijke Ordening en Milieubeheer (XI) voor het jaar 2006.

Kamerstukken II 2005-2006 30 300 XII nr. 2. Vaststelling van de begrotingsstaten van het Ministerie van Verkeer en Waterstaat (XII) voor het jaar 2006 MEMORIE VAN TOELICHTING.

Kamerstukken II 2005-2006, 30 300 A nr. 2. Vaststelling van de begroting van de uitgaven en de ontvangsten van het Infrastructuurfonds voor het jaar 2006.

Kamerstukken II 2005-2006, 30 489 nr. 2. Wijziging van de Wet milieubeheer (luchtkwaliteitseisen).

Kamerstukken II 2005-2006, 30 489 nr. 3. Memorie van Toelichting wet luchtkwaliteitseisen.

Kamerstukken II 2005-2006, 30 489 nr. 4. Wijziging van de Wet milieubeheer (luchtkwaliteitseisen).

Kamerstukken II 2005-2006, 30 489 nr. 7. Wijziging van de Wet milieubeheer (luchtkwaliteitseisen) Verslag.

Kamerstukken II 2005-2006, 30 489 nr. 8. Wijziging van de Wet milieubeheer (luchtkwaliteitseisen). Nota van Antwoord naar aanleiding van het verslag.

Kamerstukken II 2006-2007, 30 489 nr. 12. Wijziging van de Wet milieubeheer (luchtkwaliteitseisen). Lijst van vragen en antwoorden. Vastgesteld op 15 juni 2006.

Kamerstukken II 2006-2007, 30 489 nr. 25. Wijziging van de Wet milieubeheer (luchtkwaliteitseisen). VERSLAG VAN EEN WETGEVINGSOVERLEG 25 september 2006. Vastgesteld 28 september 2006

Kamerstukken II 2006-2007, 31 031 XI nr 1. Jaarverslag en slotwet ministerie van Volkshuisvesting, Ruimtelijke Ordening en Milieubeheer 2006. Aangeboden 16 mei 2007.

Kamerstukken II 2006-2007, 31 031 XII nr. 1 Jaarverslag en slotwet ministerie van Verkeer en Waterstaat 2006: jaarverslag.

Kamerstukken II 2006–2007, 30 646 nr. 2. Evaluatie Spoedwet wegverbreding. BRIEF VAN DE MINISTER VAN VERKEER EN WATERSTAAT Aan de Voorzitter van de Tweede Kamer der Staten-Generaal, Den Haag, 25 oktober 2006.

Kamerstukken II 2007-2008, 29 385 nr. 15. Aanleg en de aanpassing van hoofdinfrastructuur VERSLAG VAN EEN ALGEMEEN OVERLEG. Vastgesteld 20 december 2007.

Kamerstukken II 2007-2008, 30 175 nr. 32, p. 3. Besluit Luchtkwaliteit 2005. BRIEF VAN DE MINISTER VAN VOLKSHUISVESTING, RUIMTELIJKE ORDENING EN MILIEUBEHEER Aan de Voorzitter van de Tweede Kamer der Staten-Generaal Den Haag, 21 september 2007.

Kamerstukken II 2007-2008, 30 175 nr. 42. Besluit luchtkwaliteit 2005. VERSLAG VAN EEN ALGEMEEN OVERLEG. Vastgesteld 9 november 2007.

Kamerstukken II 2007-2008, 30 175 nr. 46. Besluit luchtkwaliteit 2005. VERSLAG VAN EEN ALGEMEEN OVERLEG 19 december 2007. Vastgesteld 29 januari 2008.

Kamerstukken II 2007-2008, 30 504 nr. 4. Tracé A4 Burgerveen – Leiden VERSLAG VAN EEN ALGEMEEN OVERLEG 6 september 2007. Vastgesteld 2 oktober 2007.

Kamerstukken II 2009-2010, 30 175 nr. 98. BRIEF VAN DE MINISTER VAN VOLKSHUISVESTING, RUIMTELIJKE ORDENING EN MILIEUBEHEER, 26 April 2010.

Staatscourant (1982a), 149, *Projectprogramma Plan Integratie Milieubeleid*, 6 August 1982.

Staatscourant (1982b), 29, *Projectopdracht Plan Integratie Milieubeleid*, 11 February 1982.

VOMIL (1972). *Urgent Memorandum on Environmental Hygiene*. Kamerstukken II 1971-1972, 11906 nr. 2.

VOMIL (1974). *Budget for the Parliamentary Year 1975*. Kamerstukken II 1974 1975, 13100 XVII nr. 2.

VOMIL (1976). *Memorandum on Ambient Environmental Standards*. Kamerstukken II 1976-1977, 14318 nrs. 1-2.

VROM (1983). *Indicative multi year Programme Air 1984-1988*. Kamerstukken II 1983-984, 18100 XI nr. 7.

VROM (1984a). *Planningsnota "Meer dan de Som der Delen"* Kamerstukken II 1983-1984, 18292 nrs. 1-2.

VROM (1984b). *Indicative multi year Programme Environmental Management 1985-1989*. Kamerstukken II 1984-1985, 18602 nrs. 1-2.

VROM (1989). *National Environmental Policy Plan (NMP), 'Choosing or Losing'*. Kamerstukken II 1988-1989, 21137 nrs. 1-2.

VROM (1992). *Evaluatie Besluiten Luchtkwaliteit*. Publikatiereeks lucht & energie, nr. 107. Den Haag: VROM.

VROM (1993). *NEPP 2, The Environment as Yardstick*. Kamerstukken II 1993-1994, 23560 nr. 2.

VROM (1998). *National Environmental Policy Plan 3*. Den Haag: VROM.

VROM (2002). *Notitie Vaste Waarden nieuwe Vormen, Milieubeleid 2002-2006*. Den Haag: VROM.

VROM (2004). *Beleidsnota Verkeersemissies, Met schonere, zuiniger en stillere voertuigen en klimaatneutrale brandstoffen op weg naar duurzaamheid*. Den Haag: VROM.

VROM (2005). Nationaal Luchtkwaliteitsplan. Den Haag: VROM.

VROM (2006). *Duurzaam VROM, Ambities, Daden en Dillemma's, maatschappelijk jaarverslag 2006*. Den Haag: VROM.

V&W (1988). *Tweede Structuurschema Verkeer en Vervoer, Deel A.* Kamerstukken II, 1988 – 1989 20922, 1/2.

V&W (1990). *Tweede Structuurschema Verkeer en Vervoer, Deel D.* Den Haag: V&W

V&W (1996). *Nota Samen Werken aan Bereikbaarheid.* Kamerstukken II 1996–1997, 25 022 nr. 1/2.

V&W (1999). *Perspectievennota Verkeer en Vervoer.* Kamerstukken II 1998–1999, 26 428 nr. 1.

V&W (2001). *Nationaal Verkeer en Vervoersplan, van A naar Beter.* Den Haag: V&W.

V&W (2006). *A4 Burgerveen – Leiden 2004, Wijzigingen Voorgenomen Tracébesluit.* Den Haag: V&W

V&W (2007). *Sneller Besluitvorming Infrastructuur,* Press release Ministry of V&W, 18 September 2007.

European policy documents

6th EAP (2002). Decision no 1600/2002/EC of the European Parliament and of the Council of 22 July 2002 laying down the sixth community environment action programme.

AEAT, (August 2005). Cost-Benefit Analysis of Policy Option Scenarios for the Clean Air for Europe program for Service Contract for carrying out cost-benefit analysis of air quality Related Issues in particular in the Clean Air for Europe (CAFÉ) Programme. AEAT/ ED48763001/ABC Scenarios. Issue 2.

Council of Ministers (2007a). Revised Addendum to I/A Item note, interinstitutional file 2005/0183(COD), Brussels, 12 June 2007.

Council of Ministers (2007b). Document 16341/07 14335/05 ENV 525 ENER 170 IND 76 TRANS 232 ENT 141 CODEC 1010. Proposal for a Directive of the European Parliament and of the Council on ambient air quality and cleaner air for Europe – Outcome of the European Parliament's second reading, Strasbourg, 10 to 13 December, on file with the author.

ENVI (2006a)2005/0184(COD). Draft report on the proposal for a DIRECTIVE OF THE EUROPEAN PARLIAMENT AND OF THE COUNCIL on ambient air quality and cleaner air for Europe 04-04, 2006.

ENVI (2006b). PE 374.011v01-00, amendments 36 – 195. Draft report on the proposal for a DIRECTIVE OF THE EUROPEAN PARLIAMENT AND OF THE COUNCIL on ambient air quality and cleaner air for Europe, 08-05 2006.

ENVI (2006c). A6-0234/2006. Report on the proposal for a DIRECTIVE OF THE EUROPEAN PARLIAMENT AND OF THE COUNCIL on ambient air quality and cleaner air for Europe 29-06, 2006.

European Commission (1993). Towards Sustainability (Fifth Environmental Action Programme) Journal of the European Communities No. C 138, 17-05 1993.

European Commission (1994). COM(1994)104 final. Proposal for a Directive on ambient air quality assessment and management. Brussels: European Commission.

European Commission (1995). COM(1995)372 final. Report from the Commission on the State of Implementation of Ambient Air Quality Directives.

European Commission (1997). COM(1997)500. Proposal for a Council Directive relating to limit values for sulphur dioxide, oxides of nitrogen, particulate matter and lead in ambient air. V

European Commission (1998). The ambient air quality Framework directive - Clean air for Europe's cities. Luxembourg: Office for Official Publications of the European Union.

European Commission (2000a). COM(2000)626. 'A review of the auto-oil II programme'.

European Commission (2000b). COM(2000)1. COMMUNICATION FROM THE COMMISSION on the precautionary principle.

European Commission (2001). COM(2001)245 final. COMMUNICATION FROM THE COMMISSION The Clean Air for Europe (CAFE) Program: Towards a Thematic Strategy for Air Quality, Brussels, 04.05.2001.

European Commission (2005a). COM(2004)845 final. Review of Council Directive 1999/30/EC relating to limit values for sulphur dioxide, nitrogen dioxide and oxides of nitrogen, particulate matter and lead in ambient air, with consideration of Council Directive 96/62/EC on ambient air quality assessment and management, 04-01 2005.

European Commission (2005b). COM(2005) 446 final. communication from the commission to the council and the European Parliament Thematic Strategy on air pollution, Brussels, 21.09.2005.

European Commission (2005c). COM(2005)447 final. Proposal for a DIRECTIVE OF THE EUROPEAN PARLIAMENT AND OF THE COUNCIL on ambient air quality and cleaner air for Europe.

European Parliament (1998). Official Journal of the European Communities, No 4-519, 1998/99 Session. Report of proceedings from 11 May to 15 May 1998.

SEC(2001)688, Commission staff Working Paper on Clean Air For Europe (CAFE) Brussels, 04/05/2001.

SEC (2004) 1713. COMMISSION STAFF WORKING DOCUMENT Annex to: The Report in support of the review of Council Directive 1999/30/EC relating to limit values for sulphur dioxide, nitrogen dioxide and oxides of nitrogen, particulate matter and lead in ambient air, with consideration of Council Directive 96/62/EC on ambient air quality assessment and management.

SEC (2005) 1133. COMMISSION STAFF WORKING PAPER Annex to: The Communication on Thematic Strategy on Air Pollution and The Directive on 'Ambient Air Quality and Cleaner Air for Europe' Impact Assessment.

Steering Group, (2001 January). *Summary Report of the Meeting of the Air Quality Steering Group*, 23 January 2001.

Steering Group, (2004 May) *9th CAFE Steering Group Meeting - Minutes Brussels*, 17-18 May 2004.

Steering Group (2005 February). *12th CAFE Steering Group Meeting - Minutes* Brussels, 21-22 February 2005.

Steering Group (2005 April). *13th CAFE Steering Group Meeting - Minutes Brussels*, 14 April 2005.

Technical Working Group on Particulate Matter (1997). Ambient Air Pollution by Particulate Matter Position Paper, Final Version, 8-04, 1997.

WG Imp (2003 January). Minutes of the 6th Meeting of the Working Group on Implementation under the CAFE program on 8 and 9. January 2003 in Paris.

WG Imp (2003 March). Minutes of the 7th Meeting of the Working Group on Implementation under the CAFE program from 27 and 28 March 2003 in Düsseldorf.

WG IMP (2004). Recommendations on the review of Council Directive 1999/30/EC, June 2004.

WG PM (2004). FINAL DRAFT SECOND POSITION PAPER ON PARTICULATE MATTER, April 2004.

WG TSP (2002 November). Minutes of the second meeting of the CAFE Working Group on Target Setting and Policy Assessment, held in Brussels on 6 November 2002, Feb, 2003 final.

WG TSP (2003 Januay) Minutes of the first meeting of the CAFE Working Group on Target Setting and Policy Assessment held in Brussels on 30 January 2003.

WG TSP (2003 September) Minutes of the Target Setting and policy assessment Working Group meeting held in Brussels on 12 September 2003 (Final).

UK POLICY DOCUMENTS

DETR (1997). A new deal for trunk roads. London: Department of the Environment, Transport and the Regions.

DETR (1998). A New Deal for Transport: Better for everyone. White Paper. London: Department of the Environment Transport and the Regions.

DOE (1995). Air Quality, Meeting the Challenge, The Governments Strategic Policy for Air Quality Management, London: Department of the environment.

Hansard 28 Jan. 1993. House of Commons Hansard Debates for 28 Jan 1993, vol 217 cc849-50W.

Hansard 10 Feb. 1993. House of Commons Hansard Debates for 10 Feb 1993, vol 218 cc647-49W

Hansard 17 Feb. 1993. House of Commons Hansard Debates for 17 Feb 1993, vol 219 c276W.

Hansard 2 March 1993. House of Commons Hansard Debates for 2 Mar 1993, vol 220 c69W, c127-28W.

Hansard 22 March 1993. House of Commons Hansard Debates for 22 Mar, 1993, vol 221 c490W.

Hansard 23 March 1993. House of Commons Hansard Debates for 23 Mar, 1993, vol 221 c536W.

Hansard 25 June 1993. House of Commons Hansard Debates for 25 June, 1993, vol 227, 624.

Hansard 1 Dec. 1993. House of Commons Hansard Debates for 1 Dec 1993, vol 233, cc. 1036/1037.

Hansard 17 Dec. 1993. House of Commons Hansard Debates for 17 Dec 1993, vol 234, cc1089-90W.

Hansard 17 March 1994. House of Commons Hansard Debates for 17 March 1994, vol 239 cc828-30W

Hansard 18 July 1994. House of Commons Hansard Debates for 18 July 1994, vol. 247, col. 10.

Hansard 20 July 1994, House of Commons Hansard Debates for 20 July 1994, vol 247 cc512-532.

Hansard 31 Oct. 1994. House of Commons Hansard Debates for 31 Oct 1994, vol 248 cc1274-318.

House of Commons (1998). House of Commons - Environmental Audit - Minutes of Evidence Memorandum submitted by the Green Minister, the Department of the Environment, Transport and the Regions. House of Commons (1998). House of Commons - Environmental Audit - Minutes of Evidence Memorandum submitted by the Green Minister, the Department of the Environment, Transport and the Regions.

NEWSPAPER AND MAGAZINE ARTICLES, DUTCH AND UK

BN/De Stem, 07-04, 2005. 'Rechter kan nu ieder bouwproject afkeuren'. Door Arthur 't Hart.

Cobouw, 22-03, 2008. 'Wetsvoorstel beperken inspraak infra-projecten'.

De Telegraaf, 15-10, 2004. 'Ons land is vieste van Europa'.

De Telegraaf, 11-05, 2005. 'Aantal Smogdoden geschat op achttienduizend'.

De Telegraaf, 28-06, 2005. 'Bouw belemmerd door vervuilde lucht'.

ENDS Report 225, October 1993. 'Framework rules on air quality promised by end of year'.

ENDS report 250, November 1995. 'Dust flies over proposed PM10 standard'.

ENDS Europe 26 april 2001. 'Integration push for EU air quality policy'.

ENDS Europe, 28 June 2006. 'Ministers agree new EU air quality limits'.

ENDS Europe, 3 July 2008. 'Netherlands requests EU air pollution delay.

Euractiv, 21-09 2005. 'Clean Air Strategy seeks balance between health and business concerns'. Website Euractiv a (last accessed 23-06 2015).

Euractiv, 27-09, 2006. 'Parliament softens clean air law'. Website Euractiv b (last accessed 23-06 2015).

Euractiv, 12 -12 2007. 'EP vote paves way for tighter air quality rules'. Website Euractiv c (last accessed 23-06 2015).

Het Financieele Dagblad, 31-05, 2005. 'Gevolgen van luchtnorm EU 'desastreus''.

Het Financieele Dagblad, 02-06, 2005. 'Planologen vrezen dat Nederland op slot gaat'.

Het Financieele Dagblad, 04-06, 2005. 'Plan ontwikkelaars om toch te kunnen bouwen'.

Het Financieele Dagblad, 28-06, 2005. 'Van Geel moet zijn karwei nog afmaken'.

Het Financieele Dagblad, 23-11, 2005. 'Werkgevers tegen luchtwet Van Geel'.

Het Parool, 28-06, 2005. 'Vuile lucht: 100.000 banen weg'.

NRC Handelsblad 11-02, 1999. 'Werkbezoek Minister Pronk aan Overschie'.

NRC Handelsblad 28-06, 2005. 'Veel banen weg door wet over luchtkwaliteit'

The Guardian, (5-07, 1994). 'Dangerous ozone levels could be regular feature of hot summer days'.

The Guardian (18-07, 1994). 'Commentary: wanted: a politician who will finally clear the air'.

The Independent, (10-10, 1993). 'Gasping for breath: Are exhaust fumes suffocating our children? Geoffrey Lean investigates the causes of the asthma epidemic'.

The Independent, (10-03, 1994). 'Pollution in air 'kills 10,000 each year''. London.

The Independent (5-07, 1994). 'Old bangers and clean air'. London.

The Independent, (17-07, 1994). 'The day Britain choked; UK suffers world's biggest asthma outbreak'. London.

Trouw (11-02, 1999). 'Pronk op bezoek in Overschie'.

Trouw (07-04, 2001). 'Gezondheid snel weg; luchtkwaliteit'.

Trouw (05-08, 2003). 'Wonen bij snelweg ongezond'.

Trouw (10-06, 2005). '18.000 doden is een voorzichtige schatting'.

de Volkskrant (25-09, 2004). 'Uitbreiding snelweg in gevaar'.

de Volkskrant (25-09, 2004). 'Klem tussen auto en gezondheid'.

de Volkskrant (01-10, 2004). 'Van Geel: Nederland gaat op slot'.

de Volkskrant (07-05, 2005). 'Recht op schone lucht wint van het verkeer'.

de Volkskrant (11-05, 2005). 'Duizenden sterven eerder door fijn stof'.

de Volkskrant (12-11, 2011). 'Wientjes: Jezus dacht ik, waar zijn we nu in beland?'

Dutch Law

Stb. 1979, 99, Wet Geluidshinder (Noise Abatement Act).

Stb. 1986, 78, Besluit Luchtkwaliteit Zwaveloxide en Zwevende Deeltjes (Air Quality Order on SO_2 and Particulates).

Stb. 1992, 551, Wet milieubeheer (Law on Environmental Management).

Stb. 2000, 393, VROM budget for the year 2000, Law of 14 September 2000, regarding changes in the budget of the Ministry of Housing Spatial Planning and Environmental Management for the year 2000.

Stb. 2001, 269, Besluit Luchtkwaliteit 2001 (Air Quality Order 2001, and Explanatory Memorandum).

Stb. 2003, 256, Spoedwet Wegverbreding (Emergency Law on Road Expansion).

Stb. 2005, 316, Besluit Luchtkwaliteit 2005 (Air Quality Order 2005).

Stb. 2007, 414, Titel 5.2 van de Wet milieubeheer, Wet van 11 oktober 2007 tot wijziging van de Wet milieubeheer (luchtkwaliteitseisen), 'Wet Luchtkwaliteit' (Air Quality Law).

Stb. 2007, 440, Besluit niet in betekenende mate bijdragen (luchtkwaliteitseisen) (Decision not to a significant extent).

Stb. 2009, 14, Besluit Gevoelige Bestemmingen (General Administrative Order on Sensitive Destinations).

Stb. 2010, 135, Crisis en Herstelwet (Crisis and Recovery Law).

Stb. 2010, 175, Wet herstructurering Raad van State (Council of State Restructuring Act).

EU Laws

(Directive 80/779/EEC). Council Directive 80/779/EEC on air quality limit values and guide values for sulphur dioxide and suspended particulates OJ L 229, 30.8.1980.

(Framework Directive). Council Directive 1996/62/EC on Ambient Air Quality Assessment and Management OJ L 296, 21.11.1996.

(Daughter Directive). Council Directive 1999/30/EC of 22 April 1999 relating to limit values for sulphur dioxide, nitrogen dioxide and oxides of nitrogen, particulate matter and lead in ambient air OJ L 163, 29.6.1999.

(CAFE Directive). Council Directive 2008/50/EC Directive 2008/50/EC of the European Parliament and of the Council of 21 May 2008 on ambient air quality and cleaner air for Europe, OJ L 152, 11.6.2008.

Court Cases

HR 05-11-1965, NJ 1966, 136, LJN AB7079, Kelderluik.

ABRvS 200308160, Barneveld Hoevelaken, 12 May 2004. Annotation: A.G.A. Nijmeijer.

ABRvS 200401178/1, Vught Ekkerswijer, 15 September 2004. Annotation: J.M.H.F. Teunissen, J.W.A. Fleuren.

ABRvS 200307780/1, Hendrik-Ido-Ambacht, 22 September 2004. Annotation: J. Struiksma.

ABRvS 200406190/2, Zuidas, 22 November 2004.

ABRvS 200409107/2, kokkelvissers, 3 December 2004.

ABRvS 200400465/1, Stadion ADO, 26 January 2005.

ABRvS 200407748/1, Wegaanpassing A 27, 13 April 2005.

ABRvS 200604266/1, Veerkaden Den Haag, 18 April 2007.

ABRvS 200602152/1, tracébesluit Burgerveen-Leiden, 25 july 2007.

ABRvS 200704973/1, 80 Kilometer, 26 March 2008.

ABRvS 200900883/1, 24 Oktoberplein, 31 March 2010.

Websites

Website Aarhus convention:
http://ec.europa.eu/environment/aarhus/

Website Airclim:
http://www.airclim.org/policy/sub6_10.php
Website Bigwobber:
http://bigwobber.nl/wp-content/uploads/osd/201504/3227.pdf
Website CAFE docs:
http://ec.europa.eu/environment/archives/cafe/general/keydocs.htm
Website COMEAP:
http://www.comeap.org.uk/documents/archive/25-archive-advisory-group-on-the-medical-aspects-of-air-pollution-episodes
Website comitology:
http://europa.eu/legislation_summaries/glossary/comitology_en.htm
Website Council of State:
https://www.raadvanstate.nl/over-de-raad-van-state/geschiedenis.html#karel
Website Den Haag Vandaag 26 april 2005:
http://www.novatv.nl/page/detail/uitzendingen/3391/Den+Haag+Vandaag%3A+luchtvervuiling#
Website Duijvestijn:
http://www.parlement.com/9291000/modulesf/glnjwpcg
Website Dutch policy documents:
https://zoek.officielebekendmakingen.nl/zoeken
Website Earth Chater:
http://www.earthcharterinaction.org/content/pages/Read-the-Charter.html
Website ENV.A.3Final report: http://www.google.nl/url?sa=t&rct=j&q=&esrc=s&source=web&cd=2&ved=0CC8QFjAB&url=http%3A%2F%2Fwww.unece.org%2Fenv%2Fpp%2Fcompliance%2FC2008-23%2FAmicus%2520brief%2FAnnexHSadeleer Report.pdf&ei=6RGMVb61NKbm7gbDhpPwBQ &usg=AFQjCNGDM2YLXweu0ga5Bn 58gVXtvcFVQQ &sig2=TsiAcaj0ERSHt5pnojYepg& bvm=bv.96782255,d.ZGU
Website Euractiv a:
http://www.euractiv.com/climate-environment/Parliament-softens-eu-clean-air-law/article-158251
Website Euractiv b:
http://www.euractiv.com/climate-environment/ep-vote-paves-way-tighter-eu-air-quality-rules/article-169059
Website Euractiv c:
http://www.euractiv.com/climate-environment/clean-air-strategy-seeks-balance-health-business-concerns/article-144604
Website European Commission codecision:
http://ec.europa.eu/codecision/stepbystep/text/index_en.htm
Website EuroParliament a:
http://www.europarl.europa.eu/oeil/popups/summary.do?id=944201&t=e&l=en.
Website EuroParliament b:
http://www.europarl.europa.eu/sides/getDoc.do?type=CRE&reference=20060925&secondRef=ITEM-013&language=EN&ring=A6-2006-0234 http://www.europarl.europa.eu/oeil/popups/ficheprocedure.do?reference=2005/0183%28COD%29&l=en

Website European Integration online:
http://eiop.or.at/eiop/texte/2009-025a.htm
Website grenzeloos:
http://www.grenzeloos.org/content/schone-lucht-een-recht
Website Greenpeace:
http://www.greenpeace.org/eastasia/publications/reports/climate-energy/2012/ranking-cities-air-action/
Website GroenLinks:
http://voorschoten.groenlinks.nl/node/73696
Website Harvard School of Public Health:
http://www.hsph.harvard.edu/news/magazine/f12-six-cities-environmental-health-air-pollution
Website Integrated assessment:
http://sedac.ciesin.columbia.edu/mva/iamcc.tg/mva-questions.html
Website International Conference on Environment and Economics:
http://sedac.ciesin.columbia.edu/entri/texts/oecd/OECD-4.02.html
Website Leefmilieu:
http://www.leefmilieu.nl/burgernetwerk/symposium4.html
Website Loonwijzer:
http://www.loonwijzer.nl/home/nieuws/nieuws-2010/tien-best-betaalde-en-meest-voorkomende-beroepen
Website Maarten Hajer:
http://www.maartenhajer.nl/?page_id=14
Website Milieudefensie:
http://www1.milieudefensie.nl/verkeer/activiteiten/luchtkwaliteit/schonelucht-juridisch
(Currently offline 20-06, 2015)
Website Milieuactueel.nl:
http://www.milieuactueel.nl/nieuws/2005/2190/milieudefensie-eist-verlaging-snelheid-op-stadssnelwegen.html
Website Milieuloket
http://www.milieuloket.nl/id/vgzdksj6rsdo/nieuws/milieudefensie_sleept_den_haag_voor?ctx=vhurdyxq5fuu&start_tab0=64
Website NRC:
http://archief.nrc.nl
Website Oxidative Stress:
http://www.oxidativestressresource.org/
Website Parliament UK:
http://www.Parliament.uk/business/publications/hansard/commons/by-date/#session=66850&year=1994&month=6&day=18
Website prelex a:
http://ec.europa.eu/prelex/detail_dossier_real.cfm?CL=en&DosId=21414
Website prelex b:
http://ec.europa.eu/prelex/detail_dossier_real.cfm?CL=en&DosId=108551

Website policy documents:
https://zoek.officielebekendmakingen.nl/zoeken,
Website REACH:
http://ec.europa.eu/environment/chemicals/index.htm
Website Rechtspraak:
http://rechtspraak.nl
Website Ridder.punt.nl:
http://ridder.punt.nl/content/2005/06/100-000-banen-in-lucht-op
Website Saneringstool:
http://www.saneringstool.nl/saneringstool.html
Website STAB:
http://www.stab.nl/jurisprudentie/tijdschriftartikel/Pages/Artikel-2012-nummer-12.aspx
Website The Economist:
http://www.economist.com/node/21546053
Website Wegenwiki (SVV2):
http://www.wegenwiki.nl/Tweede_Structuurschema_Verkeer_en_Vervoer
Website Wegenwiki (NVVP):
http://www.wegenwiki.nl/Nationaal_Verkeers-_en_Vervoersplan
Website WOB:
http://www.rijksoverheid.nl/documenten-en-publicaties/wob-verzoeken/2005/06/13/diverse-stukken-over-het-besluit-luchtkwaliteit.html
Website Working Group Imp:
http://ec.europa.eu/environment/archives/cafe/working_groups/wg_implementation.htm
Website Working Group PM:
http://ec.europa.eu/environment/archives/cafe/working_groups/wg_particulate_matter.htm call accessed 15-12 2011
Website Zembla:
http://redir.vara.nl/tv/zembla/welcome2.html?20050901/zembla

ABOUT THE AUTHOR

Tobias Arnoldussen graduated in September 2001 from the law faculty of the University of Amsterdam on a master thesis about the criminalisation of stalking. In 2002 he obtained an additional master title in philosophy (M.A.) from the same university. After living and working in Istanbul, Turkey for a period of 6 years, he returned to academic life in the Netherlands and started his PhD research on the social construction of the air quality clash at the Erasmus School of Law.

Tobias presented his research during a number of national and international conferences, published articles and several book chapters and taught courses. During his research and teaching activities, he combined insights from various disciplines such as law, sociology of law and philosophy. In 2012 Tobias was appointed as a teacher in the Faculty of Law of the University of Amsterdam at which he currently teaches a number of courses related to law and various social scientific disciplines. He is teaching and lecturing at the PPLE (Politics, Psychology, Law and Economics) College of the University of Amsterdam, an interdisciplinary honours college with an international character. There he tutors mainly international students on a broad array of topics, from the sociology of business and management to political philosophy and human rights law.

In addition Tobias works as a lecturer in the Instituut voor Wijsbegeerte, now integrated in the Internationale School Voor Wijsbegeerte. In this capacity he lectures on diverse topics in Western philosophy. His research interests concerns the fields of philosophy and sociology of law, especially environmental law. Tobias organised a PhD meeting in Rotterdam, is a member of the Dutch Flemish Association of the Social-scientific study of Law (VSR), a member of the Dutch Centre for German Idealism and was on the editorial board of Recht der Werkelijkheid, the peer reviewed journal of the VSR.

He is a keen chess player, an avid reader and enjoys the company of good friends, together with slow, home cooked food and a glass of fine red wine.

Index

M

N

O

P